Preparation of Nanomaterial Modified Electrode and Its Sensing Application

Preparation of Nanomaterial Modified Electrode and Its Sensing Application

Editors

Dong Liu
Baiqing Yuan

MDPI • Basel • Beijing • Wuhan • Barcelona • Belgrade • Manchester • Tokyo • Cluj • Tianjin

Editors
Dong Liu
Jiangsu University
China

Baiqing Yuan
Ludong University
China

Editorial Office
MDPI
St. Alban-Anlage 66
4052 Basel, Switzerland

This is a reprint of articles from the Special Issue published online in the open access journal *Nanomaterials* (ISSN 2079-4991) (available at: https://www.mdpi.com/journal/nanomaterials/special_issues/electrode_sensor).

For citation purposes, cite each article independently as indicated on the article page online and as indicated below:

LastName, A.A.; LastName, B.B.; LastName, C.C. Article Title. *Journal Name* **Year**, *Volume Number*, Page Range.

ISBN 978-3-0365-5971-1 (Hbk)
ISBN 978-3-0365-5972-8 (PDF)

Contents

About the Editors

Dong Liu

Dr. Dong Liu obtained his Ph.D. degree in analytical chemistry from Changchun Institute of Applied Chemistry (CIAC), Chinese Academy of Sciences (CAS), in 2013. Now, he is an associate professor at the School of Agricultural Engineering, Jiangsu University, China. His current research interests are focused on the synthesis of functional nanomaterials and their applications in analysis and catalysis. He has published more than 40 papers in peer-reviewed journals in e.g., Analytical Chemistry, Biosensors and Bioelectronics, and Journal of Hazardous Materials.

Baiqing Yuan

Dr. Baiqing Yuan received his Ph.D. in analytical chemistry from the Changchun Institute of Applied Chemistry, Chinese Academy of Sciences in 2010. He is currently an associate professor of chemistry in the school of chemistry and materials science with the research interests in electroanalysis and electrocatalysis. He has published more than 70 papers in peer-reviewed journals, which have been cited more than 1500 times.

nanomaterials

MDPI

Editorial

Editorial for Special Issue: "Preparation of Nanomaterial Modified Electrode and Its Sensing Application"

Dong Liu [1,*] and Baiqing Yuan [2,*]

[1] School of Agricultural Engineering, Jiangsu University, Zhenjiang 212013, China
[2] School of Chemistry and Materials Science, Ludong University, Yantai 264025, China
* Correspondence: dongliu@ujs.edu.cn (D.L.); bqyuan@ldu.edu.cn (B.Y.)

Citation: Liu, D.; Yuan, B. Editorial for Special Issue: "Preparation of Nanomaterial Modified Electrode and Its Sensing Application". *Nanomaterials* **2022**, *12*, 4010. https://doi.org/10.3390/nano12224010

Received: 8 November 2022
Accepted: 11 November 2022
Published: 15 November 2022

Electrochemical sensors have attracted enormous attention for their precision, high sensitivity, rapid response, and ease-of-use for analysis. Electrochemical sensors use an electrode as transducer element for the target. In particular, nanomaterials with excellent properties are essential for the modification of electrodes, which highlights the importance of research on the preparation of nanomaterials for electrochemical sensing. Our Special Issue includes eleven research papers, consisting of nine articles and two reviews. These reports cover the preparation of nanomaterials (magnesium phyllosilicate [1], metal–organic frameworks (MOF) [2], and covalent–organic frameworks (COF) [3]), the fabrication of electrodes with unique properties (e.g., transparency) using carbon nanomaterials [4–7] or new nanotechnologies [8,9], and applications of electrochemical sensors [10,11].

Pecheu et al. synthesized a talc-like magnesium phyllosilicate functionalized with amine groups ($TalcNH_2$) via a sol–gel process to fabricate electrochemical sensors for stripping voltametric analysis. The $TalcNH_2$-modified electrode was used for the detection of Pb^{2+} with anodic adsorptive stripping differential pulse voltammetry. The detection and quantification limits were calculated to be 7.45×10^{-8} M and 24.84×10^{-8} M, respectively, for $TalcNH_2$-based sensors [1]. This work provided an attracting nanomaterial for the fabrication of electrochemical sensors. Zhang et al. used the Co–Cu MOF precursor to synthesize porous spindle-like Cu–Co sulfide microparticles via solvothermal sulfurization. The resultant Cu–Co sulfides exhibited high catalytic activity for the electrochemical oxidation of glucose. The Co–CuS-2-based sensor exhibited enhanced analytical performance for the nonenzymatic electrochemical detection of glucose, which allowed a sensitivity of 1475.97 A mM^{-1} cm^{-2} [2]. This work provided a new method to synthesize high-performance nanomaterials from MOFs for the construction of electrochemical sensors. Li et al. synthesized a TpPa-1 COF with abundant nitrogen and oxo-functionalities by using 1,3,5-triformylphloroglucinol and phenylenediamine to develop electrochemical sensors and electrocatalysts. Pd^{2+} was employed to modify the TpPa-1 COF by taking advantage of the high adsorption affinity of TpPa-1 COF to Pd^{2+}. The Pd^{2+}-modified TpPa-1 COF exhibited high catalytic activity for the electrochemical oxidation of hydrazine and the reduction of nitrophenol. The nitrogen-doped carbon derived from TpPa-1 showed superior electrocatalytic activity for the oxidation of reduced glutathione (GSH), which can be used for GSH detection [3]. The synthesized TpPa-1 COFs can be suitable substrates and precursors for potential applications of electrochemical sensing and electrocatalysis.

Carbon nanomaterials, particularly carbon nanotubes and graphene, are among the most investigated nanomaterials for the fabrication of electrodes. Angizi et al. estimated the possibility to develop a pH-sensing platform with several graphene derivatives, and demonstrated the dependence of pH-sensing behaviors on the type and defect density of graphene derivatives. Eventually, a highly pH-sensitive platform was successfully fabricated with thermally annealed graphene oxide, which allowed for a maximum current change of 175% in a pH range from 3 to 9 [4]. Their research could offer a new way to develop micro- and nano-sized pH sensors based on graphene derivatives. Sun et al. designed a portable

E. coli detection system using the combination of immunomagnetic separation technology, a graphene-based transparent electrode, and the improved adenosine triphosphate bioluminescence technology. Such a system offered a detection range of 3.1×10^1–10^6 CFU/mL and a coefficient of variation of 3.96%, and no more than 30 min was required to complete the detection. Moreover, it allowed a high accuracy of more than 94% for *E. coli* detection in beverages and food [5]. Stanojev et al. explored the fabrication of a transparent electrode with single-walled carbon nanotubes (SWCNTs) using a combination of experimental tests and computational studies. The density functional tight-binding method was used to explore the interaction of carboxylic SWCNT with some solvents. PEI and carboxylic SWCNT were deposited on soda lime glass substrates to fabricate the thin film electrode, and its stability was tested in different solvents. By optimizing the number of depositions, transparency in the MIR range and conductivity, the electrode with suitable thickness of film enables a reduction of 20% in the MIR's transparency compared with that of the thickest SWCNT layers, while its sheet resistance can be decreased to 150–200 kW/sq [6].

Graphene oxide shows some unique advantages for electrochemical sensing. Yuan et al. constructed a micro-electrochemical sensor with a graphene oxide-modified screen-printed electrode and a polydimethylsiloxane (PDMS) micro-cell for multi-analyte detection. Their experiment demonstrated that the residual oxygen-containing functional species on graphene oxide can enhance the selectivity of sensors for different targets. Meanwhile, their investigation revealed that the attachment of bovine serum albumin can favor the diffusion of probes rather than hinder electron transfer [7].

Furthermore, many nanotechnologies have been applied to fabricate electrodes to enable them with some unique properties. Li et al. fabricated a flexible and stretchable electrode with a PDMS–Ag nanosheet composite. A low resistivity of 4.28 Wm and a low resistance variation in strain in the range of 0–50% was recorded for the resultant electrode, while the electrical conductivity of the electrode remained stable over 1000 cycles. These attracting characteristics were ascribed to the flexibility and stretchability of PDMS and high electrical conductivity of Ag nanosheets. By coupling the experiment with a theoretical model and finite element simulation, the conductive mechanism of the electrode during stretching was analyzed [8]. This work would facilitate the exploration of flexible electrodes in terms of structural design and material selection and promote the development of flexible electronic and wearable devices. Yan et al. constructed an electrochemical immunosensor using an indium tin oxide (ITO) electrode with an amino-rich nanochannels array for the detection of prostate-specific antigens (PSA). The amino-group-containing precursor was employed to grow the vertically ordered mesoporous silica nanochannel film on ITO, and the introduction of surfactant micelles caused aldehyde groups to anchor recognitive antibodies for the assembly of an immuno-recognitive interface. For detection, the specific recognition of PSA by antibody depressed the diffusion of the probe, causing the electrochemical signal to "Turn-off". Such an immunosensor offered a detection limit of 8.1 pg/mL and was used in the analysis of human serum [9]. The developed immunosensor can be also applied to develop a universal immunosensing platform for tumor marker detection.

In addition to excellent nanomaterials, strategies to assemble the sensing interface at the electrode are also vital for improving the analytical performance of electrochemical sensors. Chang et al. reviewed advances in the in-situ assembly of nanomaterials on the surface of electrodes for efficient signal readout and amplification. Furthermore, the assembly of organic molecules and biomolecules on electrodes were also discussed [10].

Recently, thanks to their unique advantages, electrochemical sensors are being widely used to monitor the important small molecules in life. Ahmad et al. reviewed the recent advances in electrochemical biosensing of H_2O_2, and a subdivision of electrochemical sensors was provided by types of nanomaterials used for fabrication. The strengths and weaknesses of the sensors were discussed by the comparison of their analytical performances. They highlighted the advantages of electrochemical sensors for H_2O_2 detection and the vital role of nanomaterials for such sensors in the early detection of cancerous cells [11].

This Special Issue has highlighted the important role of nanomaterials in fabricating high-performance electrochemical sensors. Meanwhile, the results provided by this Special Issue can favor the design and application of electrochemical sensors in the future.

Author Contributions: Writing, D.L. and B.Y.; review and editing, D.L. and B.Y.; supervision, D.L. and B.Y. All authors have read and agreed to the published version of the manuscript.

Funding: D.L. acknowledges the National Natural Science Foundation of China (NO. 61901193) and the Natural Science Foundation of Jiangsu Province (NO. BK20200104), and B.Q.Y. thanks the National Natural Science Foundation of China (NO. 22272068).

Acknowledgments: The Guest Editors would like to thank to all authors for submitting their research to the Special Issue, to the reviewers for enhancing the quality and impact of all papers, and to the editorial staff for their support during the publication.

Conflicts of Interest: The authors declare no conflict of interest.

References

1. Pecheu, C.N.; Jiokeng, S.L.Z.; Tamo, A.K.; Doungmo, G.; Doench, I.; Osorio-Madrazo, A.; Tonle, I.K.; Ngameni, E. Fabrication of an Organofunctionalized Talc-like Magnesium Phyllosilicate for the Electrochemical Sensing of Lead Ions in Water Samples. *Nanomaterials* **2022**, *12*, 2928. [CrossRef] [PubMed]
2. Zhang, D.; Zhang, X.; Bu, Y.; Zhang, J.; Zhang, R. Copper Cobalt Sulfide Structures Derived from MOF Precursors with Enhanced Electrochemical Glucose Sensing Properties. *Nanomaterials* **2022**, *12*, 1394. [CrossRef]
3. Li, G.; Yuan, B.; Chen, S.; Gan, L.; Xu, C. Covalent Organic Frameworks-TpPa-1 as an Emerging Platform for Electrochemical Sensing. *Nanomaterials* **2022**, *12*, 2953. [CrossRef]
4. Angizi, S.; Huang, X.X.; Hong, L.; Akbar, M.A.; Selvaganapathy, P.R.; Kruse, P. Defect Density-Dependent pH Response of Graphene Derivatives: Towards the Development of pH-Sensitive Graphene Oxide Devices. *Nanomaterials* **2022**, *12*, 1801. [CrossRef] [PubMed]
5. Sun, Z.; Guo, J.; Wan, W.B.; Wang, C.X. A System of Rapidly Detecting Escherichia Coli in Food Based on a Nanoprobe and Improved ATP Bioluminescence Technology. *Nanomaterials* **2022**, *12*, 2417. [CrossRef] [PubMed]
6. Stanojev, J.; Armaković, S.; Joksović, S.; Bajac, B.; Matović, J.; Srdić, V.V. Comprehensive Study of the Chemistry behind the Stability of Carboxylic SWCNT Dispersions in the Development of a Transparent Electrode. *Nanomaterials* **2022**, *12*, 1901. [CrossRef] [PubMed]
7. Yuan, B.Q.; Gan, L.J.; Li, G.; Xu, C.Y.; Liu, G. A Micro Electrochemical Sensor for Multi-Analyte Detection Based on Oxygenated Graphene Modified Screen-Printed Electrode. *Nanomaterials* **2022**, *12*, 711. [CrossRef] [PubMed]
8. Li, C.W.; Huang, K.; Yuan, T.K.; Cong, T.Z.; Fan, Z.; Pan, L.J. Fabrication and Conductive Mechanism Analysis of Stretchable Electrodes Based on PDMS-Ag Nanosheet Composite with Low Resistance, Stability, and Durability. *Nanomaterials* **2022**, *12*, 2628. [CrossRef] [PubMed]
9. Yan, L.; Xu, S.; Xi, F.N. Disposal Immunosensor for Sensitive Electrochemical Detection of Prostate-Specific Antigen Based on Amino-Rich Nanochannels Array-Modified Patterned Indium Tin Oxide Electrode. *Nanomaterials* **2022**, *12*, 3810. [CrossRef]
10. Chang, Y.; Xia, N.; Huang, Y.; Sun, Z.; Liu, L. In Situ Assembly of Nanomaterials and Molecules for the Signal Enhancement of Electrochemical Biosensors. *Nanomaterials* **2021**, *11*, 3307. [CrossRef] [PubMed]
11. Ahmad, T.; Iqbal, A.; Halim, S.A.; Uddin, J.; Khan, A.; El Deeb, S.; Al-Harrasi, A. Recent Advances in Electrochemical Sensing of Hydrogen Peroxide (H_2O_2) Released from Cancer Cells. *Nanomaterials* **2022**, *12*, 1475. [CrossRef] [PubMed]

nanomaterials

Article

Fabrication of an Organofunctionalized Talc-like Magnesium Phyllosilicate for the Electrochemical Sensing of Lead Ions in Water Samples

Chancellin Nkepdep Pecheu [1], Sherman Lesly Zambou Jiokeng [1,2,3], Arnaud Kamdem Tamo [4,5,6], Giscard Doungmo [7], Ingo Doench [4,5,6], Anayancy Osorio-Madrazo [4,5,6,*], Ignas Kenfack Tonle [1,*] and Emmanuel Ngameni [8]

1 Electrochemistry and Chemistry of Materials, Department of Chemistry, University of Dschang, Dschang P.O. Box 67, Cameroon
2 Institut für Anorganische Chemie und Strukturchemie, Heinrich-Heine-Universität Düsseldorf, 40204 Düsseldorf, Germany
3 Laboratoire de Chimie Physique et Microbiologie pour les Matériaux et l'Environnement (LCPME), UMR 7564 CNRS—Université de Lorraine, 405, rue de Vandœuvre, 54600 Villers-lès-Nancy, France
4 Laboratory for Bioinspired Materials BMBT, Institute of Microsystems Engineering IMTEK-Sensors, University of Freiburg, 79110 Freiburg, Germany
5 Freiburg Center for Interactive Materials and Bioinspired Technologies FIT, University of Freiburg, 79110 Freiburg, Germany
6 Freiburg Materials Research Center FMF, University of Freiburg, 79104 Freiburg, Germany
7 Institut für Anorganische Chemie, Christian-Albrechts-Universität zu Kiel, Max-Eyth-Str. 2, 24118 Kiel, Germany
8 Laboratory of Analytical Chemistry, Department of Chemistry, University of Yaounde 1, Yaoundé P.O. Box 812, Cameroon
* Correspondence: anayancy.osorio@imtek.uni-freiburg.de (A.O.-M.); ignas.tonle@univ-dschang.org (I.K.T.); Tel.: +49-761-203-67363 (A.O.-M.); +237-696-141-545 (I.K.T.)

Citation: Pecheu, C.N.; Jiokeng, S.L.Z.; Tamo, A.K.; Doungmo, G.; Doench, I.; Osorio-Madrazo, A.; Tonle, I.K.; Ngameni, E. Fabrication of an Organofunctionalized Talc-like Magnesium Phyllosilicate for the Electrochemical Sensing of Lead Ions in Water Samples. *Nanomaterials* **2022**, *12*, 2928. https://doi.org/10.3390/nano12172928

Academic Editors: Dong Liu and Baiqing Yuan

Received: 20 July 2022
Accepted: 22 August 2022
Published: 25 August 2022

Abstract: A talc-like magnesium phyllosilicate functionalized with amine groups (TalcNH$_2$), useful as sensor material in voltammetry stripping analysis, was synthesized by a sol–gel-based processing method. The characterizations of the resulting synthetic organoclay by scanning electron microscopy (SEM), X-ray diffraction, N_2 sorption isotherms (BET method), Fourier transform infrared spectroscopy (FTIR), CHN elemental analysis and UV–Vis diffuse reflectance spectroscopy (UV–Vis-DRS) demonstrated the effectiveness of the process used for grafting of amine functionality in the interlamellar clay. The results indicate the presence of organic moieties covalently bonded to the inorganic lattice of talc-like magnesium phyllosilicate silicon sheet, with interlayer distances of 1568.4 pm. In an effort to use a talc-like material as an electrode material without the addition of a dispersing agent and/or molecular glue, the TalcNH$_2$ material was successfully dispersed in distilled water in contrast to natural talc. Then, it was used to modify a glassy carbon electrode (GCE) by drop coating. The characterization of the resulting modified electrode by cyclic voltammetry (CV) and electrochemical impedance spectroscopy (EIS) revealed its charge selectivity ability. In addition, EIS results showed low charge transfer resistance (0.32 Ω) during the electro-oxidation of $[Fe(CN)_6]^{3-}$. Kinetics studies were also performed by EIS, which revealed that the standard heterogeneous electron transfer rate constant was (0.019 ± 0.001) cm.s^{-1}, indicating a fast direct electron transfer rate of $[Fe(CN)_6]^{3-}$ to the electrode. Using anodic adsorptive stripping differential pulse voltammetry (DPV), fast and highly sensitive determination of Pb(II) ions was achieved. The peak current of Pb^{2+} ions on TalcNH$_2$/GCE was about three-fold more important than that obtained on bare GCE. The calculated detection and quantification limits were respectively 7.45×10^{-8} M (S/N = 3) and 24.84×10^{-8} M (S/N 10), for the determination of Pb^{2+} under optimized conditions. The method was successfully used to tap water with satisfactory results. The results highlight the efficient chelation of Pb^{2+} ions by the grafted NH_2 groups and the potential of talc-like amino-functionalized magnesium phyllosilicate for application in electrochemical sensors.

Keywords: sol–gel processing; talc-like clay; amino-functionalized clay; glassy carbon electrode; Pb(II) detection; water bioremediation

1. Introduction

Talc or steatite is a clay mineral with the chemical formula $Mg_3Si_4O_{10}(OH)_2$. It consists of tetrahedral Si and octahedral Mg sheets forming the layered structure 2:1 [1,2]. Silicon dioxide (SiO_2), magnesium oxide (MgO) and H_2O are the main constituents of talc [3]. The talc surface consists of basal cleavage faces and edges. The neutral surface faces consist of tetrahedral siloxane with -Si-O-Si- bonds, giving them a non-polar and hydrophobic character, while the edges are hydrophilic due to the presence of charged ions (Mg^{2+} and OH^-) [4]. The hydrophobic behaviour of the surface layers of talc is due to oxygen atoms [5]. Talc has very interesting properties, it is organophilic, has a lamellar structure, chemical inertia, high thermal stability, low electrical conductivity, heat resistance, wide particle size distribution and high specific surface area [6,7]. Due to its hydrophobic nature, talc is highly valued for various applications such as paints, adhesives and sealants [8]. Despite its good properties, its chemical reactivity is limited when it has to undergo chemical surface modification with organic molecules, due to the rather strong attractive forces between the sheets [9]. Previous studies have shown that talc can be easily dispersed in a surfactant/polymer by absorption on its surface [10]. However, its difficult dispersion in aqueous solution limits its applications, for example, in the development of electrochemical sensors [11].

Synthetic clay materials are very popular compared to their natural counterparts due to their high purity [12,13]. Inorganic–organic materials based on magnesium silicates, similar to natural talc, have been synthesised by a sol–gel process. Under normal conditions of temperature and pressure, covalent attachment of organic functionalities occurs, thus creating homogeneous pure inorganic-organic hybrid materials with controlled porosity [14]. Ca^{2+} and Ni^{2+} ions have been extensively studied to obtain these organosilicate materials [15–18]. However, other organomodified phyllosilicates have also been synthesised, including aluminium [15], copper [19], zinc [20], and calcium [21] inside the inorganic layer. It should be noted that the magnesium and nickel-based organosilicates have a lamellar structure similar to the 2:1 trioctahedral phyllosilicates. The main objective of these synthetic materials remains their applicability, which would be multiple in case of organo-functionalisation [22]. Organoclays can be used as adsorbents, environmental barriers, polymer fillers, catalytic supports, electrochemical sensors or chemical sensors [23]. Previous work has shown their use for binding cations in aqueous solutions at the solid/liquid interface [16,19,20,24]. Their ability to extract heavy metals is a promising property to be explored in many applications [25].

Lead is a heavy metal that can have negative effects on human health, including kidney disease, cardiovascular effects, reproductive toxicity and irreversible nerve damage [26,27]. The Quantification of lead in real samples is usually performed by electrothermal atomic absorption spectrometry [28], atomic fluorescence spectrometry [29], inductively coupled plasma spectrometry [30]. As these methods are somewhat limited in selectivity and sensitivity, electrochemical methods have been widely studied as electrochemical sensors [31,32]. In this regard, several works based on chemically modified electrode materials have been proposed [33–38]. Most of these studies demonstrate organoclays for heavy metal electroanalysis due to the affinity between the organic modifier and the target species. To our knowledge, no work in the literature mentions the use of amino-functionalized magnesium phyllosilicates of the synthetic talc type as a glassy carbon electrode (GCE) modifier.

This work describes the preparation of a new amino-functionalised magnesium phyllosilicate obtained by a sol–gel process. The resulting synthesised material and natural talc were characterised by physicochemical methods. The characterised material was tested as electrochemical sensor with electrochemical characterizations of the modified electrode, which was tested for the detection of Pb^{2+} ions in aqueous solution, by means of anodic stripping differential pulse voltammetry (DPV).

2. Materials and Methods

2.1. Chemicals

The natural talc clay mineral (Nat-Talc) used in this work is a commercial sample (NICRON® 674, Luzenac America, Inc., 767 Old Yellowstone Trl, Three Forks, MT, USA). All chemicals were obtained commercially and used without further purifications. These included (3-aminopropyl)triethoxy-silane (APTES, 99%, Sigma-Aldrich Taufkirchen, Bavaria, Germany), $MgCl_2 \cdot 6H_2O$ (99%, Fluka, Buchs, Switzerland), $Pb(NO_3)_2$ (99%, Analar, Princeton, NJ, USA), $K_3[Fe(CN)_6]$ (Prolabo, Bern, Switzerland) and $Ru(NH_3)_6Cl_3$ (Alfa, Binfield, UK). $NaNO_3$ (99.99%, Prolabo, Bern, Switzerland), KCl (99.5%, Fisher Scientific International Inc., Pittsburgh, PA, USA), HCl (36%, Phillip Harris, Birmingham, England), NaCl (99.5%, Fisher Scientific International Inc., Pittsburgh, PA, USA). $Zn(NO_3)_2 \cdot 6H_2O$ (98%), $Cd(NO_3)_2 \cdot 4H_2O$ (98%), $Cu(NO_3)_2 \cdot xH_2O$ (99.99%) and $Hg(NO_3)_2 \cdot H_2O$ (99.99%) were from Sigma-Aldrich (Taufkirchen, Bavaria, Germany).

2.2. Material Characterization

In order to evaluate the morphology of our materials, an Amray 1610 Turbo scanning electron microscope (SEMTech Solutions, Inc., North Billerica, MA, USA) was used. For the measurements, the samples were deposited on a conductive strip placed on a specimen holder and coated with gold by vacuum sputtering.

Characterisation by Fourier Transform Infrared Spectroscopy (FTIR) was carried out using a Nicolet 8700 instrument equipped with a specular reflectance accessory (Smart Collector).

The crystallinity of the materials was assessed by X-ray diffraction analysis, using an STOE Stadi-p X-ray powder diffractometer (STOE & Cie GmbH, Darmstadt, Germany) operated at 40 kV and 30 mA with Cu Kα1 radiation (λ = 1.54056 Å), in transmission geometry with an IP-PSD (STOE & Cie GmbH, Darmstadt, Germany) and/or a DECTRIS® MYTHEN 1K detector (DECTRIS, Baden-Daettwil, Switzerland).

The Brunauer-Emmett-Teller (BET) specific surface areas of the samples were determined by means of N_2 adsorption-desorption at 77.13 K using a micrometrics model sorptometer (Thermo Electron Corporation, Sorptomatic Advanced Data Processing, Waltham, MA, USA). Before N_2 adsorption, the samples were degassed at 307.13 K under vacuum. The linear part of the BET equation was used to evaluate the surface area.

A CHNS Euro EA 3000 analyser (HEKAtech GmbH, Wegberg, Germany) was used to determine the chemical composition of our materials.

The optical properties of our materials were determined using a Shimadzu UV-Vis 3101PC Diffuse Reflectance Absorption Spectrophotometer (DRS) in the wavelength range 200–800 nm. $BaSO_4$ was used as the reflectance standard.

Voltammetry measurements were performed on a μ-Autolab potentiostat running GPES software and using a standard three-electrode cell (bare or modified GCEs used as working electrodes (WE), the saturated silver chloride electrode (Ag/AgCl/KCl) as reference electrode (RE), and a stainless-steel bar as auxiliary electrode). Cyclic voltammograms of $[Fe(CN)_6]^{3-}$ and $[Ru(NH_3)_6]^{3+}$ ions were recorded from −0.15 V to 0.7 V and from −0.6 V to 0.2 V respectively, in 0.1 M KCl at a scan rate of 50 mV/s, unless otherwise stated and without stirring. Using anodic stripping differential pulse voltammetry (ASDPV), the electroanalysis of Pb^{2+} ions involved two successive steps: open circuit preconcentration of the analyte under gentle agitation followed by voltammetric detection in the potential range of −0.7 V to −0.3 V after 30 s electrolysis at −0.8 V. Electrochemical impedance spectroscopy (EIS) measurements were performed on a Palmsens3 potentiostat driven by PS Trace 4.2 software. It was performed over the frequency range of 0.01 Hz to 10 kHz with a potential amplitude of 10 mV in a 0.1 M KCl solution containing 1 mM $[Fe(CN)_6]^{3-}/[Fe(CN)_6]^{4-}$.

2.2.1. Preparation of the Organofunctionalized Talc-like Magnesium Phyllosilicate

The synthetic organic/inorganic hybrid was synthesised according to a previously published procedure [13,39]. Firstly, 845.77 mg of magnesium chloride hexahydrate was dissolved at room temperature under stirring in 100 mL of distilled water. To this solution, a 1.0 mL ethanolic solution containing 1197.6 mg of (3-aminopropyl)triethoxysilane was added dropwise at room temperature. The mixture obtained corresponded to a Si/Mg molar ratio of 4/3, as in natural talc. The resulting dense, pale suspension was placed under stirring for 1 h, and 20 mL of 0.1 M NaOH was added dropwise. The suspension was aged for 24 h at room temperature, then filtered and washed with ethanol and distilled water to neutral pH. After centrifugation, the product obtained was dried under vacuum for 48 h at 50 °C and noted TalcNH$_2$.

2.2.2. Preparation of the Working Electrode

For the electrochemical characterisations, a GCE was used. This was first polished with alumina powder of different sizes, then placed in a 1:1 ethanol-water solution and ultrasonicated for 10 min to remove the remaining alumina particles. GCE modified with a TacNH$_2$ film was prepared by drop coating: 6 µL of TalcNH$_2$ suspension previously prepared by dispersing 5 mg of TalcNH$_2$ in 1 mL of water were drop coated on the active surface of the GCE (about 0.071 cm^2). The modified electrode was placed in an oven set at 110 °C for 4 min to dry. The modified electrode obtained was TalcNH$_2$/GCE.

3. Results and Discussion

3.1. Physicochemical Characterization of Organofunctionalized Clay Material

Elemental analysis carried out with a CHNS Euro EA 3000 analyser on both natural talc (Nat-Talc) and the synthesised material TalcNH$_2$ gave the experimental results shown in Table 1. The theoretical results are also presented in Table 1. During the synthesis of TalcNH$_2$, the ethoxy groups of the organosilane molecules were transformed into silanol by hydrolysis, being able to lose an H$^+$ proton in basic medium. By being negatively charged, these silanols aggregate into ordered molecular networks as do anionic surfactants [40]. The negatively charged micelles can act as a matrix and attract magnesium cations from solution to give a hybrid layered structure, in which the inorganic phase is formed by Si-O-Mg bonds. For TalcNH$_2$ (Table 1), the theoretical contents of CHN meet the experimental results, which correspond to the following formula of synthetic amine talc: $Mg_{1.8}Si_{0.9}O_{19}C_6H_{24}N_{1.8}$. From this formula, it was evident that the presence of nitrogen containing fractions and the obtained C/N molar ratio (3.33) were high and close to the calculated value (3.00), evidence that the organic fractions remained intact during the synthesis process. The high calculated value may be due to partial condensation, involving the methoxy groups of the silylating agents, which would have resulted in an increase in carbon content. The Si/Mg ratio of 0.50 for the TalcNH$_2$ material differs from the expected value of 1.33 for natural talc. Clearly, there is 2.67-fold more magnesium in the TalcNH$_2$ structure than in natural talc, which could be due to the presence of exchangeable cations between the layers or complexed by the pendant groups arranged in the lamellar cavity. The latter is the most plausible reason as the nitrogen atoms in the pendant groups can complex them [13].

Table 1. Chemical composition of Nat-Talc and TalcNH$_2$.

	Molar Ratios						
	Reaction Mixture		Product		Weight (%)		
	Si/Mg	C/N	Si/Mg	C/N	C	H	N
Nat-Talc	-	-	1.33	-	0.357	0.469	0
TalcNH$_2$	1.33	3.00	0.50	3.33	14.90 (14.57 *)	4.83 (4.89 *)	4.42 (5.10 *)

* Theoretical values.

Figure 1 displays nitrogen sorption isotherms for Nat-Talc (curve a) and $TalcNH_2$ (curve b). Both clay samples show type III nitrogen isotherms, which is characteristic of non-porous or macroporous solids. A specific surface of 14.39 m^2/g was obtained for natural talc. This low value compared to those of other families of clay minerals is due to the fact that N_2 molecules cannot access the interlayer regions of expanding clays or the structural tunnels of natural talc. The cumulative mesopore volume, cumulative micropore volume and surface area, and specific surface area of aminated synthetic talc were 0.013 cm^3/g, 0.0048 cm^3/g, 9.04 m^2/g, and 12.16 m^2/g, respectively. The low surface area compared to that of natural talc can be attributed to the high level of amino-functionalization of the hybrid material, as the pendant carboxyl groups of the (3-aminopropyl)triethoxy-silane (APTES) molecule block the access of nitrogen gas to the pores of the material [13,41].

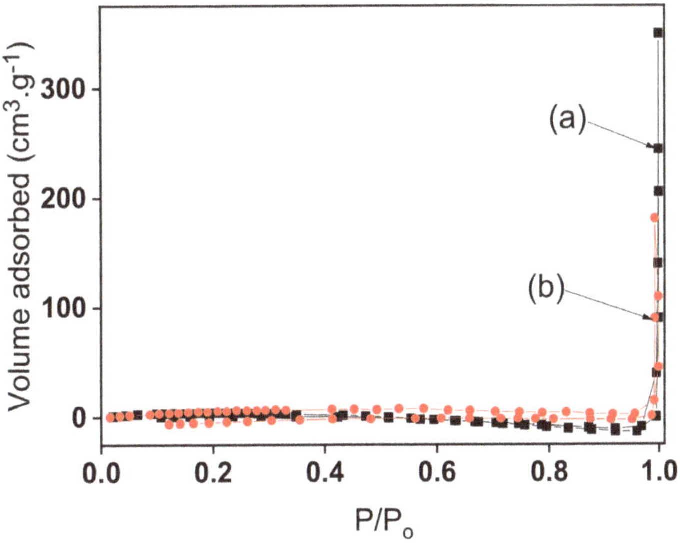

Figure 1. N_2 sorption isotherms (at 77.13 K) for (**a**) Nat-Talc, and (**b**) $TalcNH_2$ materials.

Figure 2 shows the FT-IR spectra of the raw and synthesised amine clay samples. In the spectrum of raw clay mineral (Figure 2a), a broad band at 3683 cm^{-1}, associated with (Mg-OH) vibrations, is observed [42,43]. In addition, some remarkable bands related to the inorganic structure of clay were obtained at 983 cm^{-1} and 535 cm^{-1}, due to v (Si-O-Si) and δ(Mg-OH) vibrations, respectively. Some changes were observed in the spectrum of the synthesised clay sample ($TalcNH_2$) (Figure 2b). The absorption bands at 3366 cm^{-1} υ(N-H), 2919 cm^{-1} υ(C-H), 1613 cm^{-1} δ(N-H), 1481 cm^{-1} υ(C-N), 1198 cm^{-1} υ(Si-C), and at 995 and 877 cm^{-1} associated with Si-O-Si and Si-O-Mg bonds [44–49]. Around 3600 cm^{-1}, we observed a broad band attributed to the vibration of the Mg-OH bond, and the stretching bands of water and the Si-OH bond [50–52]. The band at 1628 cm^{-1} corresponds to the bending vibration mode of water. At 535 cm^{-1}, another band is observed which is attributed to the overlapping of the Si-O and Mg(OH) vibrational modes as observed in magnesium-based and trioctahedral clays [53–55]. However, the presence of new bands observed for the synthetic amino clay corresponds to all the vibrations of the surface organosilane functions. This significant difference shows that the functionalization of the synthetic clay by the APTES molecules was effective.

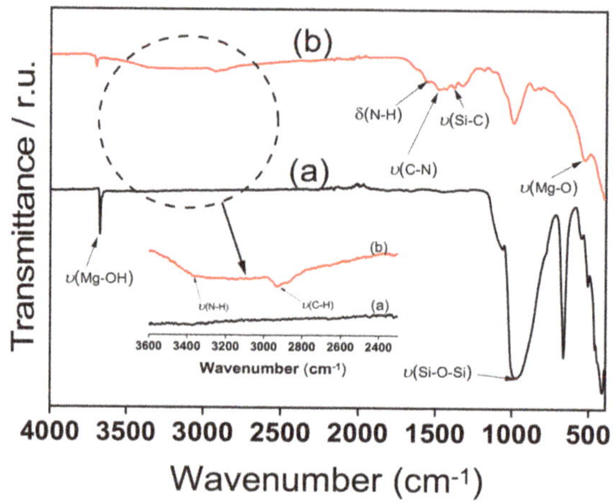

Figure 2. Fourier transform infrared (FTIR) spectra of: (**a**) Nat-Talc, and (**b**) $TalcNH_2$.

XRD characterizations were performed to compare the crystalline structures of the synthetic and the natural clay materials. XRD patterns of Nat-Talc and synthetic amino-clay ($TalcNH_2$) are shown on Figure 3. Figure 3a showed diffraction peaks at $2\theta = 9.3°$ (lamellae width/ spacing d = 950.2 pm), 19.4° and 28.7° that correspond to the primary diffractions of the (001), (020) and (003) planes of natural talc material [56]. The XRD analysis revealed the presence of small amounts of chlorite at $2\theta = 26.7°$ and of dolomite at $2\theta = 31°$, in addition to talc signals. $TalcNH_2$ material showed diffraction peaks at $2\theta = 5.63°$ (lamellae width/spacing d = 1568.4 pm) and at 21.13°, assigned to the primary diffractions of the (001) and (020) planes. The use of (3-aminopropyl)triethoxysilane (APTES) during the synthetic process led to an increase (1.65-fold) in basal spacing when comparing the natural talc value to that of the functionalized talc (950.2 pm to 1568.4 pm, respectively) (Figure 3b). This slight difference arises from the incorporation of organic chains (APTES) in the interlamellar space. This conclusion is derived from the data on the length of the organic chains. By estimating this length using a bond distance model and assuming a zigzag conformation, a value of 543 pm was obtained, for the $-(CH_2)_3NH_2$ fragments attached to the inorganic backbone. The low intensity peak at 2θ at 59.07° (156.3 pm) corresponds to a reflection in the 060 plane and is in agreement with the formation of trioctahedral layers [15].

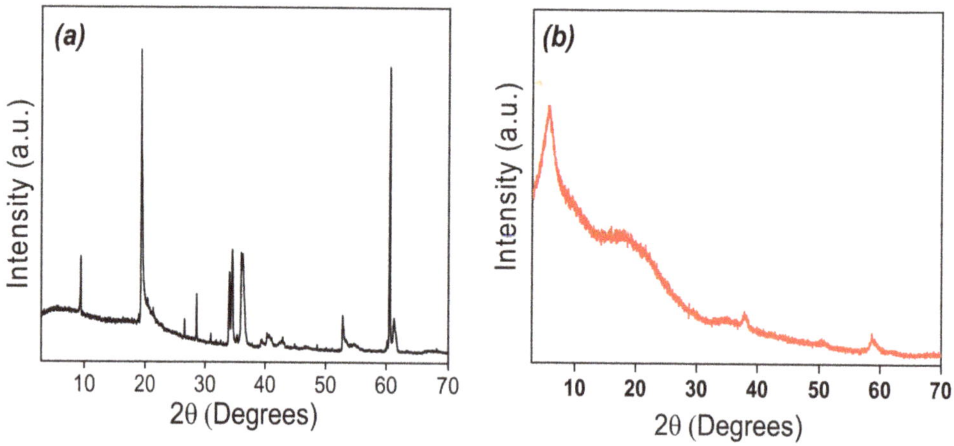

Figure 3. X-ray diffraction patterns of: (**a**) Nat-Talc, and (**b**) $TalcNH_2$.

Figure 4 shows SEM micrographs of Nat-Talc and $TalcNH_2$. The surface morphology of both materials shows a tendency towards aggregation with the microcrystalline character of talc. These observations allowed us to conclude that the structure of Nat-Talc and $TalcNH_2$ is practically the same. The raw talc sample showed a solid microporous structure. The surface of natural talc is mainly made up of aggregates or agglomerates of particles with non-uniform pores either in slits, plates or sharpened particles. After functionalization, the synthesised amine talc retained the same surface shape, with a reduction in textural microporosity probably due to the progressive diffusion of APTES molecules between its pores.

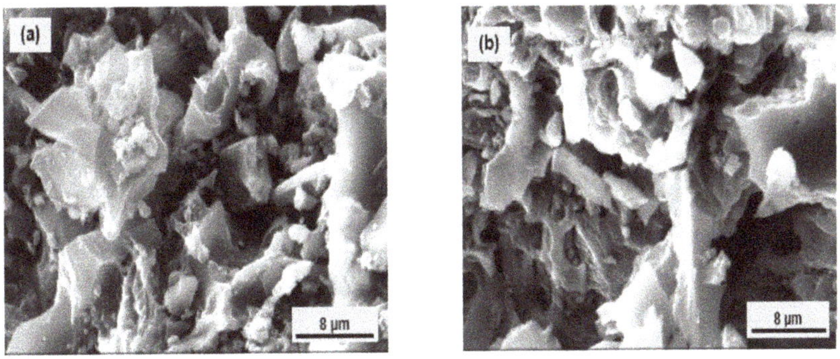

Figure 4. SEM micrographs of: (**a**) Nat-Talc, and (**b**) $TalcNH_2$.

The optical properties of raw talc and synthetic amine clay have generally been elucidated by UV-Visible Diffuse Reflectance Spectroscopy (Figure 5). It provides direct evidence of the transmission, absorption or reflection of light by a material. The UV-Vis spectra showed the strongest bands for two materials (Nat-Talc and $TalcNH_2$) around 300 nm, indicating that they reflect light in this range. The absorbance values decreased from Nat-Talc (0.01 a.u.) to $TalcNH_2$ (0.006 a.u.). The decrease in absorbance is probably related to the incorporation of organosilane into $TalcNH_2$ during the synthesis process and to the reduction of the physical surface area of the material. The light thus interacts with the amino parts of $TalcNH_2$, highlighting their energy absorption capacity.

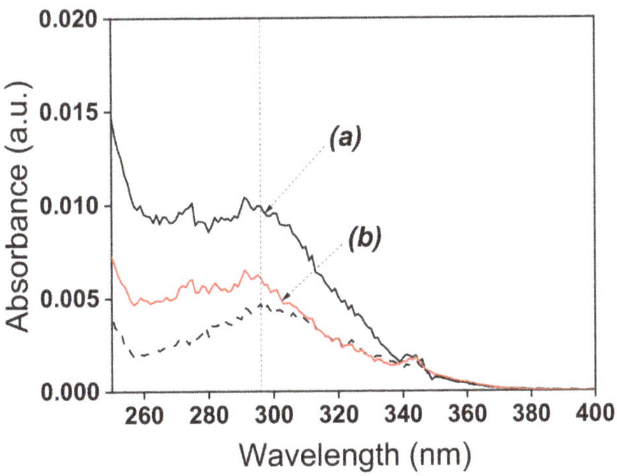

Figure 5. UV-Vis diffuse reflectance spectra of: (**a**) Talc, and (**b**) $TalcNH_2$. The dot line corresponds to the reflectance standard $BaSO_4$.

In summary, the presence of C, H, N in the $TalcNH_2$ material successfully demonstrated the covalent grafting of organosilane groups onto the chemical functional groups of natural talc, results confirmed by infrared spectroscopy showing the main characteristic bands of APTEAS embedded in the tetrahedral silica layers and by the increase in interlayer distances observed on X-ray diffractograms. Based on the literature on talc and considering the physico-chemical characteristics carried out on the virgin talc used in this work, we noticed that this mineral clay is made up of neutral layers stacked on top of each other and connected by van der Waals interactions. The basal faces of the talc do not carry any -OH functional groups or active ions, while the lateral faces carry very few -SiOH and MgOH functions. The latter two chemical functions mentioned behave as Brönsted acids and are more reactive, whereas the basal surface of talc, consisting of Si-O-Si siloxane bonds, has a low Lewis basicity. Due to the chemical composition as well as the lamellar geometry of talc, which gives talc a hydrophobic basal surface, it is very difficult to disperse natural talc in an aqueous medium, in ethanol (96%) and in dilute solutions of acids and alkali hydroxides. Although compounds such as nafion, carboxymethylcellulose (CMC) can increase the solubility of talc, allowing better adhesion of talc on a glassy carbon electrode for electroanalysis applications, our study aims at the possibility of using talc-like material without binder on a solid electrode. The nafion, a membrane and ion exchange resins, or CMC with numerous hydroxyl (-OH) functions and carboxyl groups of interest in electroanalysis [57–63], would automatically modify the electrochemical (ion exchange, charge transfer resistances) and electroanalytical properties (better accumulation of Pb^{2+} ions and consequently a better detection limit in aqueous solution) of talc, when the latter is used to prepare composites. Since these compounds do not solve the problem of hydrophobicity of talc and consequently difficulty to disperse it in water, while the synthesized amino-functionalized talc can be dispersed in aqueous solution and be stable on the glassy carbon electrode, we undertook to study its electrochemical and electroanalytical properties by comparing it to bare GCE.

3.2. *Electrochemical Characterization of Modified GCE by Cyclic Voltammetry*

The electrochemical characterization of the sensor consisting of the GCE modified with $TalcNH_2$ clay ($TalcNH_2$/GCE) was first performed in solution with neutral pH using cyclic voltammetry (CV). Negatively charged ($[Fe(CN)_6]^{3-}$) and positively charged ($[Ru(NH_3)_6]^{3+}$) redox probes were probes were used to verify and confirm the ion exchange properties of $TalcNH_2$. The analysis of $[Fe(CN)_6]^{3-}$ was carried out within a potential window ranging from −0.15 V to 0.7 V in 0.1 M KCl and the results are depicted on Figure 6a. As can be seen, the CV peak current of the first scan on $TalcNH_2$/GCE (Ipa = 4.91 µA and Ipc = 5.15 µA) is lower compared to that observed on bare GCE (Ipa = 8.07 µA and Ipc = 8.93 µA). This is due to a gradual increase of ions on the electrode surface through the binding sites or pores of the electrode material. Not all $[Fe(CN)_6]^{3-}$ ion binding sites on the electrode surface have the same accessibility. An increase in CV peak currents is then observed up to the 20th scan (Ipa = 24.5 µA and Ipc = 23.5 µA). The high peak currents (anodic and cathodic) and a strong accumulation up to the 20th scan could be due to the electrostatic attraction between the negatively charged redox system $[Fe(CN)_6]^{3-}$ and the positively charged amino ($-NH_3^+$) groups present on the organosilane fragments used in the synthesis process. Several recently published works have highlighted the good affinity that protonated amine functions have in an acid medium to interact (through electrostatic interactions) with either negatively charged molecules or those having free pairs of electrons capable of reacting with other species in solution [64–66].

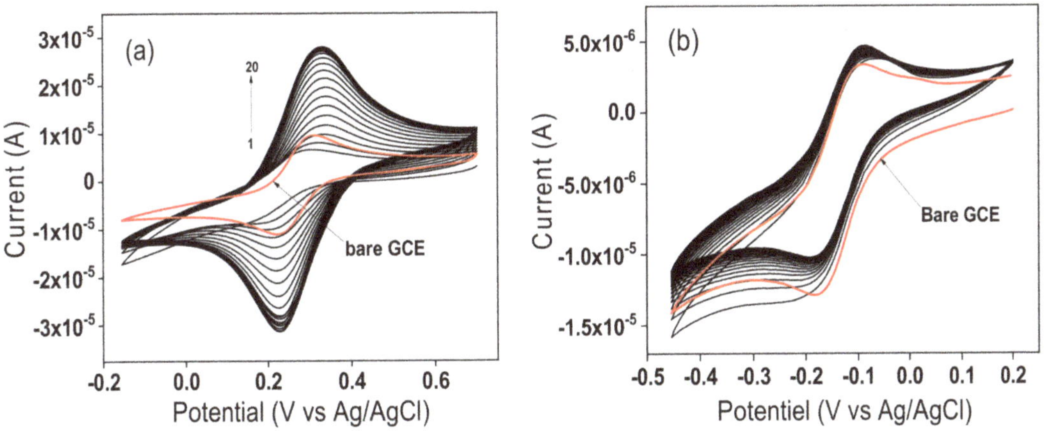

Figure 6. Multisweep cyclic voltammograms recorded in 0.1 M KCl containing (**a**) 1 mM $[Fe(CN)_6]^{3-}$ and (**b**) 1 mM $[Ru(NH_3)_6]^{3+}$ on $TalcNH_2/GCE$. The red curve in (**a**,**b**) corresponds to the probe signal recorded on the bare GCE, v = 50 mV/s.

Similarly, the analysis of $[Ru(NH_3)_6]^{3+}$ ions was carried out in the potential range from −0.6 V to 0.2 V in 0.1 M KCl, and the results are presented on Figure 6b. The comparison of the signals revealed that the oxidation and reduction peaks are more intense on the bare GCE (Ipa 5.490 μA, Ipc 6.187 μA) compared to $TalcNH_2/GCE$ (Ipa 2.110 μA, Ipc 2.164 μA). The lower peak current and non-accumulation obtained on $TalcNH_2/GCE$ could be explained by the fact that the synthesised and protonated $TalcNH_2$ film in solution acted as an electrostatic barrier, preventing the absorption of the cationic species $[Ru(NH_3)_6]^{3+}$. These electrochemical results support the results obtained during the physico-chemical characterisations which confirmed the effectiveness of the synthesis of a talc-like material and the incorporation of APTES into the structure of the synthesised product.

3.3. Determination of Electroactive Surface Area

The electrochemically active surface areas of bare GCE and $TalcNH_2/GCE$ were estimated, using the $[Fe(CN)_6]^{3-/4-}$ redox system and applying the Randles–Sevcik for a reversible Equation (1):

$$I_p = (2.69 \times 10^5) A n^{3/2} D^{1/2} C V^{1/2} \quad (1)$$

where I_p is the peak current, A the electrode electroactive area (cm^2), n the number of electrons transferred, D the diffusion coefficient of $[Fe(CN)_6]^{3-}$ in a 0.1 M KCl solution, C the concentration of $[Fe(CN)_6]^{3-}$ ($mol \cdot cm^{-3}$) and v the potential scan rate ($V \cdot s^{-1}$) [67]. Cyclic voltammetry experiments at different scan rates, as shown on Figures 7 and 8 were performed, and the obtained slopes of the Ip (peak current) vs. $v^{1/2}$ plots for the $[Fe(CN)_6]^{3-}$ oxidation process were 4.29×10^{-5} $A \cdot v^{-1/2} \cdot s^{1/2}$ for the GCE and 5.59×10^{-5} $A \cdot v^{-1/2} \cdot s^{1/2}$ for the $TalcNH_2/GCE$. From the D value for $[Fe(CN)_6]^{3-}$ equal to 7.6×10^{-6} $cm^2 \cdot s^{-1}$ [68], the corresponding electroactive areas were (0.058 ± 0.001) and (0.075 ± 0.002) cm^2 for the GCE and $TalcNH_2/GCE$, respectively. From these results, we can conclude that the electrochemical response of the $[Fe(CN)_6]^{3-}$ probe is affected by modification on the GCE, with the $TalcNH_2$ material allowed an increase in the electroactive area by 1.3 fold compared to the GCE.

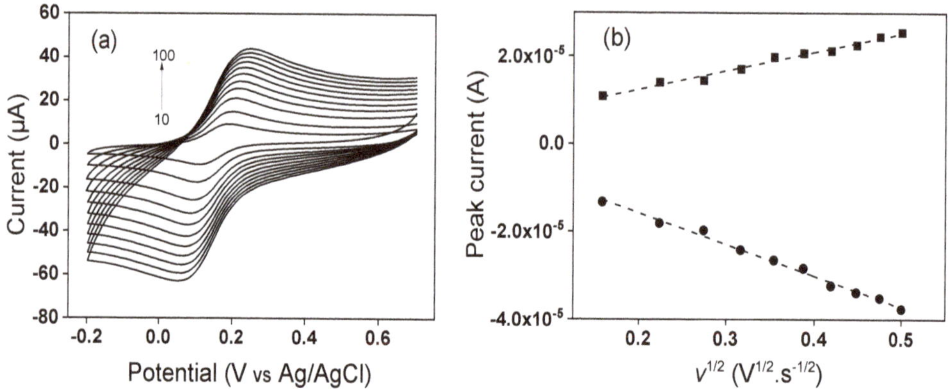

Figure 7. (**a**) Cyclic voltammograms at various scan rates (10 to 100 $mV \cdot s^{-1}$) recorded in 0.1 M KCl containing 1 mM $[Fe(CN)_6]^{3-}$ on bare GCE, and (**b**) linear plot of Ip (peak current) vs. $v^{1/2}$.

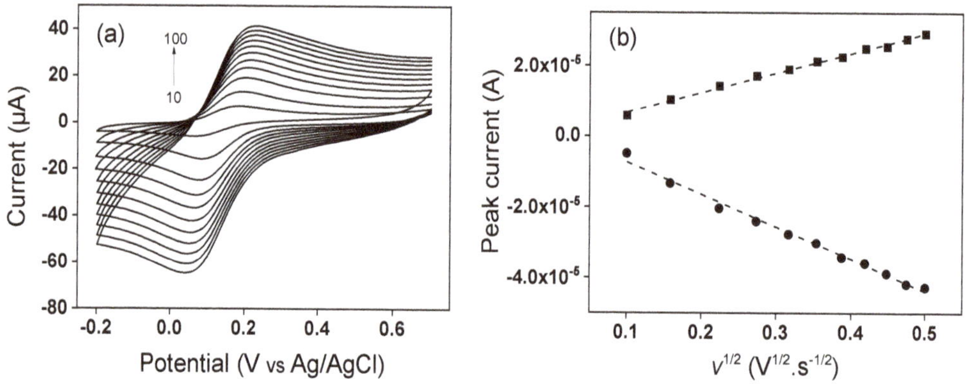

Figure 8. (**a**) Cyclic voltammograms at various scan rates (10 to 100 $mV \cdot s^{-1}$) recorded in 0.1 M KCl containing 1 mM $[Fe(CN)_6]^{3-}$ on TalcNH$_2$/GCE, and (**b**) linear plot of Ip (peak current) vs. $v^{1/2}$.

3.4. Impedance Characterization

Using electrochemical impedance spectroscopy (EIS), it is possible to elucidate the heterogeneous electron transfer properties at the electrode-solution interface. The EIS curves display semi-circular and linear areas corresponding respectively to a process limited by electron transfer and diffusion. The diameter of semicircle is equal to the charge transfer resistance (R_{ct}) [69]. Figure 9 shows the Nyquist plots recorded at bare GCE and TalcNH$_2$/GCE in 0.1 M KCl containing 1 mM $[Fe(CN)_6]^{3-/4-}$. On the bare GCE, the value of Rct was 2725.7 Ω (Figure 9a). On TalcNH$_2$/GCE, the value of Rct was decreased to 186.8 Ω (Figure 9b), which was much smaller than that of TalcNH$_2$/GCE, indicating higher electron transfer for the TalcNH$_2$-modified electrode. It was due to the presence of high conductive amino-synthetic material on GCE. As known, higher electron transfer and charge carrier density will increase the sensitivity of the target sensor. All these results indicated that Pb^{2+} ions can be successfully reduced on the surface of TalcNH$_2$/GCE. The higher electrocatalytic behavior of the TalcNH$_2$/GCE was confirmed by its lower charge transfer resistance. The EIS was also used to calculate the standard heterogeneous rate constant for the two electrodes in accordance with Equation (2) [50]:

$$k^o = \frac{RT}{F^2 R_{ct} A C} \quad (2)$$

where k° is the standard heterogeneous electron transfer rate constant ($cm \cdot s^{-1}$), R the universal gas constant ($8.314\ J \cdot K^{-1} \cdot mol^{-1}$), T the thermodynamic temperature (298.15 K), F the Faraday constant ($96485\ C \cdot mol^{-1}$), R_{ct} the electron transfer resistance (Ω), A the electrode surface area (cm^2) and C the concentration of the $[Fe(CN)_6]^{3-/4-}$ solution ($10^{-6}\ mol \cdot cm^{-3}$).

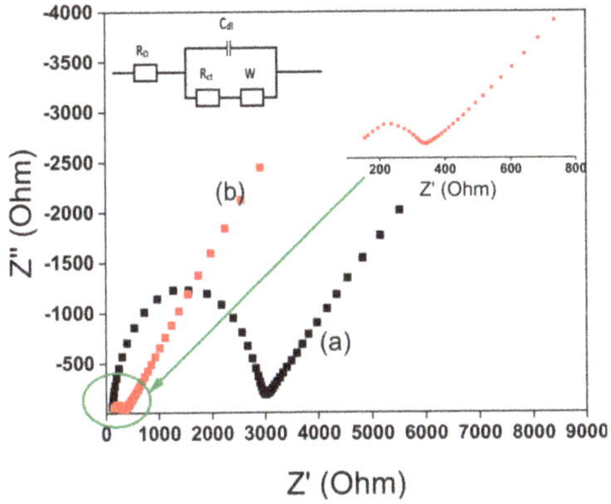

Figure 9. Electrochemical impedance spectroscopy (EIS) of bare GCE (**a**) and $TalcNH_2/GCE$ (**b**) in 0.1 M KCl containing 1 mM $K_3Fe(CN)_6/K_4Fe(CN)_6$. Frequency range: 0.01 Hz–10 kHz.

The k° values were obtained for the bare GCE ($0.0017 \pm 0.0003\ cm \cdot s^{-1}$) and $TalcNH_2/GCE$ ($0.019 \pm 0.001\ cm \cdot s^{-1}$). The k° values give information on the kinetic ease of a reaction process. A system with a low k° value will reach equilibrium faster than a system with a high k° value. The k° value is greater on $TalcNH_2/GCE$ than on GCE, indicating a faster electron transfer on this electrode, a property that is very beneficial in electrochemistry both in terms of energy savings and analysis time.

3.5. Electrochemical Behavior of Pb^{2+} Ions at $TalcNH_2/GCE$

After the characterization of the $TalcNH_2$ and the $TalcNH_2/GCE$, the ability of the modified electrode to detect Pb^{2+} ions was investigated and preliminary experiments were carried out. The comparison between the anodic stripping differential pulse voltammetry responses on a bare GCE (curve a) and $TalcNH_2/GCE$ (curve b) is shown on Figure 10. It indicated that the sensitivity was higher on the $TalcNH_2/GCE$.

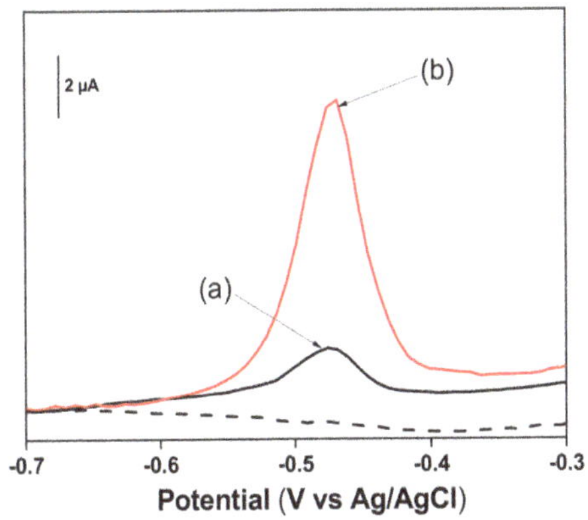

Figure 10. Anodic stripping differential pulse voltammetry (ASDPV) curves obtained on (**a**) GCE, and (**b**) $TalcNH_2/GCE$ of 5 μM Pb^{2+} in 0.1 M HCl after 30 s electrolysis at −0.8 V. Accumulation time: 1 min.

The results demonstrated that the peak height of Pb^{2+} ions on $TalcNH_2$/GCE was about 3 times more important than that obtained on bare GCE. This shows the better affinity between Pb^{2+} ions and $TalcNH_2$ synthesized clay which can be explained by the presence of amino functional groups $-NH_2$ on its surface and in its structure that can easily chelate Pb^{2+} ions.

3.6. Optimization of the Experimental Conditions for the Detection of Lead Ions atTalcNH$_2$/GCE

To achieve a best detection of Pb^{2+} ions with the $TalcNH_2$ modified GCE, parameters such as pH of the accumulation medium, the concentration of the detection medium, the deposition potential and the deposition time were optimised.

The pH of the accumulation or detection medium can influence the electrochemical response of the modified electrodes with respect to the detection of heavy metal ions [70]. The ionization of functional groups on the $TalcNH_2$ clay surface depends on the pH of the solution. At pH < pKa (the pKa of amines being around 9–10), most of these functional groups are mainly in ionized form (protonated amine) and can exchange H^+ with metal ions in solution. The effect of the pH of the accumulation medium (from 1 to 9) on the peak current was studied in the range of 1.0 to 9.0, in order to find the right pH value to define the optimal values of detection of lead ions. The results were shown in Figure 11a.

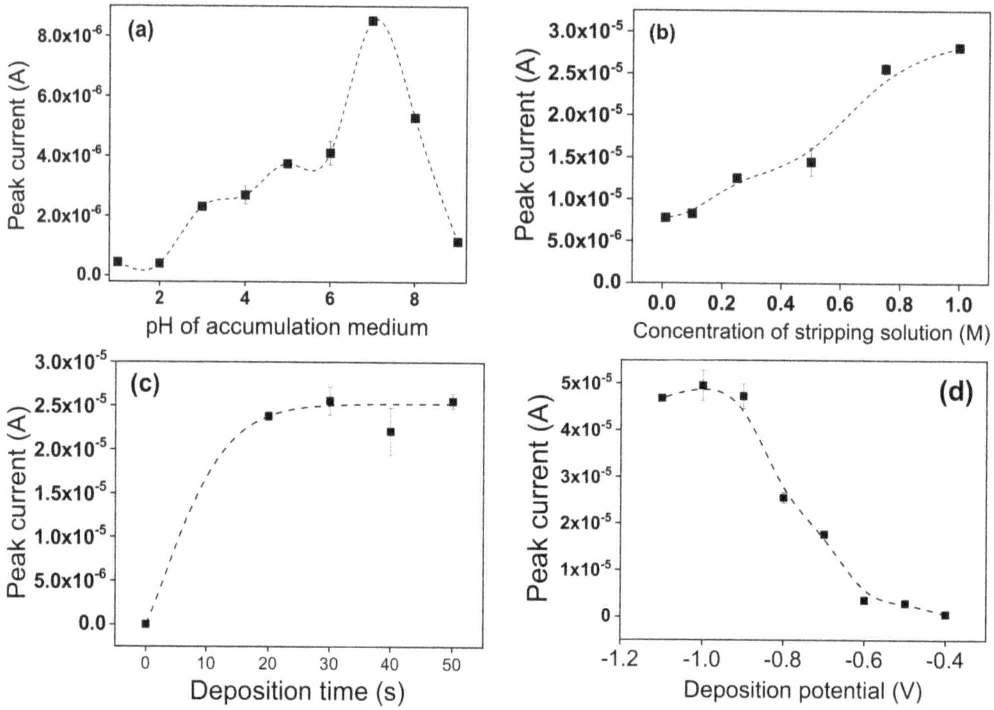

Figure 11. Influence of pH of the accumulation medium (**a**), the concentration of the stripping solution (**b**), the deposition time (**c**) and the deposition potential (**d**), on the stripping current response of $TalcNH_2$/GCE (accumulation time of 1 min in 5 μM Pb^{2+}). The experiments were conducted in triplicate.

The results (Figure 11a) showed the current peaks are low for pH values up 5.0 due to its mainly protonated state (protonated amine groups) of $TalcNH_2$ material that can prevent the fixation of lead ions (predominant in accumulation medium) on its surface via electrostatic repulsions forces. The absorption of H^+ (H^+ coming from hydrochloric acid

(HCl) used) hinders the adsorption of Pb^{2+}, as it is a stronger acid than Pb^{2+}. However, the peak currents increase when the pH increase from 1.0 to 7.0, and further decrease up to a pH value of 9.0. Yet, in the pH range 3.0–7.0, polymeric hydroxocomplexes of lead, mainly $Pb(OH)^-$ and $Pb(OH)_2$ predominate due to hydrolysis reactions [71], which inhibit the accumulation of lead ions [72]. The pH value 7 was chosen as optimal value for further studies. The pH range from 6.0 to 7.0 corresponds approximately to the beginning of the formation of the first lead monohydroxide complex $Pb(OH)^+$ and to a low concentration of H^+ in solution. $Pb(OH)^+$ can therefore be easily absorbed by $TalcNH_2$ when the concentration of H^+ becomes much lower.

The concentration of detection medium (H_3O^+, Cl^-) was studied from 0.01 M to 1 M on $TalcNH_2$/GCE and the results were presented in Figure 11b. As shown, the electrode response was less between 0.01 M and 0.1 M due to the smaller amount of H^+ ions in the solution capable of weakening the complexes formed during the accumulation stage. After 0.1 M, the electrode response increased significantly and was more quantitative at 1 M. This may be due to the increase of H^+ ions in solution. A concentration of 1 M of HCl was selected as the stripping medium for further studies. The deposition time was studied in the range of 0 to 45 s. The results in Figure 11c showed that the stripping current intensities of Pb^{2+} ions increased up to 30 s and almost reached the maximum at 30 s. Regarding sensitivity and measurement time in practice, 30 s was retained as the optimal deposition time for further investigations. To obtain the best sensitivity for $TalcNH_2$/GCE, the deposition potential was studied from −0.4 to −1.1 V. The results presented in Figure 11d showed higher peak currents between −1.1 V and −0.9 V. The electrode response significantly decreased when the deposition potential shifted from −0.9 V to −0.4 V, due to lower amount of energy required to reduce Pb^{2+} ions. In the following study, −0.9 V was used as the optimal deposition potential.

3.7. Detection Limit, Interference Study and Analytical Application of the Developed Sensor

Under the optimal conditions, Figure 12a showed that the anodic stripping peak currents of Pb^{2+} increased with concentration in the range of 0.8 μM to 2.5 μM (curve 1 to curve 6). The calibration curve (insert of Figure 12a) was linear over the studied concentration range, with the equation $Ip(A) = 3.6[Pb^{2+}](M) + (2.66 \times 10^{-6})$ ($R^2 = 0.997$). The detection limit (D_L) and quantitation limit (Q_L) were calculated from $D_L = 3\ S_b/m$ [73,74], and $Q_L = 10\ S_b/m$, where S_b is the standard deviation of the blank and m is the slope of the calibration curve. D_L and Q_L were estimated to be 7.45×10^{-8} M and 24.84×10^{-8} M, respectively, and the sensitivity of the method of 3.6 $\mu A.\mu M^{-1}$ was obtained. The performances of the sensor developed were compared with some other electrochemical sensors of Pb^{2+} ions (Table 2). Prior to the application of $TalcNH_2$/GCE as a sensing device, the reproducibility of the electrode was checked by performing five successive measurements of 5 μM Pb^{2+} on different electrodes (GCE surface was renewed between successive runs). A relative standard deviation of 5.65% was obtained, showing the stability, repeatability and reproducibility of $TalcNH_2$/GCE (Figure 12b).

Figure 13 shows the DPASV responses toward 5 μM Pb^{2+} in the presence of 5 μM Cu^{2+}, Cd^{2+}, Zn^{2+} and Hg^{2+} under optimal experimental conditions. It was observed that the presence of Cd^{2+} and Zn^{2+} does not significantly influence the signal of Pb^{2+} ions. However, a decrease in the peak current of Pb^{2+} ions was observed upon addition of Cu^{2+}, due to its high affinity for protonated amine groups on the synthesized $TalcHN_2$. A significant increase in peak current was observed when Hg^{2+} were added, due to code position of Hg(0) during the detection step. It was found that some other ions, Na^+, K^+, Cl^-, $SO_4{}^{2-}$ and $NO_3{}^-$, which were not listed in Figure 13, did not interfere with the signal of Pb^{2+}.

The applicability of $TalcNH_2$/GCE as a sensing device was evaluated for the detection of traces of lead in tap water sample. Tap water was used to prepare accumulation medium at a final Pb^{2+} concentration of 5 μM. After analysis under optimal conditions, a recovery amount of 95.65% of added Pb^{2+} ions was obtained. A high recovery obtained indicated that the $TalcNH_2$/GCE can be successfully applied to analyse water polluted by lead ions.

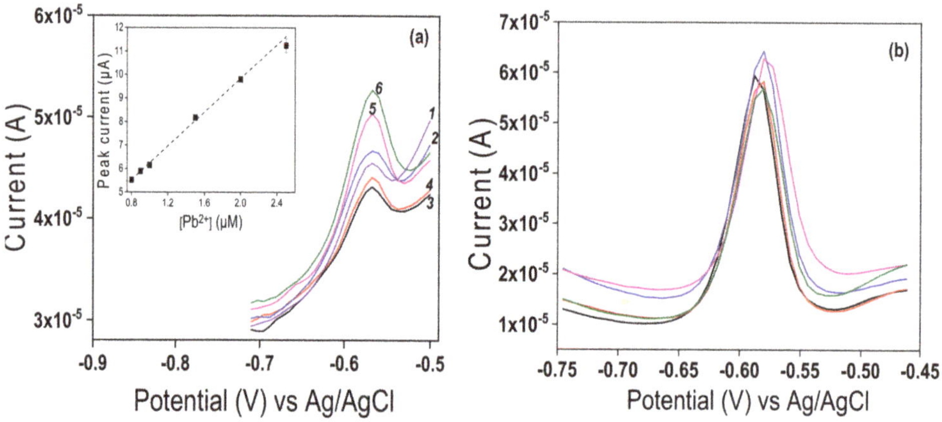

Figure 12. (**a**) Differential pulse anodic stripping voltammetry (DPASV) response of the TalcNH$_2$/GCE at different concentrations of Pb^{2+} ions under optimal conditions (insert corresponding to calibration curve). The experiments were conducted in triplicate. (**b**) Series of 5 DPASV response of 5 μM Pb^{2+}, recorded in 0.1 M HCl after 1 min accumulation in pH 6 aqueous solution.

Table 2. Comparison of the performance of some Pb^{2+} ion sensors.

Electrode	D_L (μM)	Linear Range (μM)	Reference
$Ca_{10}(PO_4)_6(OH)_2$ modified CPE *	0.0768	0.002–0.24	[75]
Nanosized HA nafion modified GCE	0.001	0.005–0.8	[76]
GCE covered by engineered MWCNT **	0.0188	0.4–80	[77]
GCE covered by zinc oxide nanofibers functionalized by L-cysteine	0.0012	0.03–0.42	[78]
CPE modified with magnetic eggshell nanocomposite and MWCNT	0.452	1.5–600	[79]
TalcNH$_2$/GCE	0.0745	0.8–2.5	This work

* CPE: Carbon paste electrode, ** MWCNT: Multi-walled carbon nanotubes.

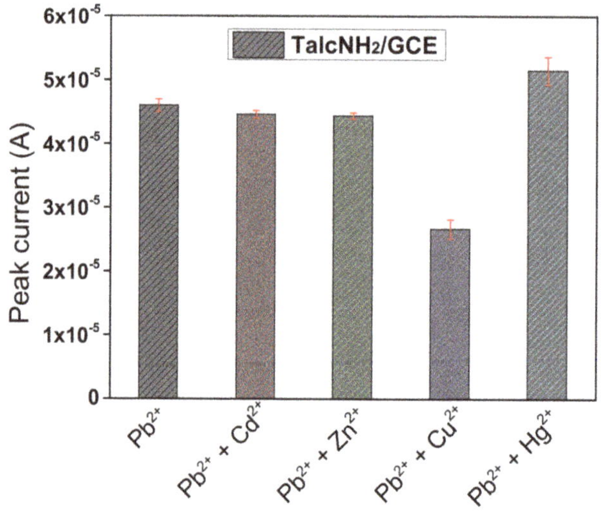

Figure 13. Interference studies of TalcNH$_2$/GCE at 5 μM Pb^{2+} ions in the presence of 5 μM Cu^{2+}, Cd^{2+}, Zn^{2+} and Hg^{2+} ions under optimal experimental conditions. Experiments were performed in triplicate.

4. Conclusions

Overall, this study has shown that it is possible to use a synthetic talc-type clay, easily functionalized with amine chelating groups as electrode material to improve the sensitivity and selectivity of the resulting modified electrodes. Indeed, a synthetic talc-like magnesium phyllosilicate amino-functionalized was obtained from the sol–gel process. Infrared spectroscopy, X-ray diffraction, CHN elemental analysis, SEM, BET method and UV–Vis diffuse reflectance spectroscopy study results confirmed the synthesis process. We have demonstrated that a $TalcNH_2$/GCE can be used for the analysis of Pb^{2+} ions by stripping voltammetry. The ability of the $TalcNH_2$ to complex Pb^{2+} ions onto the electrode was precise and accurate. After optimization of the factors affecting the pre-concentration and stripping steps, a detection limit of 7.45×10^{-8} M was obtained. The amino-functionalized talc/glassy carbon electrode prepared in this study was applied in the determination of Pb^{2+} ions in tap water with good results. This study demonstrates the potential of $TalcNH_2$ synthetic material, a hydrophilic talc-like magnesium phyllosilicate amino-functionalized to serve as electrode material for the preparation of sensing devices dedicated to the detection of heavy metals.

Author Contributions: Conceptualization, S.L.Z.J., A.O.-M. and I.K.T.; validation, S.L.Z.J., A.O.-M. and I.K.T.; investigation, C.N.P., S.L.Z.J., A.K.T., I.D. and G.D.; data curation, S.L.Z.J., C.N.P., A.K.T. and I.D.; writing-original draft preparation, C.N.P. and S.L.Z.J.; writing-review and editing, S.L.Z.J., I.D., A.K.T., A.O.-M., I.K.T. and E.N.; project administration, I.K.T., A.O.-M. and E.N.; funding acquisition, I.K.T. and A.O.-M. All authors have read and agreed to the published version of the manuscript.

Funding: This work was financially supported by the Alexander von Humboldt Foundation (Germany), and by the Emmy Noether Programme of the German Research Foundation DFG (Grant number: OS 497/6−1). The article processing charge (APC) was funded by the Baden-Wuerttemberg Ministry of Science, Research, the Arts (MWK), and the University of Freiburg in the funding program Open Access Publishing.

Data Availability Statement: Not applicable.

Acknowledgments: Wolfgang Bensch, Huayna Terraschke and Felix Tuczek (Christian-Albrechts-Universität zu Kiel, Germany) are thanked for their technical support. We thank the Emmy Noether Program of the German Research Foundation DFG (grant no. OS 497/6−1), and the Alexander von Humboldt Foundation (Germany) for funding. A.K.T. and S.L.Z.J. thank the German Academic Exchange Service (DAAD) for financial support.

Conflicts of Interest: The authors declare no conflict of interest.

References

1. Guggenheim, S.; Adams, J.M.; Bain, D.C.; Bergaya, F.; Brigatti, M.F.; Drits, V.A.; Formoso, M.L.L.; Galán, E.; Kogure, T.; Stanjek, H. Summary of recommendations of nomenclature committees relevant to clay mineralogy: Report of the Association Internationale pour l'Etude des Argiles (AIPEA) Nomenclature Committee for 2006. *Clays Clay Miner.* **2006**, *54*, 761–772. [CrossRef]
2. Wang, J.; Somasundaran, P. Adsorption and conformation of carboxymethyl cellulose at solid–liquid interfaces using spectroscopic, AFM and allied techniques. *J. Colloid Interface Sci.* **2005**, *291*, 75–83. [CrossRef] [PubMed]
3. Lugwisha, E.H.J. Properties of fired bodies made from Tanzanian talc-clay mixes for ceramic applications. *Tanzan. J. Sci.* **2009**, *32*, 69–79. [CrossRef]
4. Ahmed, M.M.; Ibrahim, G.A.; Hassan, M.M.A. Improvement of Egyptian talc quality for industrial uses by flotation process and leaching. *Int. J. Miner. Process.* **2007**, *83*, 132–145. [CrossRef]
5. Temuujin, J.; Okada, K.; Jadambaa, T.; Mackenzie, K.J.D.; Amarsanaa, J. Effect of grinding on the preparation of porous material from talc by selective leaching. *J. Mater. Sci. Lett.* **2002**, *21*, 1607–1609. [CrossRef]
6. Sanchez-Soto, P.J.; Wiewióra, A.; Avilés, M.A.; Justo, A.; Pérez-Maqueda, L.A.; Pérez-Rodríguez, J.L.; Bylina, P. Talc from Puebla de Lillo, Spain. II. Effect of dry grinding on particle size and shape. *Appl. Clay Sci.* **1997**, *12*, 297–312. [CrossRef]
7. Wallqvist, V.; Claesson, P.M.; Swerin, A.; Schoelkopf, J.; Gane, P.A.C. Influence of Wetting and Dispersing Agents on the Interaction between Talc and Hydrophobic Particles. *Langmuir* **2009**, *25*, 6909–6915. [CrossRef]
8. Da Fonseca, M.G.; Airoldi, C. New amino-inorganic hybrids from talc silylation and copper adsorption properties. *Mater. Res. Bull.* **2001**, *36*, 277–287. [CrossRef]
9. Wesołowski, M. Thermal decomposition of talc: A review. *Thermochim. Acta* **1984**, *78*, 395–421. [CrossRef]

10. Feng, B.; Peng, J.; Guo, W.; Zhang, W.; Ai, G.; Wang, H. The effect of changes in pH on the depression of talc by chitosan and the associated mechanisms. *Powder Technol.* **2018**, *325*, 58–63. [CrossRef]
11. Yi, H.; Zhao, Y.; Rao, F.; Song, S. Hydrophobic agglomeration of talc fines in aqueous suspensions. *Colloids Surf. A Physicochem. Eng. Asp.* **2018**, *538*, 327–332. [CrossRef]
12. Boyd, S.A.; Lee, J.-F.; Mortland, M.M. Attenuating organic contaminant mobility by soil modification. *Nature* **1988**, *333*, 345–347. [CrossRef]
13. Da Fonseca, M.G.; Silva, C.R.; Airoldi, C. Aminated Phyllosilicates Synthesized via a Sol–Gel Process. *Langmuir* **1999**, *15*, 5048–5055. [CrossRef]
14. Fukushima, Y.; Tani, M. An organic/inorganic hybrid layered polymer: Methacrylate–magnesium(nickel) phyllosilicate. *J. Chem. Soc. Chem. Commun.* **1995**, *63*, 241–242. [CrossRef]
15. Burkett, S.L.; Press, A.; Mann, S. Synthesis, Characterization, and Reactivity of Layered Inorganic–Organic Nanocomposites Based on 2:1 Trioctahedral Phyllosilicates. *Chem. Mater.* **1997**, *9*, 1071–1073. [CrossRef]
16. Fonseca, M.G.; Airoldi, C. Mercaptopropyl magnesium phyllosilicate—Thermodynamic data on the interaction with divalent cations in aqueous solution. *Thermochim. Acta* **2000**, *359*, 1–9. [CrossRef]
17. Fukushima, Y.; Tani, M. Synthesis of 2:1 Type 3-(Methacryloxy) propyl Magnesium (Nickel) Phyllosilicate. *Bull. Chem. Soc. Jpn.* **1996**, *69*, 3667–3671. [CrossRef]
18. Silva, C.R.; Fonseca, M.G.; Barone, J.S.; Airoldi, C. Layered Inorganic–Organic Talc-like Nanocomposites. *Chem. Mater.* **2002**, *14*, 175–179. [CrossRef]
19. Da Fonseca, M.G.; Airoldi, C. New layered inorganic–organic nanocomposites containing n-propylmercapto copper phyllosilicates. *J. Mater. Chem.* **2000**, *10*, 1457–1463. [CrossRef]
20. Da Fonseca, M.G.; da Silva Filho, E.C.; Machado Junior, R.S.A.; Arakaki, L.N.H.; Espinola, J.G.P.; Airoldi, C. Zinc phyllosilicates containing amino pendant groups. *J. Solid State Chem.* **2004**, *177*, 2316–2322. [CrossRef]
21. Minet, J.; Abramson, S.; Bresson, B.; Sanchez, C.; Montouillout, V.; Lequeux, N. New Layered Calcium Organosilicate Hybrids with Covalently Linked Organic Functionalities. *Chem. Mater.* **2004**, *16*, 3955–3962. [CrossRef]
22. Mizutani, T.; Fukushima, Y.; Okada, A.; Kamigaito, O. Synthesis of nickel and magnesium phyllosilicates with 1:1 and 2:1 layer structures. *Bull. Chem. Soc. Jpn.* **1990**, *63*, 2094–2098. [CrossRef]
23. Sasai, R.; Itoh, H.; Shindachi, I.; Shichi, T.; Takagi, K. Photochromism of Clay–Diarylethene Hybrid Materials in Optically Transparent Gelatin Films. *Chem. Mater.* **2001**, *13*, 2012–2016. [CrossRef]
24. Da Fonseca, M.G.; Silva, C.R.; Barone, J.S.; Airoldi, C. Layered hybrid nickel phyllosilicates and reactivity of the gallery space. *J. Mater. Chem.* **2000**, *10*, 789–795. [CrossRef]
25. Toeri, J.; Osorio-Madrazo, A.; Laborie, M.-P. Preparation and Chemical/Microstructural Characterization of Azacrown Ether-Crosslinked Chitosan Films. *Materials* **2017**, *10*, 400. [CrossRef]
26. Järup, L. Hazards of heavy metal contamination. *Br. Med. Bull.* **2003**, *68*, 167–182. [CrossRef]
27. Elliott, P.; Arnold, R.; Barltrop, D.; Thornton, I.; House, I.M.; Henry, J.A. Clinical lead poisoning in England: An analysis of routine sources of data. *Occup. Environ. Med.* **1999**, *56*, 820–824. [CrossRef]
28. Anthemidis, A.N.; Ioannou, K.-I.G. Development of a sequential injection dispersive liquid–liquid microextraction system for electrothermal atomic absorption spectrometry by using a hydrophobic sorbent material: Determination of lead and cadmium in natural waters. *Anal. Chim. Acta* **2010**, *668*, 35–40. [CrossRef]
29. Wan, Z.; Xu, Z.; Wang, J. Flow injection on-line solid phase extraction for ultra-trace lead screening with hydride generation atomic fluorescence spectrometry. *Analyst* **2006**, *131*, 141–147. [CrossRef] [PubMed]
30. Hsieh, H.-F.; Chang, W.-S.; Hsieh, Y.-K.; Wang, C.-F. Lead determination in whole blood by laser ablation coupled with inductively coupled plasma mass spectrometry. *Talanta* **2009**, *79*, 183–188. [CrossRef] [PubMed]
31. Arnold, M.A.; Meyerhoff, M.E. Recent Advances in the Development and Analytical Applications of Biosensing Probes. *C R C Crit. Rev. Anal. Chem.* **1988**, *20*, 149–196. [CrossRef]
32. Ngouoko, J.J.K.; Tajeu, K.Y.; Temgoua, R.C.T.; Doungmo, G.; Doench, I.; Tamo, A.K.; Kamgaing, T.; Osorio-Madrazo, A.; Tonle, I.K. Hydroxyapatite/L-Lysine Composite Coating as Glassy Carbon Electrode Modifier for the Analysis and Detection of Nile Blue A. *Materials* **2022**, *15*, 4262. [CrossRef]
33. Sun, D.; Sun, Z. Electrochemical determination of Pb2+ using a carbon nanotube/Nafion composite film-modified electrode. *J. Appl. Electrochem.* **2008**, *38*, 1223–1227. [CrossRef]
34. Yantasee, W.; Lin, Y.; Hongsirikarn, K.; Fryxell, G.E.; Addleman, R.; Timchalk, C. Electrochemical Sensors for the Detection of Lead and Other Toxic Heavy Metals: The Next Generation of Personal Exposure Biomonitors. *Environ. Health Perspect.* **2007**, *115*, 1683–1690. [CrossRef]
35. Tonlé, I.K.; Letaief, S.; Ngameni, E.; Walcarius, A.; Detellier, C. Square Wave Voltammetric Determination of Lead(II) Ions Using a Carbon Paste Electrode Modified by a Thiol-Functionalized Kaolinite. *Electroanalysis* **2011**, *23*, 245–252. [CrossRef]
36. Ngassa, G.B.P.; Tonlé, I.K.; Walcarius, A.; Ngameni, E. One-step co-intercalation of cetyltrimethylammonium and thiourea in smectite and application of the organoclay to the sensitive electrochemical detection of Pb(II). *Appl. Clay Sci.* **2014**, *99*, 297–305. [CrossRef]
37. Jiokeng, S.L.Z.; Dongmo, L.M.; Ymélé, E.; Ngameni, E.; Tonlé, I.K. Sensitive stripping voltammetry detection of Pb(II) at a glassy carbon electrode modified with an amino-functionalized attapulgite. *Sens. Actuators B Chem.* **2017**, *242*, 1027–1034. [CrossRef]

38. Guenang, L.S.; Dongmo, L.M.; Jiokeng, S.L.Z.; Kamdem, A.T.; Doungmo, G.; Tonlé, I.K.; Bassetto, V.C.; Jović, M.; Lesch, A.; Girault, H. Montmorillonite clay-modified disposable ink-jet-printed graphene electrode as a sensitive voltammetric sensor for the determination of cadmium(II) and lead(II). *SN Appl. Sci.* **2020**, *2*, 443. [CrossRef]
39. Sales, J.A.A.; Airoldi, C. Calorimetric investigation of metal ion adsorption on 3-glycidoxypropyltrimethylsiloxane + propane-1,3-diamine immobilized on silica gel. *Thermochim. Acta* **2005**, *427*, 77–83. [CrossRef]
40. Tanev, P.T.; Pinnavaia, T.J. A Neutral Templating Route to Mesoporous Molecular Sieves. *Science* **1995**, *267*, 865–867. [CrossRef]
41. Sales, J.A.A.; Petrucelli, G.C.; Oliveira, F.J.V.E.; Airoldi, C. Some features associated with organosilane groups grafted by the sol–gel process onto synthetic talc-like phyllosilicate. *J. Colloid Interface Sci.* **2006**, *297*, 95–103. [CrossRef] [PubMed]
42. Ukrainczyk, L.; Bellman, R.A.; Anderson, A.B. Template Synthesis and Characterization of Layered Al- and Mg-Silsesquioxanes. *J. Phys. Chem. B* **1997**, *101*, 531–539. [CrossRef]
43. Whilton, N.T.; Burkett, S.L.; Mann, S. Hybrid lamellar nanocomposites based on organically functionalized magnesium phyllosilicate clays with interlayer reactivity. *J. Mater. Chem.* **1998**, *8*, 1927–1932. [CrossRef]
44. Kamdem Tamo, A.; Doench, I.; Morales Helguera, A.; Hoenders, D.; Walther, A.; Madrazo, A.O. Biodegradation of Crystalline Cellulose Nanofibers by Means of Enzyme Immobilized-Alginate Beads and Microparticles. *Polymers* **2020**, *12*, 1522. [CrossRef]
45. Marquez-Bravo, S.; Doench, I.; Molina, P.; Bentley, F.E.; Tamo, A.K.; Passieux, R.; Lossada, F.; David, L.; Osorio-Madrazo, A. Functional Bionanocomposite Fibers of Chitosan Filled with Cellulose Nanofibers Obtained by Gel Spinning. *Polymers* **2021**, *13*, 1563. [CrossRef]
46. Lall, A.; Kamdem Tamo, A.; Doench, I.; David, L.; Nunes de Oliveira, P.; Gorzelanny, C.; Osorio-Madrazo, A. Nanoparticles and Colloidal Hydrogels of Chitosan–Caseinate Polyelectrolyte Complexes for Drug-Controlled Release Applications. *Int. J. Mol. Sci.* **2020**, *21*, 5602. [CrossRef]
47. Djouonkep, L.D.W.; Tamo, A.K.; Doench, I.; Selabi, N.B.S.; Ilunga, E.M.; Lenwoue, A.R.K.; Gauthier, M.; Cheng, Z.; Osorio-Madrazo, A. Synthesis of High Performance Thiophene–Aromatic Polyesters from Bio-Sourced Organic Acids and Polysaccharide-Derived Diol: Characterization and Degradability Studies. *Molecules* **2022**, *27*, 325. [CrossRef]
48. Deussi Ngaha, M.C.; Kougoum Tchieda, V.; Kamdem Tamo, A.; Doungmo, G.; Njanja, E.; Kenfack Tonle, I. Aminoalcohol-functionalization of Alkali Palm Oil Fiber and Application as Electrochemical Sensor for 2-nitrophenol Determination. *Electroanalysis* **2022**, *9*, 5222. [CrossRef]
49. Osorio-Madrazo, A.; David, L.; Peniche-Covas, C.; Rochas, C.; Putaux, J.-L.; Trombotto, S.; Alcouffe, P.; Domard, A. Fine microstructure of processed chitosan nanofibril networks preserving directional packing and high molecular weight. *Carbohydr. Polym.* **2015**, *131*, 1–8. [CrossRef]
50. Silverstein, R.M.; Webster, F.X.; Kiemle, D.J. *Spectrometric Identification of Organic Compounds*, 8th ed.; Wiley: Hoboken, NJ, USA, 2015; ISBN 978-0-470-61637-6.
51. Pavia, D.L.; Lampman, G.M.; Kriz, G.S. *Introduction to Spectroscopy: A Guide for Students of Organic Chemistry*; Saunders College: Philadelphia, PA, USA, 1979; ISBN 9780721671192.
52. Nakamoto, K. *Infrared and Raman Spectra of Inorganic and Coordination Compounds*; John Wiley & Sons, Inc.: Hoboken, NJ, USA, 2008; ISBN 9780470405840.
53. Wilson, M.J. *Clay Mineralogy: Spectroscopic and Chemical Determinative Methods*; Springer: Dordrecht, The Netherlands, 1994. ISBN 978-94-011-0727-3.
54. Fritsch, E.; Balan, E.; Petit, S.; Juillot, F. Structural, textural, and chemical controls on the OH stretching vibrations in serpentine-group minerals. *Eur. J. Miner.* **2021**, *33*, 447–462. [CrossRef]
55. Dongmo, L.M.; Guenang, L.S.; Jiokeng, S.L.Z.; Kamdem, A.T.; Doungmo, G.; Victor, B.C.; Jović, M.; Lesch, A.; Tonlé, I.K.; Girault, H. A new sensor based on an amino-montmorillonite-modified inkjet-printed graphene electrode for the voltammetric determination of gentisic acid. *Mikrochim. Acta* **2021**, *188*, 36. [CrossRef]
56. Ferrage, E.; Martin, F.; Petit, S.; Pejo-soucaille, S.; Micoud, P.; Fourty, G.; Ferret, J.; Salvi, S.; de Parseval, P.; Fortune, J.P. Evaluation of talc morphology using FTIR and H/D substitution. *Clay Miner.* **2003**, *38*, 141–150. [CrossRef]
57. Osorio-Madrazo, A.; Laborie, M.-P. Morphological and Thermal Investigations of Cellulosic Bionanocomposites. In *Biopolymer Nanocomposites*; Dufresne, A., Thomas, S., Pothen, L.A., Eds.; John Wiley & Sons, Inc.: Hoboken, NJ, USA, 2013; pp. 411–436, ISBN 9781118609958.
58. Von Palubitzki, L.; Wang, Y.; Hoffmann, S.; Vidal-Y-Sy, S.; Zobiak, B.; Failla, A.V.; Schmage, P.; John, A.; Osorio-Madrazo, A.; Bauer, A.T.; et al. Differences of the tumour cell glycocalyx affect binding of capsaicin-loaded chitosan nanocapsules. *Sci. Rep.* **2020**, *10*, 22443. [CrossRef]
59. Samyn, P.; Osorio-Madrazo, A. Native Crystalline Polysaccharide Nanofibers: Processing and Properties. In *Handbook of Nanofibers*; Barhoum, A., Bechelany, M., Makhlouf, A., Eds.; Springer International Publishing: Cham, Switzerland, 2018; pp. 1–36, ISBN 978-3-319-42789-8.
60. Lizundia, E.; Costa, C.M.; Alves, R.; Lanceros-Méndez, S. Cellulose and its derivatives for lithium ion battery separators: A review on the processing methods and properties. *Carbohydr. Polym. Technol. Appl.* **2020**, *1*, 100001. [CrossRef]
61. Abushammala, H.; Pontes, J.F.; Gomes, G.H.; Osorio-Madrazo, A.; Thiré, R.M.S.M.; Pereira, F.V.; Laborie, M.-P.G. Swelling, viscoelastic, and anatomical studies on ionic liquid-swollen Norway spruce as a screening tool toward ionosolv pulping. *Holzforschung* **2015**, *69*, 1059–1067. [CrossRef]

62. Wei, J.; Zhou, Y.; Lv, Y.; Wang, J.; Jia, C.; Liu, J.; Zhang, X.; Sun, J.; Shao, Z. Carboxymethyl Cellulose Nanofibrils with a Treelike Matrix: Preparation and Behavior of Pickering Emulsions Stabilization. *ACS Sustain. Chem. Eng.* **2019**, *7*, 12887–12896. [CrossRef]
63. Bentley, F.E.; Passieux, R.; David, L.; Osorio-Madrazo, A. Pure Chitosan Biomedical Textile Fibers from Mixtures of Low- and High-Molecular Weight Bidisperse Polymer Solutions: Processing and Understanding of Microstructure–Mechanical Properties' Relationship. *Int. J. Mol. Sci.* **2022**, *23*, 4767. [CrossRef]
64. Amine, S.; Montembault, A.; Fumagalli, M.; Osorio-Madrazo, A.; David, L. Controlled Polyelectrolyte Association of Chitosan and Carboxylated Nano-Fibrillated Cellulose by Desalting. *Polymers* **2021**, *13*, 2023. [CrossRef]
65. Doench, I.; Torres-Ramos, M.E.W.; Montembault, A.; Nunes de Oliveira, P.; Halimi, C.; Viguier, E.; Heux, L.; Siadous, R.; Thiré, R.M.S.M.; Osorio-Madrazo, A. Injectable and Gellable Chitosan Formulations Filled with Cellulose Nanofibers for Intervertebral Disc Tissue Engineering. *Polymers* **2018**, *10*, 1202. [CrossRef]
66. Kamdem Tamo, A.; Doench, I.; Walter, L.; Montembault, A.; Sudre, G.; David, L.; Morales-Helguera, A.; Selig, M.; Rolauffs, B.; Bernstein, A.; et al. Development of Bioinspired Functional Chitosan/Cellulose Nanofiber 3D Hydrogel Constructs by 3D Printing for Application in the Engineering of Mechanically Demanding Tissues. *Polymers* **2021**, *13*, 1663. [CrossRef]
67. Opallo, M.; Lesniewski, A. A review on electrodes modified with ionic liquids. *J. Electroanal. Chem.* **2011**, *656*, 2–16. [CrossRef]
68. Goyal, R.N.; Gupta, V.K.; Chatterjee, S. Voltammetric biosensors for the determination of paracetamol at carbon nanotube modified pyrolytic graphite electrode. *Sens. Actuators B Chem.* **2010**, *149*, 252–258. [CrossRef]
69. Martin Santos, A.; Wong, A.; Araújo Almeida, A.; Fatibello-Filho, O. Simultaneous determination of paracetamol and ciprofloxacin in biological fluid samples using a glassy carbon electrode modified with graphene oxide and nickel oxide nanoparticles. *Talanta* **2017**, *174*, 610–618. [CrossRef]
70. Wang, Y.; Wang, L.; Huang, W.; Zhang, T.; Hu, X.; Perman, J.A.; Ma, S. A metal–organic framework and conducting polymer based electrochemical sensor for high performance cadmium ion detection. *J. Mater. Chem. A* **2017**, *5*, 8385–8393. [CrossRef]
71. Bouwe, R.G.B.; Tonle, I.K.; Letaief, S.; Ngameni, E.; Detellier, C. Structural characterisation of 1,10-phenanthroline–montmorillonite intercalation compounds and their application as low-cost electrochemical sensors for Pb(II) detection at the sub-nanomolar level. *Appl. Clay Sci.* **2011**, *52*, 258–265. [CrossRef]
72. Wei, Y.; Gao, C.; Meng, F.-L.; Li, H.-H.; Wang, L.; Liu, J.-H.; Huang, X.-J. SnO_2/Reduced Graphene Oxide Nanocomposite for the Simultaneous Electrochemical Detection of Cadmium(II), Lead(II), Copper(II), and Mercury(II): An Interesting Favorable Mutual Interference. *J. Phys. Chem. C* **2012**, *116*, 1034–1041. [CrossRef]
73. Ghoneim, M.M.; Hassanein, A.M.; Hammam, E.; Beltagi, A.M. Simultaneous determination of Cd, Pb, Cu, Sb, Bi, Se, Zn, Mn, Ni, Co and Fe in water samples by differential pulse stripping voltammetry at a hanging mercury drop electrode. *Anal. Bioanal. Chem.* **2000**, *367*, 378–383. [CrossRef] [PubMed]
74. Ebunang, D.V.T.; Tajeu, K.Y.; Pecheu, C.N.; Jiokeng, S.L.Z.; Tamo, A.K.; Doench, I.; Osorio-Madrazo, A.; Tonle, I.K.; Ngameni, E. Amino-Functionalized Laponite Clay Material as a Sensor Modifier for the Electrochemical Detection of Quercetin. *Sensors* **2022**, *22*, 6173. [CrossRef]
75. El Mhammedi, M.A.; Achak, M.; Chtaini, A. $Ca_{10}(PO_4)_6(OH)_2$-modified carbon-paste electrode for the determination of trace lead(II) by square-wave voltammetry. *J. Hazard. Mater.* **2009**, *161*, 55–61. [CrossRef] [PubMed]
76. Pan, D.; Wang, Y.; Chen, Z.; Lou, T.; Qin, W. Nanomaterial/Ionophore-Based Electrode for Anodic Stripping Voltammetric Determination of Lead: An Electrochemical Sensing Platform toward Heavy Metals. *Anal. Chem.* **2009**, *81*, 5088–5094. [CrossRef] [PubMed]
77. Li, X.; Zhou, H.; Fu, C.; Wang, F.; Ding, Y.; Kuang, Y. A novel design of engineered multi-walled carbon nanotubes material and its improved performance in simultaneous detection of Cd(II) and Pb(II) by square wave anodic stripping voltammetry. *Sens. Actuators B Chem.* **2016**, *236*, 144–152. [CrossRef]
78. Oliveira, V.H.B.; Rechotnek, F.; da Silva, E.P.; Marques, V.d.S.; Rubira, A.F.; Silva, R.; Lourenço, S.A.; Muniz, E.C. A sensitive electrochemical sensor for Pb^{2+} ions based on ZnO nanofibers functionalized by L-cysteine. *J. Mol. Liq.* **2020**, *309*, 113041. [CrossRef]
79. Mohammadi, S.; Taher, M.A.; Beitollahi, H. Synthesis and application of a natural-based nanocomposite with carbon nanotubes for sensitive voltammetric determination of lead (II) ions. *Int. J. Environ. Anal. Chem.* **2020**, *100*, 65–81. [CrossRef]

nanomaterials

Article

Copper Cobalt Sulfide Structures Derived from MOF Precursors with Enhanced Electrochemical Glucose Sensing Properties

Daojun Zhang [1,*], Xiaobei Zhang [1,2], Yingping Bu [1,2], Jingchao Zhang [1] and Renchun Zhang [1]

1 College of Chemistry and Chemical Engineering, Anyang Normal University, Anyang 455000, China; zzuczxb@126.com (X.Z.); m18337278507@163.com (Y.B.); zjc19830618@126.com (J.Z.); rczhang@aynu.edu.cn (R.Z.)
2 College of Chemistry, Zhengzhou University, 100 Science Road, Zhengzhou 450001, China
* Correspondence: zhangdj0410@sohu.com; Tel.: +86-372-2900040

Abstract: Nonenzymatic electrochemical detection of glucose is popular because of its low price, simple operation, high sensitivity, and good reproducibility. Co-Cu MOFs precursors were synthesized via the solvothermal way at first, and a series of porous spindle-like Cu-Co sulfide microparticles were obtained by secondary solvothermal sulfurization, which maintained the morphology of the MOFs precursors. Electrochemical studies exhibit that the as-synthesized Cu-Co sulfides own excellent nonenzymatic glucose detection performances. Compared with CuS, Co (II) ion-doped CuS can improve the conductivity and electrocatalytic activity of the materials. At a potential of 0.55 V, the as-prepared Co-CuS-2 modified electrode exhibits distinguished performance for glucose detection with wide linear ranges of 0.001–3.66 mM and high sensitivity of 1475.97 $\mu A \cdot mM^{-1} \cdot cm^{-2}$, which was much higher than that of CuS- and Co-CuS-1-modified electrodes. The constructed sulfide sensors derived from MOF precursors exhibit a low detection limit and excellent anti-interference ability for glucose detection.

Keywords: copper cobalt sulfide; porous structures; nonenzymatic glucose sensing; electrocatalysts

Citation: Zhang, D.; Zhang, X.; Bu, Y.; Zhang, J.; Zhang, R. Copper Cobalt Sulfide Structures Derived from MOF Precursors with Enhanced Electrochemical Glucose Sensing Properties. *Nanomaterials* **2022**, *12*, 1394. https://doi.org/10.3390/nano12091394

Academic Editor: Jung Woo Lee

Received: 4 March 2022
Accepted: 7 April 2022
Published: 19 April 2022

1. Introduction

Currently, diabetes as a common chronic disease is already a serious threat to human health. Therefore, developing a simple and sensitive detection method for glucose is important for clinical diagnosis and diabetes management [1,2]. Compared with colorimetry, spectroscopy, and fluorescence analytical methods, nonenzymatic electrochemical glucose detection has received widespread attention due to its low cost, simple operation, and high sensitivity [3–5]. In recent years, transition-metal oxides (TMOs) and transition-metal sulfides (TMSs) have been exploited as advanced electrocatalysts to construct high performance electrochemical sensors [6–19].

Recently, among the transition-metal based electrodes materials, copper-based oxides and sulfides with various morphologies and structures have been used as electrode materials for nonenzymatic electrochemical glucose detection. For instance, Cu/Cu_2O hollow microspheres were prepared by solvothermal conditions and exhibited a high catalytic activity for glucose oxidation [20]. The glucose electrochemical sensor constructed by CuO nanorod dispersed hollow carbon fibers (CuO NR @ PCFs) [21] showed a wide linear range (0.005–0.8 mM, 0.8–8.5 mM) and a low detection limit (0.1 μM). The reported flower-like $CuCo_2O_4/C$ microspheres [22] -constructed sensor exhibited a wide linear range and low detection limit. The $Cu_xCo_{3-x}O_4$ nano-needle framework thin-film electrode reported by Xu [23] exhibited an ultrahigh sensitivity of 13,291.7 $\mu A \cdot mM^{-1} \cdot cm^{-2}$ for glucose detection. CuS nanotubes were prepared in an O/W microemulsion system at low temperature [24], and the glucose concentration could be detected by the CuS nanotube sensor with high sensitivity (7.842 $\mu A \cdot \mu M^{-1}$). Karikalan et al. synthesized S-rGO/CuS nanocomposites to construct an electrochemical glucose sensor, and the linear concentration range of the constructed sensor was 0.0001–3.88 mM and 3.88–20.17 mM, respectively, and the detection limit of 32 nM was quite low [25]. Xu et al. demonstrated the synthesis of $CuCo_2S_4$ nanosheets on flexible carbon fiber textiles (CFT) by a hydrothermal method [26]. The sensor constructed of $CuCo_2S_4$ nanosheets had a high sensitivity of 3852.7 $\mu A \cdot mM^{-1} \cdot cm^{-2}$ and a linear range up to 3.67 mM. Compared with related transition metal oxides, copper cobalt sulfides are more suitable as electrode materials for nonenzymatic glucose sensors due to the improved electrical conductivity [26].

In recent years, metal−organic frameworks (MOFs) have served as self-sacrificial precursors for preparation of porous micro-/nanostructured transition metal oxides and sulfides [27–30]. The MOF-derived TMOs and TMSs usually exhibit porous structures and high surface areas with enhanced electrocatalytic and electrochemical energy storage performances [31–35]. In this work, the shuttle-like copper cobalt sulfide structures were synthesized via MOF sacrificial templates. The electrochemical properties of copper cobalt sulfide -constructed electrodes were studied by cyclic voltammetry and the amperometric method. At a potential of 0.55 V, the linear range of Co-CuS-2 modified electrode was 0.001–3.66 mM with a detection limit of 0.1 μM, and the sensitivity of the electrode was 1475.97 $\mu A \cdot mM^{-1} \cdot cm^{-2}$. The results indicate that the sensor owns good electrochemical sensing performance for glucose and has a potential application in glucose detection.

2. Experimental Section

2.1. Chemicals

Polyvinylpyrrolidone (PVP) and 2,5-dihydroxyterephthalic acid (H_4dobdc) were purchased from Shanghai Macklin Biochemical Co., Ltd. (Shanghai, China); $Cu(NO_3)_2 \cdot 4H_2O$, $Co(NO_3)_2 \cdot 6H_2O$, glucose (Glu), ascorbic acid (AA), and NaOH were purchased from Sinopharm Chemical Reagent Co., Ltd. (Shanghai, China); Thioacetamide (TAA), uric acid (UA), dopamine (DA), sodium chloride (NaCl), glutathione (GSH), and sucrose (Suc) were purchased from Aladdin Industrial Corporation (Shanghai, China); Ethylene glycol, ethanol, and N,N-dimethylformamide (DMF) were purchased from Tianjin Fuyu Fine Chemical Co., Ltd. (Tianjin, China). All chemicals and solvents were used without further purification.

2.2. Preparation of Spindle-like Cu-Co Sulfide Microparticles

The solvent system of Cu-Co MOF precursors synthesis is similar to that of our previously reported paper [36]. For the preparation of Co-Cu MOF precursors, $Cu(NO_3)_2 \cdot 4H_2O$, $Co(NO_3)_2 \cdot 6H_2O$, 2,5-dihydroxyterephthalic acid (H_4dobdc), and polyvinylpyrrolidone (PVP) were added in DMF/ethanol/water mixed solvent according to the molar ratio of Co/Cu of 8:2, 7:3 and 0:1, the mixture was heated at 100 °C for 12 h, and then MOF precursors were isolated and washed twice with DMF and water. Cu-Co sulfides were synthesized via an effective sulfurization treatment of MOF precursors with thioacetamide (TAA), the samples with Co/Cu ratio from high to low denoted as Co-CuS-1 and Co-CuS-2, respectively. The MOF precursors were redispersed into 3 mL of ethylene glycol. Then, 0.0043 g of TAA was added and fully stirred, the mixture was transferred into a 25 mL stainless-steel Teflon-lined autoclave and reacted at 110 °C for 12 h, and the black powder was obtained and washed three times with ethanol and water.

2.3. Materials Characterization

Powder X-ray diffraction (PXRD) analyses of the as-prepared samples were conducted on a PANalytical X'Pert PRO MPD system with $Cu_{K\alpha}$ radiation (λ = 1.5418 Å) and operated at 40 kV and 40 mA. The morphologies and compositions were analyzed by scanning electron microscopy (SEM) with energy dispersive spectroscopy (EDS) on a Hitachi SU-8010 instrument and X-ray photo-electron spectroscopy (XPS) using Thermo-Scientific system. The specific surface areas of the samples were acquired by N_2 adsorption/desorption isotherms measured on a Gemini VII 2390 analyzer at 77 K.

2.4. Electrode Preparation and Measurement

All electrochemical tests were conducted on a CHI660E electrochemical workstation with a typical three-electrode system. First, 2 mg Co-Cu sulfides was dispersed in 1.0 mL distilled H_2O via ultrasound 30 min. A glassy carbon electrode (GCE) with diameter 3 mm was polished with alumina slurries and washed with ultrapure H_2O. Afterward, 5 μL of the suspension was covered onto the GCE surface to obtain Co-CuS/GCE. The modified electrode was used as the working electrode, Ag/AgCl was used as the reference electrode, and Pt wire was used as the counter electrode.

3. Results and Discussion

The morphologies of Co-Cu MOF precursors were analyzed by scanning electron microscopy (SEM) technique and are shown in Figure 1, which exhibited a spindle-like structure with a well-distributed and smooth surface. Figure 2 shows the SEM images of derived samples of CuS (Figure 2a,b), Co-CuS-1 (Figure 2d) and Co-CuS-2 (Figure 2e). After effective sulfurization treatment of MOF precursors with TAA, the derived products can largely retain the morphology of MOF precursors; however, the surface of all samples seems rough and porous. According to the EDS mapping images of a single shuttle-like CuS particle (Figure 2c), the Cu and S elements are evenly distributed. Figures S1 and 2f show the EDS mapping images of Co-CuS-1 and Co-CuS-2, respectively. It can be seen from the images that there is a distribution of Co, Cu and S elements, which indicates that Co element is doped into CuS microparticles. The phase of as-synthesized sulfides was checked by the XRD patterns and are exhibited in Figure 3a; all the positions of the peaks are consistent with the standard card number JCPDS No.065-3588 of the hexagonal phase CuS, and no other impurity peaks appear in the patterns. It can be deduced that partial-doped cobalt ions into CuS do not change its crystal structure. Figure 3b shows the N_2 adsorption isotherm of the Co-CuS-2 sample and the corresponding pore size distribution. Co-CuS-2 is the type IV adsorption isotherm, which belongs to the typical mesoporous structure. The specific surface area calculated by the BET method is 16.3 $m^2\ g^{-1}$, and the average pore size is 31.72 nm. The element content of the Co-CuS-1 and Co-CuS-2 samples were further characterized by EDS, which indicated that the Co/Cu ratios are close to the stoichiometric ratio of raw materials (Figure 4a). XPS technique was further used to analyze the surface valance state of Co and Cu in the corresponding sulfide. Figure S2 provides the XPS survey spectra of the as-synthesized samples. The high revolution spectra are shown in Figure 4b–d. The high revolution spectra of Cu in the three samples are similar, the binding energy of the two peaks located at 931.4 and 951.3 eV was attributed to Cu $2p_{3/2}$ and Cu $2p_{1/2}$, respectively [25]. The fitted peaks of 931.4 and 951.2 eV indicate the existence of Cu^+, and the peaks at 932.4 and 953.1 eV correspond to Cu^{2+}. The high resolution Co 2p spectra in Co-CuS-1 are fitted with two doublet peaks centered at 780.5 and 796.4 eV, which correspond to Co^{3+}. The peaks at 781.9 and 797.5 eV correspond to Co^{2+}. Figure 4d shows the S 2p spectra, with the peaks centered at 161.2 and 162.4 eV for S $2p_{3/2}$ and S $2p_{1/2}$, respectively. For Co-CuS-2, the peak of 163.6 eV increased dramatically, which can be ascribed to a Metal-S bond at a low coordination environment and contributed to an increase in the intrinsic conductivity [37,38].

Figure 1. SEM images of (**a**) Cu-MOF, (**b**) CuCo-MOF-1, and (**c**) CuCo-MOF-2 precursors.

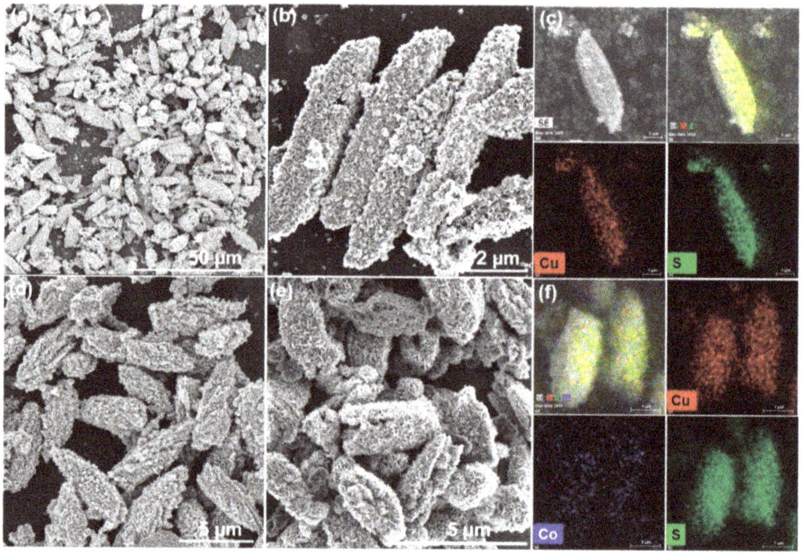

Figure 2. SEM images of copper cobalt sulfides (**a**,**b**) CuS, (**d**) Co-CuS-1, (**e**) Co-CuS-2, and the EDX mapping element distribution of (**c**) CuS, (**f**) Co-CuS-2.

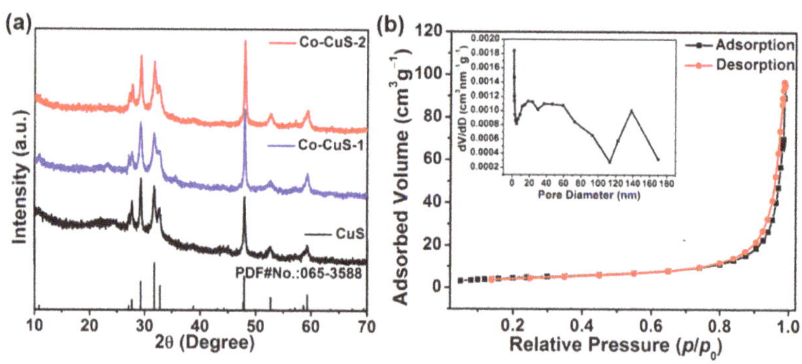

Figure 3. (**a**) XRD patterns of the samples with different proportions; (**b**) N_2 adsorption isotherm and pore distribution of Co-CuS-2.

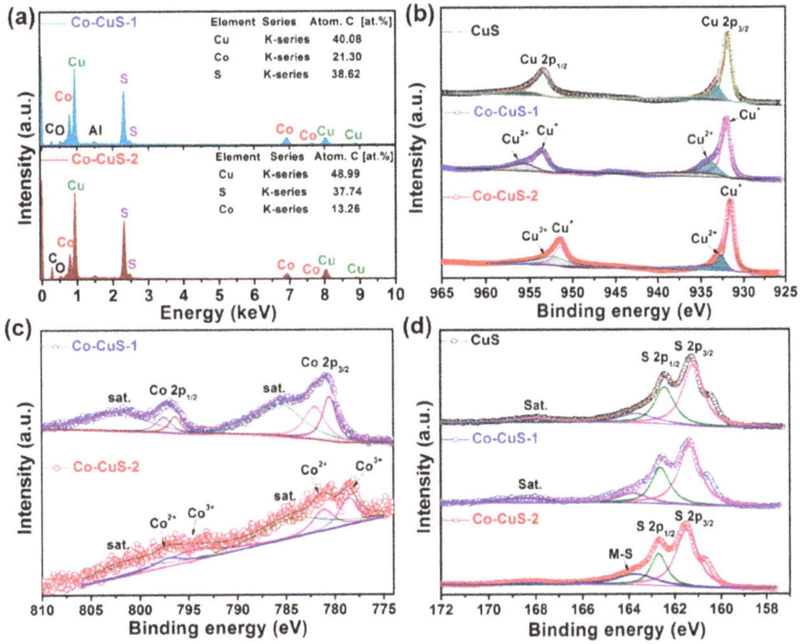

Figure 4. (**a**) EDS of CuS-1 and Co-CuS-2 samples. High resolution XPS spectra of Cu (**b**), Co (**c**), and S (**d**) for CuS, Co-CuS-1, and Co-CuS-2 samples.

Figure 5a shows the CV curves of the bare electrode and copper cobalt sulfide-modified electrodes in the 0.1 M NaOH electrolyte containing 1 mM Glu. As shown in Figure 5a, the bare electrode has almost no response, CuS- and Co-CuS-1-modified electrodes have weak redox peaks, and the Co-CuS-2-modified electrode has a pair of obvious redox peaks at 0.45/0.60 V (vs. Ag/AgCl), indicating that the Co-CuS-2 sample exhibits the best response to glucose. As shown in Figure 5b–d, the CV curves of CuS, Co-CuS-1, and Co-CuS-2 at different glucose concentrations further show that the redox peaks of the Co-CuS-2 electrode is the strongest, indicating that this material has the best electrocatalytic performance for glucose among the three samples. The possible oxidation mechanism of glucose may be described in the following three steps [22,23,26]:

$$\text{CuCo-S} + OH^- + H_2O \rightarrow \text{CuSOH} + \text{CoSOH} + e^- \quad (1)$$

$$\text{CoSOH} + OH^- \rightarrow \text{CoSO} + H_2O + e^- \quad (2)$$

$$\text{CuOSH} + \text{CoSO} + \text{glucose} \rightarrow \text{CuS} + \text{CoSOH} + \text{glucolactone} \quad (3)$$

CuSOH and CoSO intermediator might be formed through electrooxidation at alkaline conditions. The formed CuSOH and CoSO adsorbed glucose molecules and subsequently oxidated to gluconolactone in an alkaline medium.

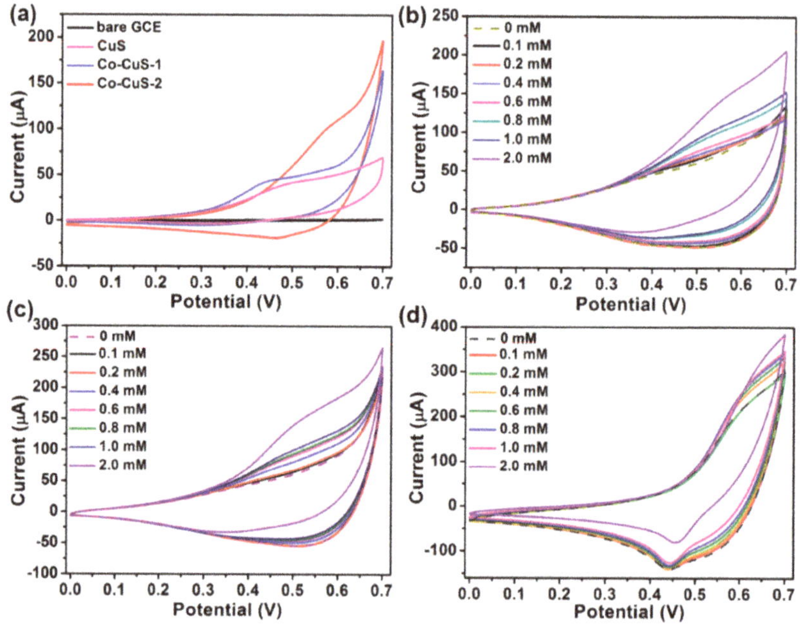

Figure 5. (**a**) CV curves of CuS, Co-CuS-1, Co-CuS-2 and bare electrode in 0.1 M NaOH electrolyte containing 1 mM Glu (20 mV·s^{-1}). CV curves in NaOH solution with different glucose concentration: (**b**) CuS, (**c**) Co-CuS-1, and (**d**) Co-CuS-2 (100 mV·s^{-1}).

In order to further acquire the kinetic information of glucose electrocatalytic oxidation of the CuS, Co-CuS-1, and Co-CuS-2 electrodes, the CV curves of three samples at scan rates varying from 20 to 180 mV·s^{-1} were studied. Figure 6a,c,e shows the CV curves of CuS, Co-CuS-1, and Co-CuS-2 in 0.1 M NaOH solution containing 1 mM glucose at different scan rates. The peak current of the three samples increases steadily with the increase in scan rate. As seen in Figure 6b,d, the peak current (anodic and cathodic) of the CuS and Co-CuS-1 electrodes increases linearly with the scan rate, and the fitted linear equations are $I_{pa} = 0.626v + 31.172$, $I_{pc} = -0.489v + 12.852$, and $I_{pa} = 0.634v + 28.708$, $I_{pc} = -0.717v + 17.782$ respectively, indicating that CuS and Co-CuS-1 electrodes are adsorption-controlled processes for oxidation glucose. Figure 6f shows that both the anodic and cathodic peak current has a linear relationship with the square root of the scan rate, indicating that Co-CuS-2 is a diffusion-controlled process. The fitted linear equations are $I_{pa} = 21.176v^{1/2} + 47.474$, $I_{pc} = -15.716v^{1/2} + 8.971$, respectively, and this may be attributed to the large surface area and good conductivity; thus, it is conducive to glucose detection.

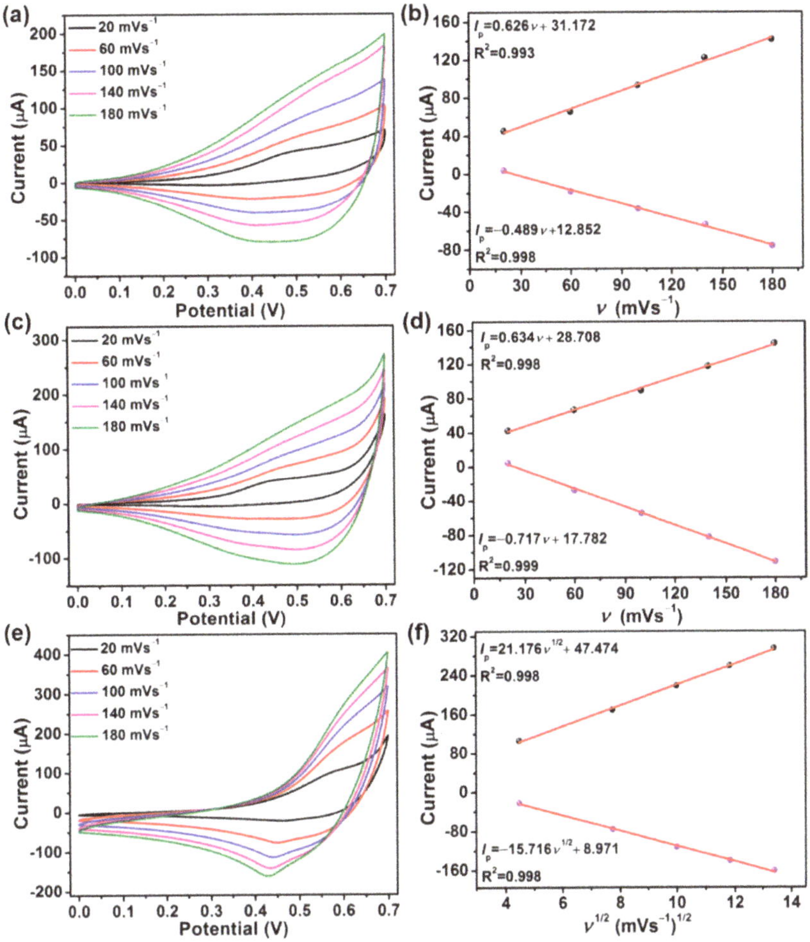

Figure 6. CV curves at different scan rates (20–180 mV s^{-1}) in a 1 mM glucose solution (**a**) CuS, (**c**) Co-CuS-1, (**e**) Co-CuS-2, and the corresponding linear calibration curves, (**b**) CuS, (**d**) Co-CuS-1, (**f**) Co-CuS-2, respectively.

In order to systematically study the effect of working potential on the electrocatalytic oxidation of glucose for CuS-, Co-CuS-1-, and Co-CuS-2-modified electrodes, the current−time (I–t) curves at different potentials were measured. Figure 7a shows the current response (0.1 M NaOH) of the CuS electrode via a continuously increasing glucose concentration at 0.5, 0.55 and 0.6 V. The CuS electrode exhibits the highest amperometric response at 0.6 V. The corresponding calibration curve at 0.6 V is also shown in Figure 7b. The linear range of the CuS electrode is 0.002–2.16 mM, the sensitivity is 905.42 $\mu A \cdot mM^{-1} \cdot cm^{-2}$, and the limit of detection (LOD) of 0.9 μM is calculated based on $3\sigma/s$, where σ is the standard deviation of the blank, and s is the slope of the calibration curve [12,39].

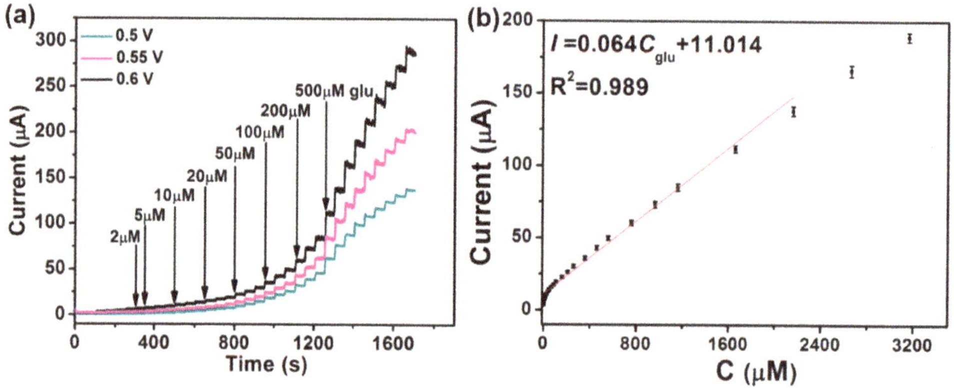

Figure 7. (**a**) Current−time curves of the CuS electrode with the continuous addition of glucose solution at 0.5, 0.55 and 0.6 V (0.1 M NaOH), respectively. (**b**) The calibration curves fitted from the current responses at 0.6 V.

Figure 8a,b shows the *I*–t curves of Co-CuS-1 and Co-CuS-2 at different potentials, which shows the best glucose electrocatalytic performance at a potential of 0.55 V after Co doping. Thus, the electrocatalytic properties of CuS, Co-CuS-1 and Co-CuS-2 at 0.55 V were compared and are shown in Figure 8c. It can be seen from the image that the amperometric response of Co-CuS-2 is the highest. Figure 8d shows the calibration curve of CuS at 0.55 V; the linear range is 0.002–2.66 mM, the sensitivity is 686.13 $\mu A \cdot mM^{-1} \cdot cm^{-2}$, and the detection limit is 0.4 μM (3σ/s). Figure 8e shows the calibration curve of Co-CuS-1; the linear range is 0.001–3.16 mM, the sensitivity increases to 1206.75 $\mu A \cdot mM^{-1} \cdot cm^{-2}$, and the detection limit decreases to 0.3 μM (3σ/s). Figure 8f shows the calibration curve of Co-CuS-2 at 0.55 V; the linear range extends from 0.001 mM to 3.66 mM, the sensitivity is 1475.97 $\mu A \cdot mM^{-1} \cdot cm^{-2}$, and the detection limit is 0.1 μM (3σ/s). Table 1 shows the details. As compared, at a potential of 0.55 V, Co-CuS-2 has the widest linear range, the highest sensitivity and lowest LOD among the three electrodes. The sensitivity of the Co-CuS-2 sensor is higher than that of the Cu/Cu_2O hollow microspheres [20], $CuCo_2O_4/C$ microspheres [22], hierarchical Co_3O_4 film [39], CuO_x-CoO_x/graphene [40], Octahedral Cu_2O [41], and CuO microspheres [42]; however, it is lower than that of the $CuCo_2S_4$ nanosheets [26] and $NiCo_2O_4$ hollow nanorods [43]. The comprehensive performance of Co-CuS-2/GCE is equivalent to or better than that of previously reported electrochemical glucose sensors (Table 2).

Table 1. Electrochemical sensing properties of Cu-Co sulfides with different doping ratios.

Electrode Material	Potential (V)	Linear Range (mM)	Detection Limit (μM)	Sensitivity ($\mu A \cdot mM^{-1} \cdot cm^{-2}$)
CuS	0.55 0.60	0.002–2.66 0.002–2.16	0.4 0.9	686.13 905.42
Co-CuS-1	0.55	0.001–3.16	0.3	1206.75
Co-CuS-2	0.55	0.001–3.66	0.1	1475.97

Table 2. Comparison of the Co-CuS-2 electrode with some reported sensors for glucose detection.

Electrode Material	Potential (V)	Linear Range (mM)	Detection Limit (μM)	Sensitivity ($\mu A \cdot mM^{-1} \cdot cm^{-2}$)	Ref.
Co-CuS-2	0.55	0.001–3.66	0.1	1475.97	This work
Cu/Cu_2O hollow microspheres	0.45	0.22–10.89	0.05	33.63 $\mu A \cdot mM^{-1}$	20

Table 2. *Cont.*

Electrode Material	Potential (V)	Linear Range (mM)	Detection Limit (μM)	Sensitivity (μA·mM^{-1}·cm^{-2})	Ref.
CuO NR @ PCFs	0.60	0.005–0.8 0.8–8.5	0.1	608	21
$CuCo_2O_4$/C microspheres	0.60	0.005–8	1.5	707.71	22
CuS nanotube	0.20	0.05–5	–	7.842 μA·mM^{-1}	24
$CuCo_2S_4$/carbon fiber textile	0.35	up to 3.67	1.01	3852.7	26
Co_3O_4 porous film	0.6	up to 3.0	1	366.03	39
CuO_x-CoO_x/ graphene	0.50 (vs. SCE)	0.005–0.57	0.5	507	40
Octahedral Cu_2O	0.60	0.3–4.1	128	241	41
CuO microspheres	0.45 (vs. SCE)	0.001–4	0.5	349.6	42
$NiCo_2O_4$ hollow nanorods	0.60	0.0003–1	0.16	1685.1	43

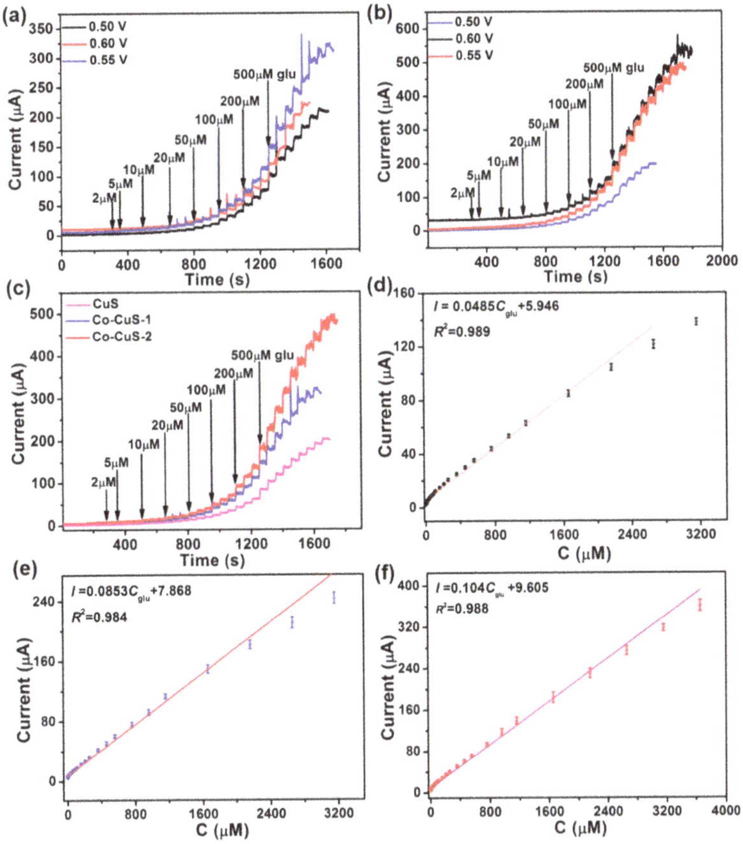

Figure 8. Current response of continuously increasing glucose concentration at 0.5, 0.55 and 0.60 V (0.1 M NaOH): (**a**) Co-CuS-1, (**b**) Co-CuS-2, (**c**) current response of CuS, Co-CuS-1 and Co-CuS-2 at 0.55 V. The corresponding calibration curves at 0.55 V: (**d**) CuS, (**e**) Co-CuS-1, (**f**) Co-CuS-2.

As shown in Figure 9a, all Nyquist diagrams contain the semicircular part at high frequency and the oblique line at low frequency. The tilt line is related to the diffusion limit step, and the R_{ct} of the electrode surface can be equal to the radius of the semicircular part. The radius of Co-CuS-2 is the smallest, indicating that the conductivity of Co-CuS-2 is the largest, which is one of the reasons for its best electrocatalytic performance for glucose. The repeatability of CuS, Co-CuS-1 and Co-CuS-2 electrodes were tested and are shown in Figure 9b. The current response of adding 200 μM Glu 13 times in 0.1 M NaOH solution at a potential of 0.55 V shows that the current response of Co-CuS-2 is high, the step change is almost unchanged, and the calculated RSD is 4.19%. In order to evaluate the selectivity of the constructed electrodes to glucose detection, as shown in Figure 9c, three different sulfide electrodes showed good anti-interference performance at a working potential of 0.55 V. The specific operation is to add 200 μM glucose (Glu), 20 μM ascorbic acid (AA), uric acid (UA), dopamine (DA), sodium chloride (NaCl), glutathione (GSH), sucrose (Suc) and 200 μM glucose (Glu) in 0.1 M NaOH supporting solution. It can be seen from the *I*–t curves that the response current for glucose of the Co-CuS-2-modified electrode remains unchanged after adding interfering substances, while the response current of interferents is almost negligible, indicating that the electrode has good selectivity for glucose detection. Figure 9d shows the stability of the as-synthesized sample-modified electrodes. After adding 200 μM Glu solution, the current response lasts for 3500 s. The results show that the retention rates of Co-CuS-2, Co-CuS-1 and CuS are 93%, 76% and 84%, respectively, indicating that Co-CuS-2 has the best stability. Therefore, the Co-CuS-2-modified electrode has good reproducibility, selectivity, and stability for the detection of glucose.

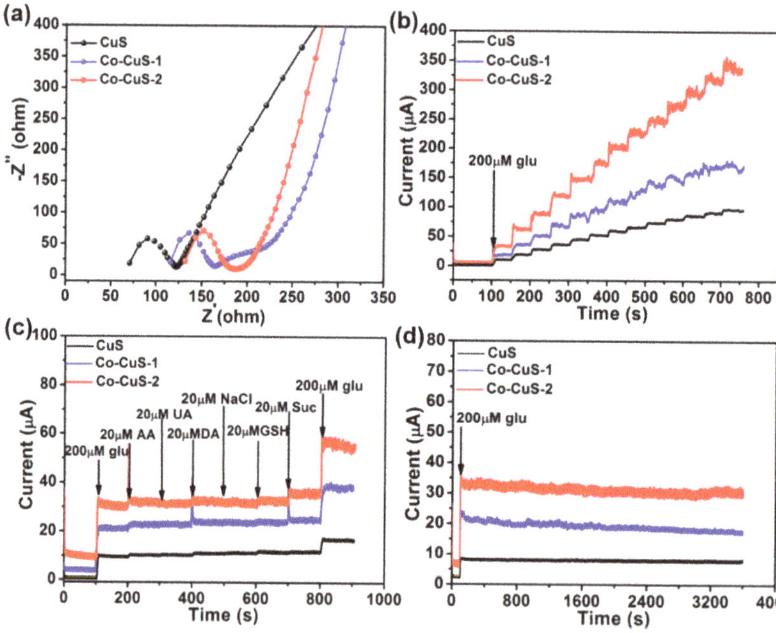

Figure 9. (**a**) Impedance diagram of CuS, Co-CuS-1 and Co-CuS-2 samples, (**b**) reproducibility, (**c**) anti-interference, and (**d**) stability.

4. Conclusions

Co/Cu MOFs precursors were synthesized by the one-step solvothermal method, and a series of porous spindle-like Cu-Co sulfide microparticles were obtained by secondary solvothermal sulfurization, which maintained the morphology of the MOFs precursors. The porous structure of these materials was conducive to the diffusion of electrolytes and analytes. Compared with CuS, Co (II) ion doping can improve the conductivity and electrocatalytic activity of the materials. At a potential of 0.55 V, the linear range of the Co-CuS-2 electrode for glucose detection was 0.001–3.66 mM with an LOD of 0.1 μM, and the sensitivity was 1475.97 $\mu A \cdot mM^{-1} \cdot cm^{-2}$, which was much better than that of the CuS and Co-CuS-1 samples. This work provides an effective strategy for glucose detection and opens up a new way to improve the electrochemical performance of non-enzyme electrochemical sensors.

Supplementary Materials: The following supporting information can be downloaded at: https://www.mdpi.com/article/10.3390/nano12091394/s1, Figure S1: EDX mapping images of Co-CuS-1 sample; Figure S2: XPS survey spectra of (a) CuS-1, (b) Co-CuS-1 and (c) Co-CuS-2 samples.

Author Contributions: Conceptualization, D.Z.; Data curation, X.Z., Y.B. and J.Z.; Formal Analysis, D.Z.; Funding acquisition, D.Z.; Investigation, X.Z. and D.Z.; Methodology, D.Z.; Project administration, D.Z. and R.Z.; Supervision, D.Z.; Writing—original draft, X.Z. and D.Z.; Writing—review & editing, X.Z., D.Z. and J.Z. All authors have read and agreed to the published version of the manuscript.

Funding: This work was supported by the National Science Foundation of China (no. 21603004, U1604119, 21501006), the Science and Technology Research Project of Henan Province (222102240096), and the Program for Innovative Research Team of Science and Technology at the University of Henan Province (18IRTSTHN006).

Institutional Review Board Statement: Not applicable.

Informed Consent Statement: Not applicable.

Data Availability Statement: Not applicable.

Conflicts of Interest: The authors declare no conflict of interest.

References

1. Heller, A.; Feldman, B. Electrochemical Glucose Sensors and Their Applications in Diabetes Management. *Chem. Rev.* **2018**, *108*, 2482–2505. [CrossRef] [PubMed]
2. Steiner, M.; Duerkop, A.; Wolfbeis, O. Optical methods for sensing glucose. *Chem. Soc. Rev.* **2011**, *40*, 4805–4839. [CrossRef] [PubMed]
3. Wang, G.; He, X.; Wang, L.; Gu, A.; Huang, Y.; Fang, B.; Geng, B.; Zhang, X. Non-enzymatic electrochemical sensing of glucose. *Microchim. Acta* **2013**, *180*, 161–186. [CrossRef]
4. Dhara, K.; Mahapatra, D. Electrochemical nonenzymatic sensing of glucose using advanced nanomaterials. *Microchim. Acta* **2018**, *185*, 49. [CrossRef] [PubMed]
5. Dong, Q.; Ryu, H.; Lei, Y. Metal oxide based non-enzymatic electrochemical sensors for glucose detection. *Electrochim. Acta* **2021**, *370*, 137744. [CrossRef]
6. Zhang, Y.; Zhou, Q.; Zhu, J.; Yan, Q.; Dou, S.; Sun, W. Nanostructured Metal Chalcogenides for Energy Storage and Electrocatalysis. *Adv. Funct. Mater.* **2017**, *27*, 1702317. [CrossRef]
7. Guo, Y.; Park, T.; Yi, J.; Henzie, J.; Kim, J.; Wang, Z.; Jiang, B.; Bando, Y.; Sugahara, Y.; Tang, J.; et al. Nanoarchitectonics for Transition-Metal- Sulfide-Based Electrocatalysts for Water Splitting. *Adv. Mater.* **2019**, *31*, 1807134. [CrossRef]
8. Yu, X.; Lou, X. Mixed Metal Sulfides for Electrochemical Energy Storage and Conversion. *Adv. Energy Mater.* **2018**, *8*, 1701592. [CrossRef]
9. Rui, X.; Tan, H.; Yan, Q. Nanostructured metal sulfides for energy storage. *Nanoscale* **2014**, *6*, 9889–9924. [CrossRef]
10. Liu, Y.; Li, X.; Shen, W.; Dai, Y.; Kou, W.; Zheng, W.; Jiang, X.; He, G. Multishelled Transition Metal-Based Microspheres: Synthesis and Applications for Batteries and Supercapacitors. *Small* **2019**, *15*, 1804737. [CrossRef]
11. Kulkarni, P.; Nataraj, S.; Geetha Balakrishna, R.; Nagarajua, D.; Reddy, M. Nanostructured binary and ternary metal sulfides: Synthesis methods and their application in energy conversion and storage devices. *J. Mater. Chem. A* **2017**, *5*, 22040–22094. [CrossRef]
12. Yang, P.; Wang, X.; Ge, C.; Fu, X.; Liu, X.; Chai, H.; Guo, X.; Yao, H.; Zhang, Y.; Chen, K. Fabrication of CuO nanosheets-built microtubes via Kirkendall effect for non-enzymatic glucose sensor. *Appl. Surf. Sci.* **2019**, *494*, 484–491. [CrossRef]
13. Li, M.; Han, C.; Zhang, Y.F.; Bo, X.J.; Guo, L.P. Facile synthesis of ultrafine Co_3O_4 nanocrystals embedded carbon matrices with specific skeletal structures as efficient non-enzymatic glucose sensors. *Anal. Chim. Acta* **2015**, *861*, 25–35. [CrossRef] [PubMed]
14. Kang, M.; Zhou, H.; Zhao, N.; Lv, B. Porous Co_3O_4 nanoplates as an efficient electromaterial for non-enzymatic glucose sensing. *CrystEngComm* **2020**, *22*, 35–43. [CrossRef]

15. Zhang, Y.Y.; Wang, L.; Yu, J.; Yang, H.; Pan, G.X.; Miao, L.F.; Song, Y.H. Three-dimensional macroporous carbon supported hierarchical ZnO-NiO nanosheets for electrochemical glucose sensing. *J. Alloys Compd.* **2017**, *698*, 800–806. [CrossRef]
16. Xiao, X.; Zhang, X.; Zhang, Z.; You, J.; Liu, S.; Wang, Y. Macro-/meso-porous $NiCo_2O_4$ synthesized by template-free solution combustion to enhance the performance of a nonenzymatic amperometric glucose sensor. *Microchimi. Acta* **2020**, *187*, 64. [CrossRef]
17. Yang, Z.; Bai, X.; Zhu, S.; Qi, C. Synthesis of porous Co_3S_4 for enhanced voltammetric nonenzymatic determination of glucose. *Microchim. Acta* **2020**, *187*, 98. [CrossRef]
18. Huo, H.; Zhao, Y.; Xu, C. 3D Ni_3S_2 nanosheet arrays supported on Ni foam for high-performance supercapacitor and nonenzymatic glucose detection. *J. Mater. Chem. A* **2014**, *2*, 15111–15117. [CrossRef]
19. Guo, Q.; Wu, T.; Liu, L.; He, Y.; Liu, D.; You, T. Hierarchically porous $NiCo_2S_4$ nanowires anchored on flexible electrospun graphitic nanofiber for high-performance glucose biosensing. *J. Alloys Compd.* **2020**, *819*, 153376. [CrossRef]
20. Wang, A.; Feng, J.; Li, Z.; Liao, Q.; Wang, Z.; Chen, J. Solvothermal synthesis of Cu/Cu_2O hollow microspheres for non-enzymatic amperometric glucose sensing. *CrystEngComm* **2012**, *14*, 1289–1295. [CrossRef]
21. Li, M.; Zhao, Z.; Liu, X.; Xiong, Y.; Han, C.; Zhang, Y.; Bo, X.; Guo, L. Novel bamboo leaf shaped CuO nanorod@hollow carbon fibers derived from plant biomass for efficient and nonenzymatic glucose detection. *Analyst* **2015**, *140*, 6412–6420. [CrossRef] [PubMed]
22. Yin, H.; Zhu, J.; Chen, J.; Gong, J.; Nie, Q. Hierarchical $CuCo_2O_4$/C microspheres assembled with nanoparticle-stacked nanosheets for sensitive non-enzymatic glucose detection. *J. Mater. Sci.* **2018**, *53*, 11951–11961. [CrossRef]
23. Xu, H.; Han, F.; Xia, C.; Wang, S.; Zhuiykov, S.; Zheng, G. Spinel sub-stoichiometric $Cu_xCo_yO_4$ nano-wire framework thin-film electrode for enhanced electrochemical non-enzymatic sensing of glucose. *Electrochim. Acta* **2020**, *331*, 135295. [CrossRef]
24. Zhang, X.; Wang, G.; Gu, A.; Wei, Y.; Fang, B. CuS nanotubes for ultrasensitive nonenzymatic glucose sensors. *Chem. Commun.* **2008**, 5945–5947. [CrossRef] [PubMed]
25. Karikalan, N.; Karthik, R.; Chen, S.; Karuppiah, C.; Elangovan, A. Sonochemical synthesis of sulfur doped reduced Graphene oxide supported CuS nanoparticles for the non-enzymatic glucose sensor applications. *Sci. Rep.* **2017**, *7*, 2494. [CrossRef] [PubMed]
26. Xu, W.; Lu, J.; Huo, W.; Li, J.; Wang, X.; Zhang, C.; Gu, X.; Hu, C. Direct growth of $CuCo_2S_4$ nanosheets on carbon fiber textile with enhanced electrochemical pseudocapacitive properties and electrocatalytic properties towards glucose oxidation. *Nanoscale* **2018**, *10*, 14304. [CrossRef]
27. Zhan, W.; Sun, L.; Han, X. Recent Progress on Engineering Highly Efficient Porous Semiconductor Photocatalysts Derived from Metal–Organic Frameworks. *Nano-Micro Lett.* **2019**, *11*, 1. [CrossRef]
28. Li, Y.; Xu, Y.; Yang, W.; Shen, W.; Xue, H.; Pang, H. MOF-Derived Metal Oxide Composites for Advanced Electrochemical Energy Storage. *Small* **2018**, *14*, 1704435. [CrossRef]
29. Salunkhe, R.; Kaneti, Y.; Yamauchi, Y. Metal−Organic Framework- Derived Nanoporous Metal Oxides toward Supercapacitor Applications: Progress and Prospects. *ACS Nano* **2017**, *11*, 5293–5308. [CrossRef]
30. Cai, Z.; Wang, Z.; Kim, J.; Yamauchi, Y. Hollow Functional Materials Derived from Metal–Organic Frameworks: Synthetic Strategies, Conversion Mechanisms, and Electrochemical Applications. *Adv. Mater.* **2019**, *31*, 1804903. [CrossRef]
31. Long, L.; Liu, X.J.; Chen, L.L.; Wang, S.Y.; Liu, M.C.; Jia, J.B. MOF-derived 3D leaf-like CuCo oxide arrays as an efficient catalyst for highly sensitive glucose detection. *Electrochim. Acta* **2019**, *308*, 243–252. [CrossRef]
32. Luo, Y.; Wang, Q.; Li, J.; Xu, F.; Sun, L.; Bu, Y.; Zou, Y.; Kraatz, H.; Rosei, F. Tunable hierarchical surfaces of CuO derived from metal–organic frameworks for non-enzymatic glucose sensing. *Inorg. Chem. Front.* **2020**, *7*, 1512–1525. [CrossRef]
33. Shu, Y.; Yan, Y.; Chen, J.; Xu, Q.; Pang, H.; Hu, X. Ni and NiO Nanoparticles Decorated Metal−Organic Framework Nanosheets: Facile Synthesis and High-Performance Nonenzymatic Glucose Detection in Human Serum. *ACS Appl. Mater. Interfaces* **2017**, *9*, 22342–22349. [CrossRef] [PubMed]
34. Kim, K.; Kim, S.; Lee, H.; Park, Y.; Bae, Y.; Kim, H. Electrochemically derived CuO nanorod from copper-based metal-organic framework for non-enzymatic detection of glucose. *Appl. Surf. Sci.* **2019**, *479*, 720–726. [CrossRef]
35. Xiao, Z.; Bao, Y.; Li, Z.; Huai, X.; Wang, M.; Liu, P.; Wang, L. Construction of Hollow Cobalt−Nickel Phosphate Nanocages through a Controllable Etching Strategy for High Supercapacitor Performances. *ACS Appl. Energy Mater.* **2019**, *2*, 1086–1092. [CrossRef]
36. Zhang, D.J.; Zhang, J.C.; Shi, H.Z.; Guo, X.L.; Guo, Y.Y.; Zhang, R.C.; Yuan, B.Q. Redox-active microsized metal-organic framework for efficient nonenzymatic H_2O_2 sensing. *Sens. Actuators B* **2015**, *221*, 224–229. [CrossRef]
37. Kang, L.; Huang, C.; Zhang, J.; Zhang, M.Y.; Zhang, N.; Liu, S.D.; Ye, Y.; Luo, C.; Gong, Z.W.; Wang, C.L.; et al. Effect of fluorine doping and sulfur vacancies of $CuCo_2S_4$ on its electrochemical performance in supercapacitors. *Chem. Eng. J.* **2020**, *390*, 124643. [CrossRef]
38. Han, X.Z.; Qin, Y.; Luo, J.X.; Zhang, F.Z.; Lei, X.D. Polygonal CuS Nanoprisms Fabricated by Grinding Reaction for Advanced Quasi-Solid-State Asymmetry Supercapacitors. *ACS Appl. Energy Mater.* **2021**, *4*, 12631–12640. [CrossRef]
39. Fan, S.; Zhao, M.; Ding, L.; Liang, J.; Chen, J.; Li, Y.; Chen, S. Synthesis of 3D hierarchical porous Co_3O_4 film by eggshell membrane for non-enzymatic glucose detection. *J. Electroanal. Chem.* **2016**, *775*, 52–57. [CrossRef]
40. Li, S.; Hou, L.; Yuan, B.; Chang, M.; Ma, Y.; Du, J. Enzyme-free glucose sensor using a glassy carbon electrode modified with reduced graphene oxide decorated with mixed copper and cobalt oxides. *Microchim. Acta* **2016**, *183*, 1813–1821. [CrossRef]

41. Li, Y.; Zhong, Y.; Zhang, Y.; Weng, W.; Li, S. Carbon quantum dots/octahedral Cu_2O nanocomposites for non-enzymatic glucose and hydrogen peroxide amperometric sensor. *Sens. Actuators B* **2015**, *206*, 735–743. [CrossRef]
42. Liu, X.; Yang, Y.; Liu, R.; Shi, Z.; Ma, L.; Wei, M. Synthesis of porous CuO microspheres assembled from (001) facetexposed nanocrystals with excellent glucose-sensing performance. *J. Alloys Compd.* **2017**, *718*, 304–310. [CrossRef]
43. Yang, J.; Cho, M.; Lee, Y. Synthesis of hierarchical $NiCo_2O_4$ hollow nanorods via sacrificial-template accelerate hydrolysis for electrochemical glucose oxidation. *Biosens. Bioelectron.* **2016**, *75*, 15–22. [CrossRef] [PubMed]

nanomaterials

MDPI

Article

Covalent Organic Frameworks-TpPa-1 as an Emerging Platform for Electrochemical Sensing

Gang Li, Baiqing Yuan *, Sidi Chen, Liju Gan and Chunying Xu

School of Chemistry and Materials Science, Ludong University, Yantai 264025, China
* Correspondence: baiqingyuan1981@126.com or bqyuan@ldu.edu.cn

Abstract: Covalent organic frameworks (COFs) are a new type of metal-free porous architecture with a well-designed pore structure and high stability. Here an efficient electrochemical sensing platform was demonstrated based on COFs TpPa-1 constructed by 1,3,5-triformylphloroglucinol (Tp) with p-phenylenediamine (Pa-1), which possesses abundant nitrogen and oxo-functionalities. COFs TpPa-1 exhibited good water dispersibility and strong adsorption affinities for Pd^{2+} and thus was used as loading support to modify Pd^{2+}. The Pd^{2+}-modified COFs TpPa-1 electrode (Pd^{2+}/COFs) showed high electrocatalytic activity for both hydrazine oxidation reaction and nitrophenol reduction reaction. In addition, TpPa-1-derived nitrogen-doped carbon presented high activity for the electro-oxidation of reduced glutathione (GSH), and sensitive electrochemical detection of GSH was achieved. The presented COFs TpPa-1 can be utilized as a precursor as well as support for anchoring electro-active molecules and nanoparticles, which will be useful for electrochemical sensing and electrocatalysis.

Keywords: covalent organic frameworks; electrochemical sensor; hydrazine; nitrophenol; nitrogen doped carbon; reduced glutathione

Citation: Li, G.; Yuan, B.; Chen, S.; Gan, L.; Xu, C. Covalent Organic Frameworks-TpPa-1 as an Emerging Platform for Electrochemical Sensing. *Nanomaterials* **2022**, *12*, 2953. https://doi.org/10.3390/nano12172953

Academic Editor: Adriano Sacco

Received: 2 August 2022
Accepted: 22 August 2022
Published: 26 August 2022

1. Introduction

The synthesis and construction of efficient electrode materials to gain high sensitivity and selectivity is crucial for electroanalysis. The successful development of nonenzymatic electrochemical sensors depends critically on electrode materials with excellent electrocatalytic properties because many analytes are electrochemically inactive when using traditional electrodes. Up to now, different categories of materials, including noble metals, metal oxides, and carbons (such as CNT and graphene), have been applied in electrochemical sensing [1]. Recently, porous materials have demonstrated advantages in constructing electrochemical sensors due to their high specific surface areas and powerful adsorption and loading capability [2].

Metal–organic frameworks (MOFs) and covalent organic frameworks (COFs), as two emerging families of crystalline porous materials, have been explored in electroanalysis and electrocatalysis [3,4]. Compared to conventional porous materials such as zeolites, porous oxides, and carbons, MOFs and COFs show some intriguing features for electrochemical applications, including inter-connected porosity, tunable intra-framework chemical functionality, and high surface area [5]. Wang et al. synthesized graphene aerogel (GA) and metal–organic framework (MOF) composites via the in situ growth of the MOF UiO-66-NH_2 crystal on the GA matrix. The interface showed high sensitivity and selectivity for the electrochemical sensing of multiple heavy-metal ions. GA was employed as the backbone for UiO-66-NH_2, which also improved the electrical conductivity of the composites by accelerating the electron transfer in the matrix. UiO-66-NH_2 provided a binding site for heavy-metal ions due to the interaction between hydrophilic species and metal cations [6]. Xu et al. developed a facile furazolidone sensor based on COFs [7]. In this work, 1,3,5-tris-(4-amino-phenyl) benzene (TAPB) and terephthaldicarboxaldehyde (TPA) were used as monomers to synthesize the COF, which was then self-assembled on the surface of the NH_2-CNTs via a simple and rapid one-pot strategy at room temperature. The prepared COF@NH_2-CNT composites exhibited a high surface area and excellent conductivity, thus achieving high sensitive detection of furazolidone. However, most MOFs are not chemically stable in water. If the MOF is unstable in the aqueous buffer used for electroanalysis, the MOF coating modified on the electrode surface may suffer from dissolving or converting into MOF-derived metal hydroxides or oxides [8]. Unlike MOFs, COFs are regularly integrated into metal-free crystalline organic structures through strong covalent bonds between the organic building blocks, leading to more robust frameworks and relatively lower densities. Many COFs possess the sp^2 carbon networks with a high degree of p-conjunction and thus lead to at least a modest electronic conductivity, which facilitates the electron transfer rate. Furthermore, the presence of electronegative atoms (N and O) and π bonds within COFs can allow more specific target analyte recognition through the formation of hydrogen bonds or π-π bonds, thus improving the sensitivity and selectivity of COF-based electrochemical sensors [7]. These unique characteristics of COFs make them promising materials for electrochemical sensing.

COFs-derived porous carbons are also considered ideal platforms for electrochemical sensing due to their high conductivity, porosity, chemical and thermal stability, high electrocatalytic activity, and powerful supporting performances for anchoring molecules and nanomaterials. Most intrinsic electrocatalytic activities are contributed to the oxo-functionalities and edge-plane-like sites (defects) present on the surface of carbon materials [9]. COFs are often rich with nonmetal elements such as N, B, S, and P, which can induce heteroatom doping into derived carbons. Heteroatom doping forms charge delocalization and changes in the electronic structure due to the difference in atomic size, bond length, and coordination between the carbon atoms and dopants, leading to abundant defect sites [10]. N-doping can increase the surface polarity and wettability of the electrodes and also improve the electrocatalytic activity and adsorption [11,12]. Here COFs TpPa-1 constructed by 1,3,5-triformylphloroglucinol (Tp) with p-phenylenediamine (Pa-1) possess abundant nitrogen and oxo-functionalities, which was demonstrated for Pd^{2+} immobilization and fabricating nitrogen-doped porous carbon. The presented electrocatalysts showed high responses for hydrazine oxidation, nitrophenol reduction, and glutathione (GSH) oxidation.

2. Experimental

2.1. Chemicals and Solutions

GSH, palladium chloride ($PdCl_2$), hydrazine (wt. 80%), graphite powder, liquid paraffin, p-nitrophenol (PNP), and o-nitrophenol (ONP) were purchased from Sigma-Aldrich (St. Louis, MO, USA). COFs-TpPa-1 was purchased from Nanjing XFNano Materials Tech Co., Ltd. (Nanjing, China) The pore size of COFs-TpPa-1 is in the range of 1.5–1.8 nm with a BET specific surface area of ~1360 m^2/g. All of the other chemicals were of analytical reagent grade and used without further purification.

2.2. Apparatus

The morphologies were characterized by scanning electron microscopy (SEM) (Hitachi SU8010, Tokyo, Japan) and transmission electron microscopy (TEM) (JEM-2100, JEOL, Tokyo, Japan). The X-ray photoelectron spectroscopy (XPS) was recorded on a Thermo ESCALAB 250 Xi spectrometer (Waltham, MA, USA) fitted with a monochromatic Al Kα X-ray source. X-ray diffraction patterns were collected by a Bruker D8 (Billerica, MA, USA) advance powder X ray Cu Ka radiation diffractometer. All of the electrochemical experiments were conducted on a CHI 750E electrochemical workstation with a conventional three electrode system consisting of a modified working electrode, platinum coil auxiliary electrode, and Ag/AgCl (saturated KCl) reference electrode. Electrochemical impedance spectra (EIS) were carried out in 0.1 M KCl containing 5.0 mM Fe $(CN)_6{}^{3-}/{}^{4-}$ in the frequency range of 1 MHz to 0.1 Hz at 0.25 V.

2.3. Electrode Preparation and Modification

2.3.1. Preparation of Pd^{2+}/COFs/GCE

Prior to modification, the glassy carbon electrode (GCE) (3 mm diameter, 0.07 cm^2) was successively polished with 1, 0.3, and 0.05 μm alumina paste to a mirror finish and then rinsed with deionized water followed by ultrasonic treatment in water and ethanol, respectively. The COFs-modified GCE (COFs/GCE) was prepared by dropping 10.0 μL COFs aqueous suspension (2.0 mg/mL) on a cleaned GCE and dried at room temperature. In order to anchor Pd^{2+}, the prepared COFs/GCE was dipped in 10 mM $PdCl_2$ solution for 30 min, which was then washed with water and left to dry at room temperature (Pd^{2+}/COFs/GCE). Pd^{2+}/GCE was prepared by soaking the GCE in Pd^{2+} solution by electroless deposition. The electroactive surface areas of the different electrodes could be calculated by the Randles–Sevcik equation using the CV responses of the corresponding electrodes in 0.1 M KCl solution in the presence of 5.0 mM $K_3Fe(CN)_6$ at different scan rates at 298 K (Figures S1–S3).

2.3.2. Preparation of PCs/CPE

The nitrogen-doped porous carbon materials (PCs) were prepared by calcination. A number of COFs were calcined at 800 °C (heating rate of 10 °C min^{-1}) for 80 min under nitrogen flow in the tubular furnace, and black carbonized materials were obtained after cooling to room temperature naturally. To prepare the PCs modified electrode, 3.0 mg PCs were ultrasonically dispersed in 1.0 mL water for 1 h. The carbon paste electrode (CPE) was prepared by thoroughly mixing 1 mg of the graphite powder and 300 μL of liquid paraffin using an agate mortar and pestle for 40 min. Then, a suitable amount of the above mixture was filled firmly into one glass tube. The electrical contact was made by inserting a copper wire into the paste in the inner hole of the tube. Finally, 5.0 μL of 3.0 mg/mL PC suspension was dropped onto the freshly prepared CPE surface and dried at room temperature (PCs/CPE).

3. Results and Discussion

3.1. Electrochemical Sensing of Hydrazine

3.1.1. Characterization of Pd^{2+}/COFs/GCE

The surface morphology and elemental composition of Pd^{2+}/COFs/GCE were investigated by SEM and XPS. As shown in Figure 1, the SEM image of Pd^{2+}/COFs/GCE shows the rough and inter-connected porous surface. The morphology is mainly composed of complicated long strips, indicating that COFs were modified on the GCE surface. XPS was also carried out to explore the surface composition and chemical states (Figure 2). The characteristic peaks of C, N, O, and Pd elements are found in the XPS survey spectrum (Figure S4). The O1s XPS spectrum exhibits four peaks centered at 530.79, 531.96, 533.18, and 535.29 eV (Figure 2a), which can be attributed to C−OH, C=O, C−O, or H−O, and chemisorbed oxygen in carboxylic groups or water (H-O-H), respectively [13]. The abundant oxo functionalities present on COFs can improve their dispersion in water, which is beneficial for the modification of COFs on the electrode surface. The N1s XPS spectra show two characteristic peaks at 400.06 and 403.53 eV (Figure 2c), which are assigned to the C=N and C-N groups, respectively [14]. Figure 2b displays the XPS Pd3d spectra of Pd^{2+}/COFs, including two peaks at 338.10 and 343.40 eV, corresponding to the peak of the low energy band (Pd3d 3/2) and high energy band (Pd3d 5/2), respectively. The Pd^{2+} species might be attached to COFs via N coordination [15]. The results confirm the existence of Pd on the composite surface, and the Pd^{2+} is successfully loaded on the COFs. The XPS data showed that the atomic amount of Pd is 0.16%. In order to investigate the stability of COFs, the XRD patterns of COFs, Pd^{2+}/COFs, and Pd^{2+}/COFs soaked in 0.1 M NaOH were performed (Figure 3). The characteristic peaks of these samples observed in XRD patterns were in accord with the reported literature [15,16] and remained unchanged, indicating that Pd^{2+} modification and NaOH solution soaking did not change the crystalline structure. The results showed that COF-TpPa-1 possessed good chemical stability. It was also found that COF-TpPa-1 was still stable even though soaked in 1.0 mol/L HNO_3, 6.0 mol/L NaOH, and HCl for three days [16].

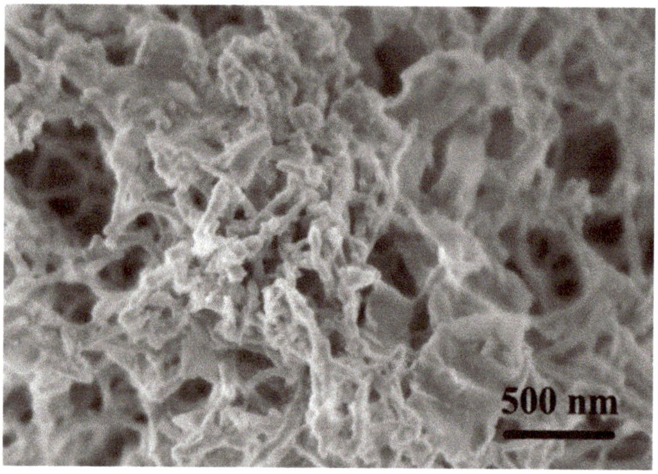

Figure 1. SEM image of Pd^{2+}/COFs/GCE.

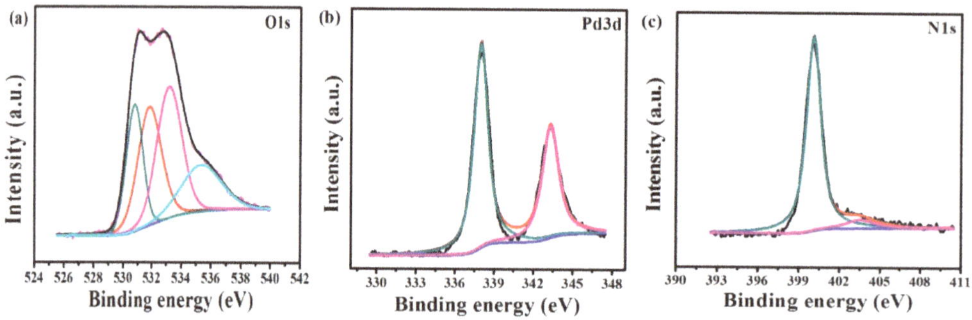

Figure 2. XPS (**a**) O1s, (**b**) Pd3d and (**c**) N1s spectra of Pd2+/COFs.

3.1.2. Electrochemical Oxidation of Hydrazine

Figure 4 shows the CVs of COFs/GCE, Pd^{2+}/GCE, and Pd^{2+}/COFs/GCE in 0.1 M NaOH in the absence and presence of 5.0 mM hydrazine at a scan rate of 100 mV s^{-1}. At COFs/GCE, the hydrazine oxidation began at the potential of 0.25 V (Figure 4a) and reached the oxidation peak at about 0.75 V. However, for Pd^{2+}/GCE, two oxidation peaks were observed at −0.25 and 0.75 V, respectively (Figure 4b). Pd^{2+} was reduced to Pd metal by electroless deposition when GCE was immersed in a Pd^{2+} solution [17]. The first oxidation wave was ascribed to the electrocatalytic activity of Pd, while the second one originated from the bare GCE. Pd^{2+}/COFs/GCE showed three oxidation peaks at −0.50, −0.33, and 0 V (Figure 4c), and no oxidation peak was found at about 0.75 V. This indicates that Pd^{2+}/COFs/GCE demonstrated the lowest oxidation over-potential among these modified electrodes, suggesting that the combination of Pd^{2+} and COFs formed new catalytic sites and thus enhanced the oxidation of hydrazine. The different electrodes were also investigated by EIS (Figure S5). The electron transfer rate of different electrodes follows the order: Bare GCE > Pd^{2+}/COFs/GCE > COFs/GCE. The higher charge transfer resistance of COFs/GCE is attributed to the lower electrical conductivity.

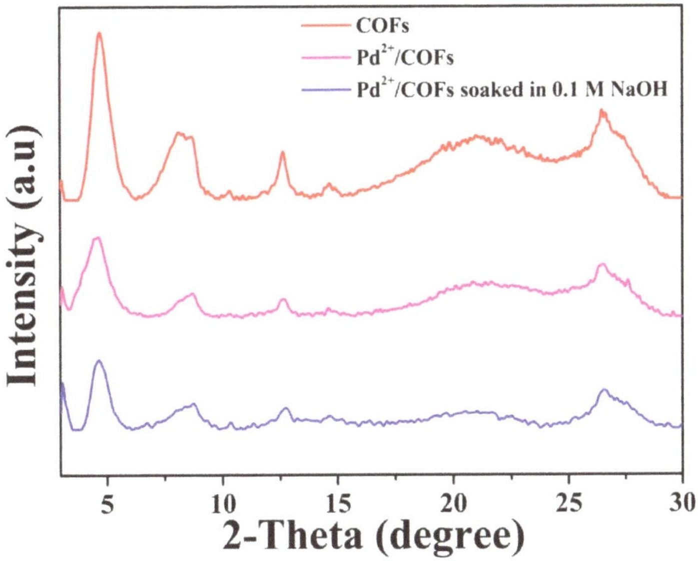

Figure 3. XRD patterns of COFs, Pd^{2+}/COFs and Pd^{2+}/COFs soaked in 0.1 M NaOH for 24 h.

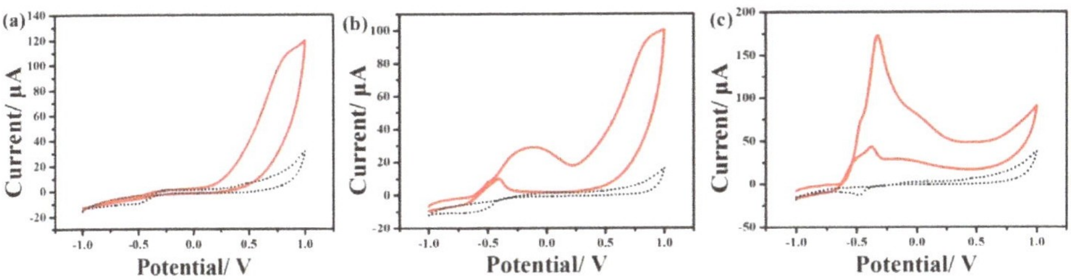

Figure 4. CVs of COFs/GCE (**a**), Pd^{2+}/GCE (**b**), Pd^{2+}/COFs/GCE (**c**) in the absence (dotted line) and presence (solid line) of 5 mM hydrazine at a scan rate of 100 mV s^{-1}, respectively. (Electrolyte: 0.1 M NaOH).

3.1.3. Analytical Performances for Hydrazine

Figure 5 shows the amperometric current–time response of Pd^{2+}/COFs/GCE to the successive additions of hydrazine in 0.1 M NaOH. Upon each successive addition of different concentrations of hydrazine, a stair-shaped slot was raised rapidly and then reached a steady-state oxidation current within 2 s. The inset (a) of Figure 5 shows a linear response over a concentration range from 0.5 μM to 1.6 mM with a sensitivity of 10 μA μM^{-1} cm^{-2} and a detection limit of 0.2 μM (S/N = 3). Table S1 summarizes the sensing performances of hydrazine with Pd^{2+}/COFs/GCE and other modified electrodes presented in the literature. The results show that Pd^{2+}/COFs/GCE demonstrated an excellent analytical performance for hydrazine, owing to the high electrocatalytic activity.

In order to explore the selectivity of Pd^{2+}/COFs/GCE, the interference test was also examined at Pd^{2+}/COFs/GCE with 50.0 μM hydrazine in 0.1 M NaOH containing 0.5 mM of the possible interference substances such as KCl, NaCl, Na_2SO_4, NaBr, $CaCl_2$, NaAc, glucose, $(NH_4)_2SO_4$, $NaNO_3$, and $NaNO_2$ (Figure S6). It was found that 100-fold interferents almost did not interfere with the determination of hydrazine. Moreover, a remarkable current response with the addition of hydrazine was observed again after adding the mentioned interference species. The results indicate that our proposed Pd^{2+}/COFs/GCE sensor displays excellent selectivity towards the electrochemical oxidation of hydrazine.

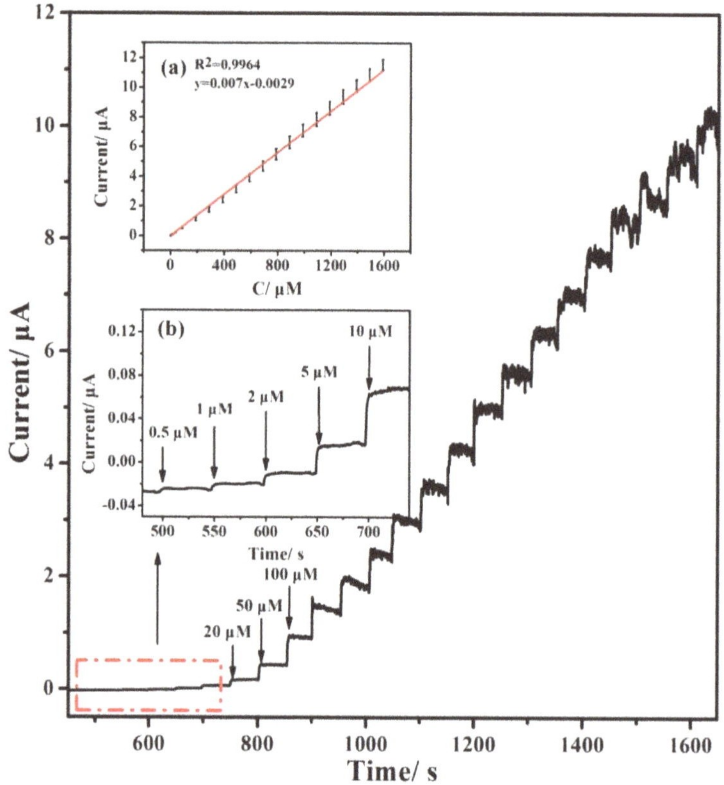

Figure 5. Amperometric responses of the Pd^{2+}/COFs/GCE to successive additions of various concentrations of hydrazine in 0.1 M NaOH at −0.1 V. (Inset: (**a**) the corresponding calibration plot, (**b**) the amplified response curve.).

The long-term stability of the sensor was further tested by a continuous operation. After 1000s of continuous i-t running for 10.0 μM hydrazine, the electrode still retained 89% of its initial value. In addition, ten successive analyses of 10.0 μM hydrazine using the same electrode showed a relative standard deviation (RSD) value of 4.5%. A reproducible experiment was also evaluated by using five parallel electrodes for the same solution, and the RSD of the current response was found to be 6.4%, suggesting an acceptable reproducibility.

3.1.4. Sample Analysis of Hydrazine

To confirm the practical applicability of the proposed sensor, the validity of Pd^{2+}/COFs/GCE for the determination of hydrazine spiked in real samples was tested by a standard addition method (Figure S7). The prepared sensor presented acceptable recoveries of hydrazine in lake water samples (Table S2).

3.2. Electrochemical Sensing of ONP and PNP

3.2.1. Electrochemical Behaviors of ONP and PNP

The electrochemical behaviors of ONP and PNP were investigated at bare GCE (a), COFs/GCE (b), and Pd^{2+}/COFs/GCE (c) by CV at a scan rate of 50 mV s^{-1} in 0.1 M PBS (pH = 7.0) (Figure 6). The CVs of these electrodes in 0.1 M PBS are shown in Figure S8. For PNP, three redox peaks were observed, with one main irreversible cathodic peak located at −0.79 V (R_1) and a pair of reversible redox peaks centered at about 0.08 V (R_2) and 0.14 V (O_1) at bare GCE and COFs/GCE. Pd^{2+}/COFs/GCE also exhibited the same redox peaks, but the peak of R_1 shifted to −0.56 V. However, at bare GCE and COFs/GCE, upon the addition of ONP, four redox peaks, including the three peaks and an additional oxidation peak (O_2) appeared at −0.68 (R_1), −0.30 (R_2), −0.25 (O_1), and −0.23 V (O_2), respectively. Pd^{2+}/COFs/GCE demonstrated the same redox peaks but a much lower R_1 peak (−0.54 V). This indicated that Pd^{2+}/COFs/GCE presented superior electrocatalytic activity towards the R_1 reduction process of ONP and PNP, which was ascribed to the combination of the high catalytic performance of Pd^{2+} and efficient supporting effect of COFs.

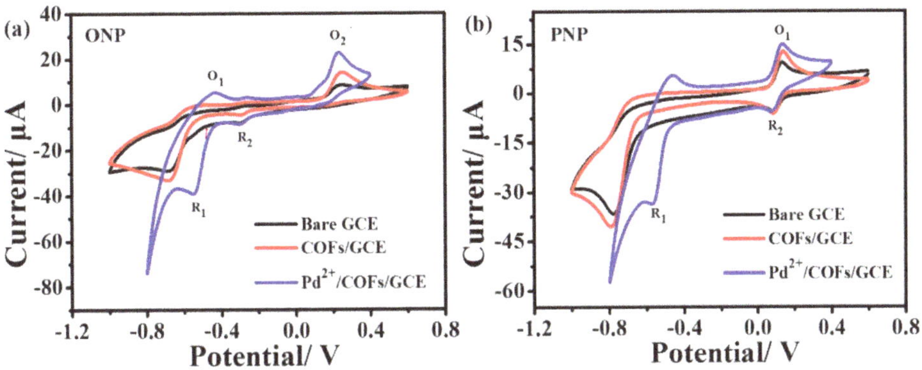

Figure 6. CVs of bare GCE, COFs/GCE, Pd^{2+}/COFs/GCE in the presence of 500 μM nitrophenol ((**a**): ONP, (**b**): PNP) at a scan rate of 50 mV s^{-1} in 0.1 M PBS (pH = 7.0).

When the potential was scanned negatively, both ONP and PNP could be reduced to hydroxylaminophenol through a four-electron and four-proton transfer process (R_1). Reversible transformation (O_1/R_2) between hydroxylaminophenol and nitrosophenol then occurred with the potential scanning back. In addition, the phenolic group of ONP gave further oxidation through a proton and an electron transfer process (O_2). The detailed reaction mechanism was proposed as Scheme 1 [18]:

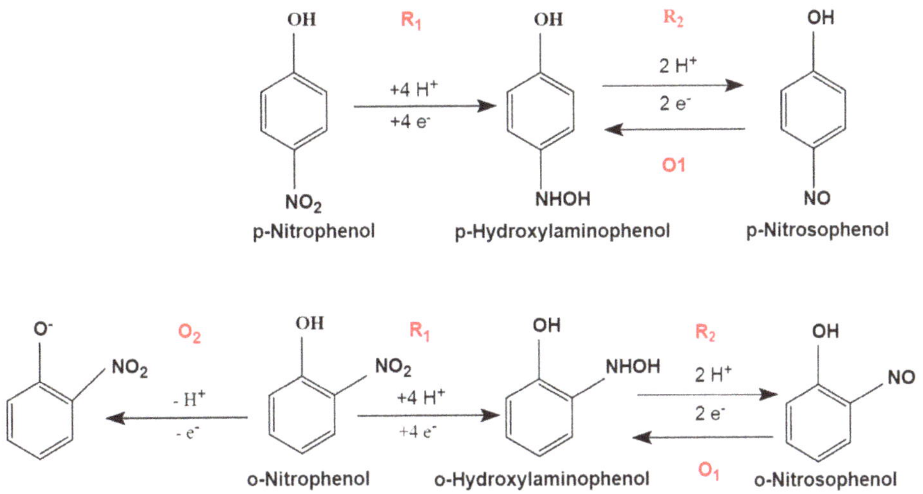

Scheme 1. Electrochemical reaction mechanism of PNP and ONP at the Pd^{2+}/COFs/GCE.

To further investigate the electrochemical performance of Pd^{2+}/COFs/GCE, the effect of the scan rate on the redox peak current (I_{pc}) was studied (Figure S9). The results showed that the peak currents of ONP and PNP were quadratic functions of the square root of the scan rate ($v^{1/2}$), which suggested that the reactions of ONP and PNP on the surface of Pd^{2+}/COFs/GCE were mixed controlled processes (surface-controlled at lower scan rates but diffusion-controlled at higher scan rates).

Next, the effect of the pH on the cathodic peak current and potential (R_1) of ONP and PNP (500 μM) was studied by DPV on the Pd^{2+}/COFs/GCE in 0.1 M PBS (Figure S10). The reduction peak current of both ONP and PNP increased firstly and then decreased and reached the maximum at pH = 7.0. In the basic solution, the protonation of PNP and ONP decreases, and the electron density of the nitro group increases due to the decrease in proton concentration, thus hampering the electrochemical reduction of PNP and ONP [19]. Therefore, pH 7.0 was selected for further electrochemical experiments.

3.2.2. Analytical Performances for ONP and PNP

Under optimal conditions, the DPV response of Pd^{2+}/COFs/GCE for different concentrations of ONP and PNP in 0.1 M PBS (pH = 7.0) was investigated, respectively (Figure 7a,b). The reduction peak current increased gradually with the concentration of ONP and PNP and exhibited a log–log linear relationship in the range of 5.0 μM–2.0 mM. The detection limit of ONP and PNP was estimated to be 1.6 and 1.3 μM (S/N = 3), respectively. Pd^{2+}/COFs/GCE was also applied for the simultaneous detection of ONP and PNP. Figure 7c clearly shows the well-separated reduction peak for ONP and PNP. The calibration curves reveal the favorable linear relationships between the logarithm of peak current and the logarithm of concentrations (5.0 μM–2.0 mM), with 1.8 and 0.9 μM for ONP and PNP, respectively. The electrochemical sensing performances based on Pd^{2+}/COFs/GCE and other reported sensors are summarized in Table S3. The results show that the Pd^{2+}/COFs/GCE sensor exhibits comparable performances with the reported methods for the detection of ONP and PNP.

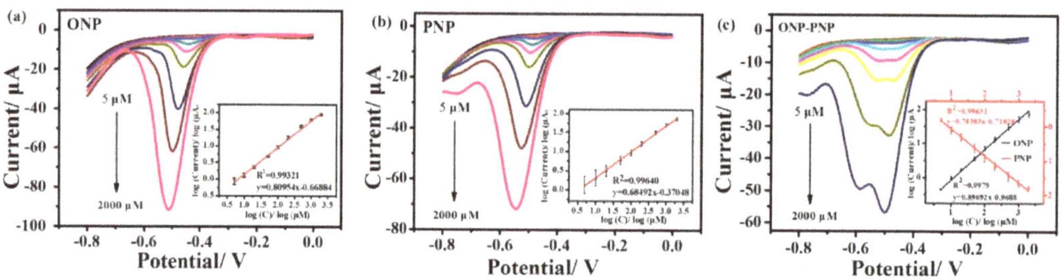

Figure 7. DPV of Pd^{2+}/COFs/GCE with individual addition of (**a**) ONP and (**b**) PNP (5, 10, 20, 50, 100, 200, 500, 1000, and 2000 μM) into 0.1 M PBS (pH = 7.0), respectively. Inset: Linear relationship between logarithm of peak current and logarithm of concentrations of (**a**) ONP and (**b**) PNP. DPV of Pd^{2+}/COFs/GCE with simultaneous addition of the same concentration of (**c**) ONP and PNP (5–2000 μM) into 0.1 M PBS (pH = 7.0). Inset: Linear relationship between logarithm of peak current and logarithm of concentrations of (**c**) ONP and PNP.

The antifouling property of different electrodes was investigated by successive CVs in the ONP and PNP solution, respectively (Figures S11–S13). The results showed that COFs/GCE exhibited the best antifouling property because of the hydrophilicity of COFs. For ONP, Pd^{2+}/COFs/GCE showed improved antifouling properties than GCE in the first six cycles. However, for PNP, both Pd^{2+}/COFs/GCE and bare GCE exhibited lower antifouling properties.

3.3. Electrochemical Sensing of GSH

3.3.1. Characterization of PC

The morphology of the PCs was characterized by SEM and TEM images. The SEM image of PCs shows the overlapping structure of stratified lamellae (Figure S14a). As shown in Figure S14b, the flake morphology can be observed in the TEM image of PCs. Furthermore, XPS characterization was employed to investigate the elemental compositions and chemical states of PCs. Figure 8 shows the XPS spectra of PCs containing C, N, and O elements. The C 1s spectra suggest the coexistence of C-C (283.60 eV), C=C (284.50 eV), C=O (287.70 eV), and π-π (292.30 eV) in PCs (Figure 8a) [20]. The N1s XPS spectra show two characteristic peaks at 397.10 and 400.00 eV (Figure 8b), which are attributed to the C=N and C-N groups, respectively [10]. Figure 8c displays the O 1s XPS spectrum of PCs, including two peaks at 531.30 and 532.20 eV, representing the C=O and C-O bonds, respectively [21]. The XPS spectra suggest that a large number of functional groups are present on the surface of the PCs.

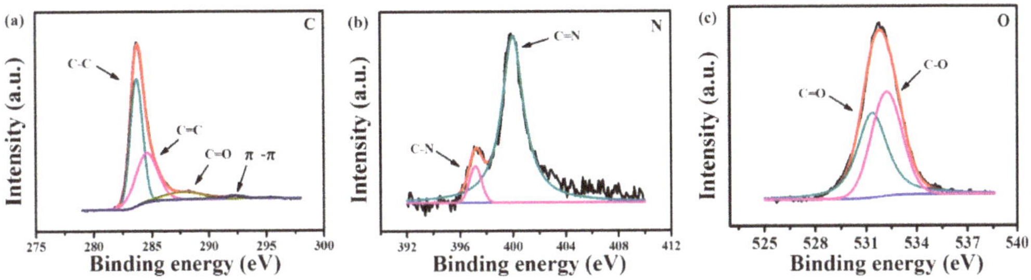

Figure 8. XPS (**a**) C, (**b**) N and (**c**) O spectra of PCs.

3.3.2. Electrochemical Oxidation of GSH

GSH is an important antioxidant in protecting cells from oxidative damage by oxidizing to form glutathione disulfifide (GSSG), and the balance of GSH and GSSG acts as an indicator of many human illnesses such as several cancers, Parkinson's disease, Alzheimer's, and HIV [22]. Metal and non-metal electrocatalysts have been explored for the non-enzymatic detection of biological thiols [23–25]. Non-metal electrocatalysts consist of redox-active organic molecules [23,26] and carbon materials, including carbon nanotube [27], ordered mesoporous carbon [28], graphene oxide [29], graphene oxide nanoribbons [30], and derived carbon [31]. In fact, the oxo-functionalities present on these carbons contribute to the electrocatalytic activity towards thiols. Recently, nitrogen-doped carbon has proven effective in not only improving the electrical conductivity but also promoting defect sites in the carbon network for better catalytic activity towards GSH [32]. Here, the N-doped PCs with abundant oxo-functionalities were explored for the electro-oxidation of GSH. Figure 9 shows the CV response for the electrochemical oxidation of GSH at PCs/CPE in 0.1 M PBS (pH 7.0) at a scan rate of 50 mV s^{-1}. As can be seen, in the absence of GSH, no anodic peak was observed on the PCs/CPE. However, in the presence of GSH, it could be noticed that a remarkable anodic current emerged at about 0.10 V, and the anodic peak was observed at 0.25 V, which could be ascribed to some oxygen-containing functional groups on PCs promoting the oxidation of GSH [26]. Furthermore, there was a very small anodic peak at 0.28 V, and this could be attributed to the edge plane-defective sites on PCs [33]. Furthermore, no cathodic current peak appeared in the CV curves, confirming that the GSH oxidation was a fully irreversible process. The results suggest that the N-doping and the oxo-functionalities present on PCs enhance the rate of electron transfer and exhibit efficient electrocatalytic activity for the oxidation of GSH.

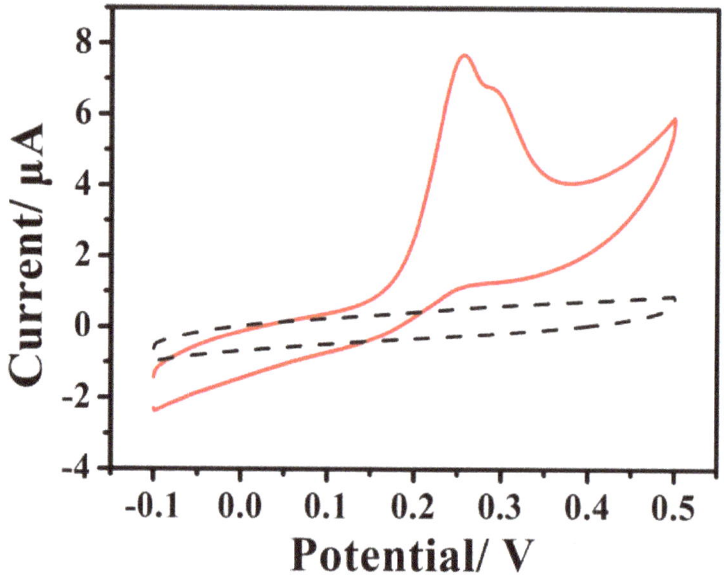

Figure 9. CV of PCs/CPE in the absence (dotted line) and presence (solid line) of 5 mM GSH in 0.1 M PBS (pH = 7.0). Scan rate: 50 mV s^{-1}.

The pH value is a crucial factor for the electrochemical oxidation of GSH, which was examined in 0.1 M PBS with various pH values containing 5.0 mM GSH on the PCs/CPE (Figure S15). The anodic peak current of GSH increased firstly and then decreased as the solution pH value increased from 3.0 to 5.0 and reached a maximum at pH = 4.0. Accordingly, pH = 4.0 is selected as the optimal condition for further detection. GSH is an important tripeptide consisting of L-cysteine, L-glutamine, and glycine with pKa1 = 2.12 (carboxylic acid of glutaminic acid), pKa2 = 3.59 (carboxylic acid of glycine), pKa3 = 8.75 (thiol of cysteine), and pKa4 = 9.65 (ammonium of glutaminic acid) [34]. Gilbertson et al. investigated the interaction and catalytic oxidation mechanisms between GO and GSH. It was found that the synergism between the adjacent epoxide and hydroxyl groups on the GO surface contributed to the catalytic oxidation of GSH [35]. It was speculated that the pH value of the solution is related to both the activity of oxo-functionalities and the ionization of GSH.

3.3.3. Analytical Performances for GSH

Under the optimal conditions, the amperometric responses of the PCs/CPE were conducted by successive injections of various concentrations of GSH into the stirred 0.1 M PBS (pH = 4.0) solution (Figure 10). Upon each addition of 200 μM GSH, a well stair-shaped amperometric response was raised quickly and reached a steady-state oxidation current within 5 s. The inset (a) of Figure 10 shows the response current of the PCs/CPE to various concentrations of GSH, and it is clear that the current increased rapidly with the addition of GSH. The enhancement of response current is ascribed to the excellent electrocatalytic property of the PCs/CPE. Figure 10b displays the linear correlation between the GSH oxidation current and concentration ranging from 5.0 μM to 1.2 mM and 1.2 to 6.4 mM, respectively. The limit of detection was calculated to be 2 μM (S/N = 3). Nevertheless, as the GSH concentration increased gradually, the sensitivity of the prepared sensor slightly decreased owing to the binding of the sulfur moiety to the electrode surface caused by the oxidation of GSH [26].

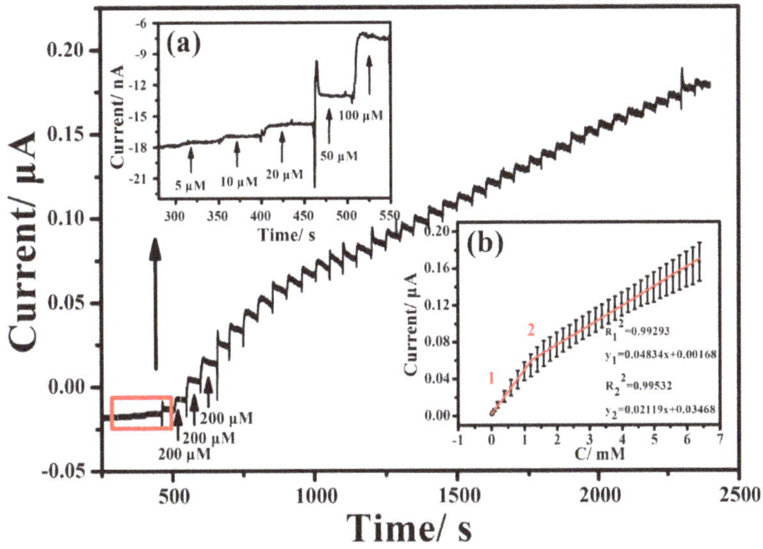

Figure 10. Amperometric responses of the PCs/CPE to various concentrations of GSH in 0.1 M PBS (pH = 4.0) at 0.25 V. (Inset: (**a**) the amplified response curve, (**b**) the calibration curve for the different concentrations of GSH).

The current responses of the PCs/CPE to various concentrations of GSH in the static 0.1 M PBS (pH = 4.0) were also conducted using the amperometric technique. The inset of Figure S16 shows that the amperometric current increased linearly with the concentrations of GSH in a wide range from 5.0 μM to 10.0 mM. The limit of detection was also calculated to be 1.0 μM (S/N = 3). Table S4 summarizes the electrochemical sensing performances of the PCs/CPE and other reported sensors. The results indicate that the PCs/CPE exhibits excellent performances compared with other reported electrodes for the detection of GSH.

3.3.4. Sample Analysis of GSH

To further assess the practical applications of the proposed PCs/CPE biosensor, the validity of PCs/CPE for the determination of GSH in spiked ample was investigated by both successive and non-successive addition methods (Figures S17 and S18). The recoveries of GSH ranged from 92% to 99% and 92% to 107%, respectively (Tables S5 and S6).

4. Conclusions

In this work, COFs TpPa-1 was explored as an efficient supporting material and precursor to construct electrocatalytic interfaces for electrochemical sensing, and environmental pollutants and biological thiol were sensitively detected. COFs TpPa-1 possesses abundant oxo-functionalities and nitrogen elements, which endow good water dispersibility, powerful adsorption ability, and enhanced electrical conductivity. The results show that COFs TpPa-1 will be a potential material in synthesizing functional materials for electrocatalysis and electroanalysis.

Supplementary Materials: The following supporting information can be downloaded at: https://www.mdpi.com/article/10.3390/nano12172953/s1, Figure S1: The CVs of COFs/GCE in 0.1 M KCl solution in the presence of 5 mM $K_3Fe(CN)_6$ at different scan rates; Figure S2: The CVs of Pd^{2+}/COFs/GCE in 0.1 M KCl solution in the presence of 5 mM $K_3Fe(CN)_6$ at different scan rates; Figure S3: The CVs of PCs/GCE in 0.1 M KCl solution in the presence of 5 mM $K_3Fe(CN)_6$ at different scan rates; Figure S4: XPS spectra of Pd^{2+}/COFs; Figure S5: EIS of PCs /GCE, Bare GCE, COFs/GCE, and Pd^{2+}/COFs/GCE in 0.1 M KCl solution in the presence of 5 mM $K_3Fe(CN)_6$; Figure S6: Amperometric curve of Pd^{2+}/COFs/GCE with 50 μM hydrazine and 500 μM of interferences such as KCl, NaCl, Na_2SO_4, NaBr, $CaCl_2$, NaAc, glucose, $(NH_4)_2SO_4$, $NaNO_3$, and $NaNO_2$. (Electrolyte: 0.1 M NaOH); Figure S7: Amperometric curve of Pd^{2+}/COFs/GCE for the detection of hydrazine spiked in water samples at −0.1 V. (Electrolyte: 0.1 M NaOH); Figure S8: CVs of Bare GCE, COFs/GCE, and Pd^{2+}/COFs/GCE in 0.1 M PBS at a scan rate of 50 mV s^{-1} (pH = 7.0); Figure S9: CVs of 500 μM of (a) ONP and (b) PNP on Pd^{2+}/COFs/GCE with different scan rates (20–200 mV s^{-1}) in 0.1 M PBS (PH = 7.0). (Inset: the corresponding calibration plot.); Figure S10: DPV curves of 500 μM of (a) ONP and (b) PNP on Pd^{2+}/COFs/GCE in 0.1 M PBS (pH = 7.0) with different pH values (1–5: 6.5, 7.0, 7.5, 8.0, and 8.5). Plots of I_{pc} vs. pH for (c) ONP and (d) PNP; Figure S11: Successive CVs of Bare GCE (a), COFs/GCE (b), and Pd^{2+}/COFs/GCE (c) in the presence of 500 μM nitrophenol ONP at a scan rate of 50 mV s^{-1} in 0.1 M PBS (pH = 7); Figure S12: Successive CVs of Bare GCE (a), COFs/GCE (b), and Pd^{2+}/COFs/GCE (c) in the presence of 500 μM nitrophenol PNP at a scan rate of 50 mV s^{-1} in 0.1 M PBS (pH = 7.0); Figure S13: Normalized peak current of bare GCE, COFs/GCE, and Pd^{2+}/COFs/GCE in the presence of 500 μM nitrophenol (a: ONP, b: PNP) at a scan rate of 50 mV s^{-1} in 0.1 M PBS (pH = 7.0) for ten cycles; Figure S14: SEM (a) and TEM (b) images of PCs; Figure S15: CVs of 5 mM of GSH on (a) PCs/CPE in 0.1 M PBS with different pH values (1–5: 3.0, 3.5, 4.0, 4.5 and 5.0). Plots of I_{pc} vs. pH for (b) GSH; Figure S16: Amperometric responses of PCs/CPE to additions of various concentrations of GSH in 0.1 M PBS (pH = 4.0) at 0.25 V. (Inset: the corresponding calibration plot.); Figure S17: Amperometric responses of PCs/CPE to successive additions of various concentrations of GSH spiked in serum in 0.1 M PBS (pH = 4.0) at 0.25 V; Figure S18: Amperometric responses of PCs/CPE for the detection of GSH spiked in human serum in 0.1 M PBS (pH = 4.0) at 0.25 V; Table S1: Comparison of the Pd^{2+}/COFs/GCE for different parameters with previously reported hydrazine electrodes; Table S2: Determination of hydrazine spiked in water samples; Table S3: Comparison of different modified GCE for detection of ONP and PNP; Table S4; Comparison of different modified electrodes for electrochemical sensing of GSH; Table S5; Determination of GSH spiked in human serum by successive additions;Table S6; Determination of GSH spiked in human serum by a non-successive addition method. References [36–68] were cited in the Supplementary Materials.

Author Contributions: Conceptualization, B.Y.; Data curation, B.Y., G.L. and C.X.; Formal Analysis, B.Y. and G.L.; Funding acquisition, B.Y.; Investigation, G.L., S.C. and L.G.; Methodology, B.Y.; Project administration, B.Y. and C.X.; Supervision, B.Y.; Validation, B.Y.; Visualization, B.Y. and G.L.; Writing—original draft, B.Y. and G.L.; Writing—review & editing, B.Y. and C.X. All authors have read and agreed to the published version of the manuscript.

Funding: This research was funded by the Natural Science Foundation of Shandong Province, China (No.ZR2020MB058).

Institutional Review Board Statement: Not applicable.

Informed Consent Statement: Not applicable.

Data Availability Statement: The authors confirm that the data supporting the findings of this study are available within the article [and/or] its supplementary materials.

Conflicts of Interest: The authors declare no conflict of interest.

References

1. Liu, H.Y.; Weng, L.Y.; Yang, C. A review on nanomaterial-based electrochemical sensors for H_2O_2, H_2S and NO inside cells or released by cells. *Microchim. Acta* **2017**, *184*, 1267–1283. [CrossRef]
2. Liu, C.S.; Li, J.J.; Pang, H. Metal-organic framework-based materials as an emerging platform for advanced electrochemical sensing. *Coordin. Chem. Rev.* **2020**, *410*, 213222. [CrossRef]
3. Yan, Y.; He, T.; Zhao, B.; Qi, K.; Liu, H.F.; Xia, B.Y. Metal/covalent-organic frameworks-based electrocatalysts for water splitting. *J. Mater. Chem. A* **2018**, *6*, 15905–15926. [CrossRef]
4. Yuan, B.Q.; Liu, D.; Yin, H.J.; Zhang, D.J. Materials for Electroanalysis Based on Advanced Frameworks. *Front. Chem.* **2021**, *3*, 638338. [CrossRef]
5. Chuang, C.H.; Kung, C.W. Metal-organic frameworks toward electrochemical sensors: Challenges and opportunities. *Electroanalysis* **2020**, *32*, 1885–1895. [CrossRef]
6. Lu, M.X.; Deng, Y.J.; Luo, Y.; Lv, J.P.; Li, T.B.; Xu, J.; Chen, S.W.; Wang, J.Y. Graphene aerogel–metal–organic framework-based electrochemical method for simultaneous detection of multiple heavy-metal ions. *Anal. Chem.* **2019**, *91*, 888–895. [CrossRef] [PubMed]

7. Sun, Y.F.; Waterhouse, G.I.; Xu, L.H.; Qiao, X.G.; Xu, Z.X. Three-dimensional electrochemical sensor with covalent organic framework decorated carbon nanotubes signal amplification for the detection of furazolidone. *Sens. Actuators B Chem.* **2020**, *321*, 128501. [CrossRef]
8. He, W.H.; Ifraemov, R.; Raslin, A.; Hod, I. Room-Temperature Electrochemical Conversion of Metal-Organic Frameworks into Porous Amorphous Metal Sulfides with Tailored Composition and Hydrogen Evolution Activity. *Adv. Funct. Mater.* **2018**, *28*, 1707244. [CrossRef]
9. Beitollahi, H.; Movahedifar, F.; Tajik, S.; Jahani, S. A review on the effects of introducing CNTs in the modification process of electrochemical sensors. *Electroanalysis* **2019**, *31*, 1195–1203. [CrossRef]
10. Hu, X.W.; Long, Y.; Fan, M.Y.; Yuan, M.; Zhao, H.; Ma, J.T.; Dong, Z.P. Two-dimensional covalent organic frameworks as self-template derived nitrogen-doped carbon nanosheets for eco-friendly metal-free catalysis. *Appl. Catal. B Environ.* **2019**, *244*, 25–35. [CrossRef]
11. Yang, L.; Cai, Z.; Hao, L.; Ran, L.L.; Xu, X.; Dai, Y.; Pan, S.Y.; Jing, B.J.; Zhou, J.L. Increase of structural defects by N doping in MoS_2 cross-linked with N-doped CNTs/carbon for enhancing charge transfer in oxygen reduction. *Electrochim. Acta* **2018**, *283*, 448–458. [CrossRef]
12. Sun, F.; Liu, X.Y.; Wu, H.B.; Wang, L.J.; Gao, J.H.; Li, H.X.; Lu, Y.F. In situ high-level nitrogen doping into carbon nanospheres and boosting of capacitive charge storage in both anode and cathode for a high-energy 4.5 V full-carbon lithium-ion capacitor. *Nano Lett.* **2018**, *18*, 3368–3376. [CrossRef] [PubMed]
13. Wang, J.; Yin, D.G.; Guo, X.D.; Luo, Z.Y.; Tao, L.Y.; Ren, J.J.; Zhang, Y. Fabrication of a Covalent Organic Framework-Based Heterojunction via Coupling with ZnAgInS Nanosphere with High Photocatalytic Activity. *Langmuir* **2022**, *38*, 4680–4691. [CrossRef]
14. Zhong, X.; Lu, Z.P.; Liang, W.; Hu, B.W. The magnetic covalent organic framework as a platform for high-performance extraction of Cr (VI) and bisphenol a from aqueous solution. *J. Hazard. Mater.* **2020**, *393*, 122353. [CrossRef] [PubMed]
15. Wang, F.N.; Zhang, J.X.; Shao, Y.H.; Jiang, H.; Liu, Y.F.; Chen, R.Z. Pd nanoparticles loaded on two-dimensional covalent organic frameworks with enhanced catalytic performance for phenol hydrogenation. *Ind. Eng. Chem. Res.* **2020**, *59*, 18489–18499. [CrossRef]
16. Li, Z.D.; Zhang, H.Q.; Xiong, X.H.; Luo, F. U (VI) adsorption onto covalent organic frameworks-TpPa-1. *J. Solid State Chem.* **2019**, *277*, 484–492. [CrossRef]
17. Choi, H.C.; Shim, M.; Bangsaruntip, S. Spontaneous reduction of metal ions on the sidewalls of carbon nanotubes. *J. Am. Chem. Soc.* **2002**, *124*, 9058–9059. [CrossRef]
18. Wang, Q.Z.; Li, R.; Zhao, Y.J.; Zhe, T.T.; Bu, T.; Liu, Y.N.; Sun, X.Y.; Hu, H.F.; Zhang, M.; Zheng, X.H. Surface morphology-controllable magnetic covalent organic frameworks: A novel electrocatalyst for simultaneously high-performance detection of p-nitrophenol and o-nitrophenol. *Talanta* **2020**, *219*, 121255. [CrossRef]
19. Afzali, F.; Zavar, M.H.A.; Rounaghi, G.; Ashraf, N.J.E.A. Gold digital versatile disc platform modified with nano-porous mercury/gold amalgam as a solid-state disposable electrochemical sensor for detection of para-nitrophenol. *Electrochim. Acta* **2016**, *209*, 654–660. [CrossRef]
20. Chen, X.; Wang, X.; Fang, D. A review on C1s XPS-spectra for some kinds of carbon materials. *Fuller. Nanotub. Carbon Nanostruct.* **2020**, *28*, 1048–1058. [CrossRef]
21. Cao, S.; Yang, J.X.; Li, J.; Shi, K.; Li, X.K. Preparation of oxygen-rich hierarchical porous carbon for supercapacitors through the co-carbonization of pitch and biomass. *Diamond Relat. Mater.* **2019**, *96*, 118–125. [CrossRef]
22. Townsend, D.M.; Tew, K.D.; Tapiero, H.J.B. The importance of glutathione in human disease. *Biomed. Pharmacother.* **2003**, *57*, 145–155. [CrossRef]
23. Harfield, J.C.; Batchelor-McAuley, C.; Compton, R.G. Electrochemical determination of glutathione: A review. *Analyst* **2012**, *137*, 2285–2296. [CrossRef] [PubMed]
24. Hanko, M.; Švorc, L'.; Planková, A.; Mikuš, P. Overview and recent advances in electrochemical sensing of glutathione—A review. *Anal. Chim. Acta* **2019**, *1062*, 1–27. [CrossRef] [PubMed]
25. Nesakumar, N.; Berchmans, S.; Alwarappan, S. Chemically modified carbon based electrodes for the detection of reduced glutathione. *Sens. Actuators B Chem.* **2018**, *264*, 448–466. [CrossRef]
26. Yuan, B.Q.; Xu, C.Y.; Zhang, R.C.; Lv, D.H.; Li, S.J.; Zhang, D.J.; Liu, L.; Fernandez, C. Bioelectronics, Glassy carbon electrode modified with 7, 7, 8, 8-tetracyanoquinodimethane and graphene oxide triggered a synergistic effect: Low-potential amperometric detection of reduced glutathione. *Biosens. Bioelectron.* **2017**, *96*, 1–7. [CrossRef]
27. Lee, P.; Compton, R.G. Selective electrochemical detection of thiol biomarkers in saliva using multiwalled carbon nanotube screen-printed electrodes. *Sens. Actuators B Chem.* **2015**, *209*, 983–988. [CrossRef]
28. Zhou, M.; Ding, J.; Guo, L.P.; Shang, Q.K. Electrochemical behavior of L-cysteine and its detection at ordered mesoporous carbon-modified glassy carbon electrode. *Anal. Chem.* **2007**, *79*, 5328–5335. [CrossRef]
29. Yuan, B.Q.; Zeng, X.Y.; Xu, C.Y.; Liu, L.; Ma, Y.H.; Zhang, D.J.; Fan, Y. Electrochemical modification of graphene oxide bearing different types of oxygen functional species for the electro-catalytic oxidation of reduced glutathione. *Sens. Actuators B Chem.* **2013**, *184*, 15–20. [CrossRef]
30. Wu, S.; Lan, X.Q.; Huang, F.F.; Luo, Z.Z.; Ju, H.X.; Meng, C.G.; Duan, C.Y. Selective electrochemical detection of cysteine in complex serum by graphene nanoribbon. *Biosens. Bioelectron.* **2012**, *32*, 293–296. [CrossRef]

31. Zhang, T.; Li, H.D.; Liu, M.X.; Zhou, H.; Zhang, Z.C.; Yu, C.; Wang, C.Y.; Wang, G.X. Improved the specificity of peroxidase-like carbonized polydopamine nanotubes with high nitrogen doping for glutathione detection. *Sens. Actuators B Chem.* **2021**, *341*, 129987. [CrossRef]
32. Wang, Y.; de Carvalho, N.A.; Tan, S.S.; Gilbertson, L.M. Leveraging electrochemistry to uncover the role of nitrogen in the biological reactivity of nitrogen-doped graphene. *Environ. Sci. Nano* **2019**, *6*, 3525–3538. [CrossRef]
33. Ndamanisha, J.C.; Bai, J.; Qi, B.; Guo, L.P. Application of electrochemical properties of ordered mesoporous carbon to the determination of glutathione and cysteine. *Anal. Biochem.* **2009**, *386*, 79–84. [CrossRef] [PubMed]
34. Gotoh, T.; Iguchi, H.; Kikuchi, K.I. Separation of glutathione and its related amino acids by nanofiltration. *Biochem. Eng. J.* **2004**, *19*, 165–170. [CrossRef]
35. Wang, Y.; Basdogan, Y.; Zhang, T.Y.; Lankone, R.S.; Wallace, A.N.; Fairbrother, D.H.; Keith, J.A.; Gilbertson, L.M. Interfaces, Unveiling the synergistic role of oxygen functional groups in the graphene-mediated oxidation of glutathione. *ACS Appl. Mater. Interfaces* **2020**, *12*, 45753–45762. [CrossRef]
36. Wang, L.; Teng, Q.Q.; Sun, X.T.; Chen, Y.T.; Wang, Y.M.; Wang, H.; Zhang, Y.F. Facile synthesis of metal-organic frameworks/ordered mesoporous carbon composites with enhanced electrocatalytic ability for hydrazine. *J. Colloid Interface Sci.* **2018**, *512*, 127–133. [CrossRef]
37. Wang, L.; Meng, T.J.; Jia, H.X.; Feng, Y.; Gong, T.; Wang, H.; Zhang, Y.F. Electrochemical study of hydrazine oxidation by leaf-shaped copper oxide loaded on highly ordered mesoporous carbon composite. *J Colloid Interface Sci.* **2019**, *549*, 98–104. [CrossRef]
38. Gharani, M.; Bahari, A.; Ghasemi, S. Preparation of MoS_2-reduced graphene oxide/Au nanohybrid for electrochemical sensing of hydrazine. *J. Mater. Sci. Mater. Electron.* **2021**, *32*, 7765–7777. [CrossRef]
39. Yao, Y.Q.; Han, X.H.; Yang, X.H.; Zhao, J.; Chai, C.P. Detection of Hydrazine at MXene/ZIF-8 Nanocomposite Modified Electrode. *Chin. J. Chem.* **2021**, *39*, 330–336. [CrossRef]
40. Zhang, X.J.; Zheng, J.B. Amperometric hydrazine sensor based on the use of a gold nanoparticle-modified nanocomposite consisting of porous polydopamine, multiwalled carbon nanotubes and reduced graphene oxide. *Mikrochim. Acta.* **2020**, *187*, 89. [CrossRef]
41. Vishnu, N.; Kumar, A.S.; Badhulika, S. Selective in-situ derivatization of intrinsic nickel to nickel hexacyanoferrate on carbon nanotube and its application for electrochemical sensing of hydrazine. *J. Electroanal. Chem.* **2019**, *837*, 60–66. [CrossRef]
42. Li, S.; Feng, W.S.; Gao, X.H.; Guo, A.M.; Li, H.J. Copper-based materials derived from metal-organic frameworks for electrochemical sensing of hydrazine. *Micro Nano Lett.* **2021**, *16*, 478–483. [CrossRef]
43. Asadi, F.; Azizi, S.N.; Ghasemi, S. Preparation of Ag nanoparticles on nano cobalt-based metal organic framework (ZIF-67) as catalyst support for electrochemical determination of hydrazine. *J. Mater. Sci. Mater. Electron.* **2019**, *30*, 5410–5420. [CrossRef]
44. Avanes, A.; Hasanzadeh-Karamjavan, M.; Shokri-Jarcheloo, G. Electrocatalytic oxidation and amperometric determination of hydrazine using a carbon paste electrode modified with beta-nickel hydroxide nanoplatelets. *Mikrochim Acta* **2019**, *186*, 441. [CrossRef] [PubMed]
45. Duan, C.Q.; Dong, Y.H.; Sheng, Q.L.; Zheng, J.B. A high-performance non-enzymatic electrochemical hydrazine sensor based on $NiCo_2S_4$ porous sphere. *Talanta* **2019**, *198*, 23–29. [CrossRef]
46. Ghasemi, S.; Hosseini, S.R.; Hasanpoor, F.; Nabipour, S. Amperometric hydrazine sensor based on the use of Pt-Pd nanoparticles placed on reduced graphene oxide nanosheets. *Mikrochim. Acta* **2019**, *186*, 601. [CrossRef]
47. Maleki, A.; Rezaee, R.; Daraei, H.; Shahmoradi, B.; Amini, N. Fabrication of a sensitive electrochemical sensor to environmental pollutant of hydrazine in real water samples based on synergistic catalysis of Ag@C core–shell and polyalizarin yellow R. *J. Alloys Compd.* **2018**, *763*, 997–1004. [CrossRef]
48. Shahid, M.M.; Rameshkumar, P.; Basirunc, W.J.; Wijayantha, U.; Chiu, W.S.; Khiew, P.S.; Huang, N.M. An electrochemical sensing platform of cobalt oxide@gold nanocubes interleaved reduced graphene oxide for the selective determination of hydrazine. *Electrochim. Acta* **2018**, *259*, 606–616. [CrossRef]
49. Mousavi-Majd, A.; Ghasemi, S.; Hosseini, S.R. Zeolitic imidazolate framework derived porous ZnO/Co_3O_4 incorporated with gold nanoparticles as ternary nanohybrid for determination of hydrazine. *J. Alloys Compd.* **2022**, *896*, 162922. [CrossRef]
50. Vinoth, S.; Sampathkumar, P.; Giribabu, K.; Pandikumar, A. Ultrasonically assisted synthesis of barium stannate incorporated graphitic carbon nitride nanocomposite and its analytical performance in electrochemical sensing of 4-nitrophenol, Ultrasonics Sonochemistry. *Ultrason. Sonochem.* **2020**, *62*, 104855. [CrossRef]
51. Zhou, Y.Y.; Zhao, J.; Li, S.H.; Guo, M.J.; Fan, Z. An electrochemical sensor for the detection of p-nitrophenol based on a cyclodextrin-decorated gold nanoparticle-mesoporous carbon hybrid. *Analyst* **2019**, *144*, 4400–4406. [CrossRef] [PubMed]
52. Ansari, A.A.; Alam, M.; Ali, M.A. Nanostructured CeO_2: Ag platform for electrochemically sensitive detection of nitrophenol. *Colloids Surf. A.* **2021**, *613*, 126116. [CrossRef]
53. Suresh, R.; Giribabu, K.; Manigandan, R.; Kumar, S.P.; Munusamy, S.; Muthamizh, S.; Narayanan, V. Polyaniline Nanorods: Synthesis, Characterization, and Application for the Determination ofpara-Nitrophenol. *Anal. Lett.* **2015**, *49*, 269–281. [CrossRef]
54. Alam, M.M.; Asiri, A.M.; Rahman, M.M. Electrochemical detection of 2-Nitrophenol using a glassy carbon electrode modified with BaO Nanorods. *Chem. Asian J.* **2021**, *16*, 1475–1485. [CrossRef] [PubMed]
55. Nejati, K.; Asadpour-Zeynali, K.; Rezvani, Z.; Peyghami, R. Determination of 2-nitrophenol by electrochemical synthesized Mg/Fe layered double hydroxide sensor. *Int. J. Electrochem. Sci.* **2014**, *9*, 5222–5234.

56. Yin, H.S.; Zhou, Y.L.; Ai, S.Y.; Cui, L.; Zhu, L.S. Electrochemical determination of 2-Nitrophenol in Water Samples using Mg-Al-SDS Hydrotalcite-Like Clay modified glassy carbon electrode. *Electroanalysis* **2010**, *22*, 1136–1142. [CrossRef]
57. Li, J.H.; He, L.Z.; Jiang, J.B.; Xu, Z.F.; Liu, M.Q.; Liu, X.; Tong, H.X.; Liu, Z.; Qian, D. Facile syntheses of bimetallic Prussian blue analogues (KxM $[Fe(CN)_6]$•nH_2O, M = Ni, Co, and Mn) for electrochemical determination of toxic 2-nitrophenol. *Electrochim. Acta* **2020**, *353*, 136579. [CrossRef]
58. Chu, L.; Han, L.; Zhang, X.L. Electrochemical simultaneous determination of nitrophenol isomers at nano-gold modified glassy carbon electrode. *J. Appl. Electrochem.* **2011**, *41*, 687–694. [CrossRef]
59. Zhang, J.J.; Cui, S.Q.; Ding, Y.P.; Yang, X.X.; Guo, K.; Zhao, J.T. Bioelectronics, Two-dimensional mesoporous $ZnCo_2O_4$ nanosheets as a novel electrocatalyst for detection of o-nitrophenol and p-nitrophenol. *Biosens. Bioelectron.* **2018**, *112*, 177–185. [CrossRef]
60. Hryniewicz, B.M.; Orth, E.S.; Vidotti, M. Enzymeless PEDOT-based electrochemical sensor for the detection of nitrophenols and organophosphates. *Sens. Actuators B Chem.* **2018**, *257*, 570–578. [CrossRef]
61. Wu, W.T.; Chen, X.; Jiao, Y.T.; Fan, W.T.; Liu, Y.L.; Huang, W.H. Versatile Construction of Biomimetic Nanosensors for Electrochemical Monitoring of Intracellular Glutathione. *Electroanal. Chem.* **2022**, *134*, e202115820.
62. Zhao, L.Z.; Zhao, L.; Miao, Y.; Zhang, C.X. Selective electrochemical determination of glutathione from the leakage of intracellular GSH contents in HeLa cells following doxorubicin-induced cell apoptosis. *Electrochim. Acta* **2016**, *206*, 86–98. [CrossRef]
63. Lee, P.T.; Goncalves, L.M.; Compton, R.G. Electrochemical determination of free and total glutathione in human saliva samples. *Sens. Actuators B Chem.* **2015**, *221*, 962–968. [CrossRef]
64. Yuan, B.Q.; Zhang, R.C.; Jiao, X.X.; Li, J.; Shi, H.Z.; Zhang, D.J. Amperometric determination of reduced glutathione with a new Co-based metal-organic coordination polymer modified electrode. *Electrochem. Commun.* **2014**, *40*, 92–95. [CrossRef]
65. Kannappan, S.; Bhat, L.R.; Nesakumar, N.; Babu, K.J.; Kulandaisamy, A.J.; Rayappan, J.B.B. Design and development of a non-enzymatic electrochemical biosensor for the detection of Glutathione. *Electroanalysis* **2022**, *34*, 1–12. [CrossRef]
66. Abbas, M.N.; Saeed, A.A.; Ali, M.B.; Errachid, A.; Zine, N.; Baraket, A.; Singh, B. Biosensor for the oxidative stress biomarker glutathione based on SAM of cobalt phthalocyanine on a thioctic acid modified gold electrode. *J. Solid State Electrochem.* **2019**, *23*, 1129–1144. [CrossRef]
67. Xu, H.Y.; Xiao, J.J.; Liu, B.; Griveau, S.; Bedioui, F. Enhanced electrochemical sensing of thiols based on cobalt phthalocyanine immobilized on nitrogen-doped graphene. *Biosens. Bioelectron.* **2015**, *66*, 438–444. [CrossRef]
68. Saranya, S.; Geetha, B.; Deepa, P.N. Simultaneous detection of glutathione, threonine, and glycine at electrodeposited RuHCF/rGO–modified electrode. *Ionics* **2019**, *25*, 5537–5550. [CrossRef]

nanomaterials

Article

Defect Density-Dependent pH Response of Graphene Derivatives: Towards the Development of pH-Sensitive Graphene Oxide Devices

Shayan Angizi [1], Xianxuan Huang [1], Lea Hong [1], Md Ali Akbar [1], P. Ravi Selvaganapathy [2] and Peter Kruse [1,*]

[1] Department of Chemistry and Chemical Biology, McMaster University, 1280 Main Street West, Hamilton, ON L8S 4M1, Canada; angizis@mcmaster.ca (S.A.); huangx87@mcmaster.ca (X.H.); hongl1@mcmaster.ca (L.H.); akbarm7@mcmaster.ca (M.A.A.)
[2] Department of Mechanical Engineering, McMaster University, 1280 Main Street West, Hamilton, ON L8S 4L7, Canada; selvaga@mcmaster.ca
* Correspondence: pkruse@mcmaster.ca

Abstract: In this study, we demonstrate that a highly pH-sensitive substrate could be fabricated by controlling the type and defect density of graphene derivatives. Nanomaterials from single-layer graphene resembling a defect-free structure to few-layer graphene and graphene oxide with high defect density were used to demonstrate the pH-sensing mechanisms of graphene. We show the presence of three competing mechanisms of pH sensitivity, including the availability of functional groups, the electrochemical double layer, and the ion trapping that determines the overall pH response. The graphene surface was selectively functionalized with hydroxyl, amine, and carboxyl groups to understand the role and density of the graphene pH-sensitive functional groups. Later, we establish the development of highly pH-sensitive graphene oxide by controlling its defect density. This research opens a new avenue for integrating micro–nano-sized pH sensors based on graphene derivatives into next-generation sensing platforms.

Keywords: pH; graphene; graphene oxide; defect; surface functionalization

Citation: Angizi, S.; Huang, X.; Hong, L.; Akbar, M.A.; Selvaganapathy, P.R.; Kruse, P. Defect Density-Dependent pH Response of Graphene Derivatives: Towards the Development of pH-Sensitive Graphene Oxide Devices. *Nanomaterials* **2022**, *12*, 1801. https://doi.org/10.3390/nano12111801

Academic Editor: Marco Cannas

Received: 30 April 2022
Accepted: 20 May 2022
Published: 25 May 2022

1. Introduction

The pH measurement in an aqueous solution is a universal need and is widely utilized in various applications, from biology [1] to wastewater treatment [2]. The standard techniques for pH detections are often based on electrochemistry (potentiometry) using a porous glass electrode [3,4]. Later, the development of ion-selective membranes introduced a new category of sensing devices, such as field-effect transistors [5,6]. The ion-selective membranes transfer the charge to a conductive active layer under varying gate potentials [7]. Other pH detection methods, including conductometric [8,9] and optical [10,11], operate based on the pH sensitivity of an indicator dye, where adding an organic redox-active indicator is essential. Nevertheless, commercial pH measurements have a number of drawbacks. The porous glass electrodes require high maintenance and are prone to performance loss at low ionic strength or high temperature [12]. The ion-selective membranes are subject to degradation and loss of sensitivity over long-term operation [13,14]. The optical measurements require the introduction of undesirable chemical species in the reaction solution [15,16]. Therefore, developing reagent-free pH-sensitive platforms has remained a relevant topic.

Since the first successful isolation of monolayer graphene from graphite [17], there has been an increasing trend towards integrating these atomically thin structures into many sensing applications, including pH detection units. This broad range of graphene sensing applications is mainly due to three of its characteristics: superior conductivity, high specific surface area, and modulable surface chemistry [18–20]. Single-layer graphene

(SLG) consists of an sp^2-hybridized covalently bonded carbon network, while vertically stacked consecutive layers of graphene are held together by van der Waals forces [21]. Since the electronic properties of graphene greatly depend on the presence of delocalized π electrons, any minor disruption of the surface infinite symmetry causes changes in its electrical properties [22,23]. Therefore, defects involving sp^3 carbon atoms or in-plane lattice asymmetries play a vital role in determining graphene's ultimate electrical and chemical properties [24]. Accordingly, the defect engineering of graphene can lead to the generation of other graphene derivatives, such as graphene oxide (GO) and reduced graphene oxide (r-GO).

The sensing performance of graphene derivatives in contact with an aqueous environment is profoundly reliant on the electrolyte parameters, including pH, oxidation-reduction potential, ionic strength, dissolved oxygen, and temperature [25]. Therefore, the concept of pH sensors based on graphene derivatives can be fully developed only if the aqueous electrolyte's impact on graphene is entirely investigated. Moreover, the pH-sensing mechanism of graphene has been shown to be defect-dependent [26], but there is still no clarity on the role of defects in the pH response of graphene devices.

This lack of understanding originates because "ideal" defect-free graphene tends to be hydrophobic [23], making it challenging to elucidate the structure of the aqueous solution/graphene interface [27]. However, oxygen-based impurities enhance the surface local charges and decrease the liquid/surface contact angle [28–30]. Therefore, the defectivity level, type of defects, and their configurations (edge or plane) on graphene play a vital role in the graphene–solution interface. Despite a few studies exploring the possibility of developing a pH-sensing platform based on graphene derivatives [28,31–35], there is still a gap in understanding their response to pH at high defect levels. Thus far, there exist only a few reports on the application of GO in pH sensing, primarily due to three main limiting factors in the development of such devices: (i) the water dispersibility of the GO due to its hydrophilic nature and high negative zeta potential [36], (ii) its low surface conductance and semiconductive nature, (iii) the lack of understanding of the pH detection mechanisms of GO [37]. In 2020, potentiometric and chemiresistive measurements of hydrothermally reduced GO revealed potential sensitivities of 66 mV/pH (pH 2–12) and 44 mV/pH (pH 4–7), respectively. Accordingly, the reported working range of the chemiresistor was limited to acidic pH, and the higher pH sensitivities required a three-electrode design [38,39]. Although the presence of oxygen functional groups was shown to be responsible for such detection, no explanation for the limited linear range and mechanism has been proposed. In a different study, a sensitivity of 51 mV/pH was obtained for a pH of 2–10 using potentiometry [40] and FETs [41]. The authors successfully demonstrated the reversibility of the charge transfer on the GO surface upon exposure to pH. While the nature of the response has remained unknown, this study has pushed the boundaries toward mono- and diprotic acid potentiometric titration units based on GO.

Herein, we demonstrate the possibility of developing a pH-sensing platform based on various graphene derivatives, including single-layer graphene (SLG), few-layer graphene (FLG), and GO, focusing on the mechanisms by which each of these structures responds to pH. Furthermore, we examine the defect density-dependent pH response of graphene by varying the -COOH, -OH, and $-NH_2$ concentrations. The results fill the existing gap in understanding the GO pH detection mechanism and creates foundations for developing reagent-free GO pH chemiresistive sensors with a high sensitivity and reproducibility. The results of this study open up a new window into the development of graphene derivative-based pH sensing devices.

2. Materials and Methods

2.1. Single-Layer Graphene Transfer Process

The SLG coated with PMMA was purchased from Graphenea Inc. To fabricate the sensor, the PMMA/SLG/Cu samples were cut into 1 cm $\times$ 1 cm squares and then placed into 200 mL 0.1 M Ammonium persulfate (APS, purchased from VWR) to etch off the copper film. After about 30 min, the PMMA/SLG samples were transferred to 500 mL DI water to rinse off the remaining APS and transferred on a glass slide (VWR). The samples were first heated to 100 °C for 30 min and then annealed for 2 h at 600 °C under N_2 using a three-heat zone tube furnace (Lindberg Thermodyne 21100). The samples were then immersed in acetone overnight (16 h) to remove the PMMA. The annealing step is essential to achieve minimum PMMA residue on graphene.

2.2. Synthesis of FLG and Sensor Fabrication

Synthesis: The synthesis process of FLG has already been reported elsewhere [42]. Briefly, 40 mg of graphite powder (Alfa Aesar, Haverhill, MA, USA, 99.99%) is mixed with 4.5 mL and 10.5 mL of IPA and DI water, respectively. The mixture is then sonicated for 6 h using an Elmasonic P60H ultrasonic cleaner, 100% power; sweep mode at 37 kHz. Then, the products are centrifuged (Eppendorf MiniSpin Plus microcentrifuge) for 5 min at 14,000 rpm (13,149× *g*). Afterward, the supernatant products are collected and centrifuged again for 15 min at the same speed. Lastly, the precipitated products are ordered and used for sensor fabrication.

Sensor fabrication: The frosted sides of the glass slides (VWR VistaVision) were initially rinsed with methanol (Fisher Scientific Canada, HPLC) and patterned by two parallel rectangles drawn by a 9B pencil (Figure 1a). Then, the FLG suspension was airbrushed (e NEO-Iwata CN Gravity Feed Dual Action Brush #N4500) using a nitrogen gas directly on the surface, preheated to 150 °C, until the resistance was measured in the range of 5–10 kΩ (Figure 1b). The copper tape was attached as sensor contacts to the pencil-drawn rectangle/airbrushed sample. To avoid direct exposure of copper tape to aqueous solutions, the contacts were covered by parafilm (Parafilm "M", VWR). Increasing the temperature to 70 °C (above the melting point of parafilm) leads to a more uniform coverage (Figure 1c). An SEM image of an FLG film deposited for a sensor and the I-V curve characteristics of a typical device are shown in Figure 1d,e, respectively.

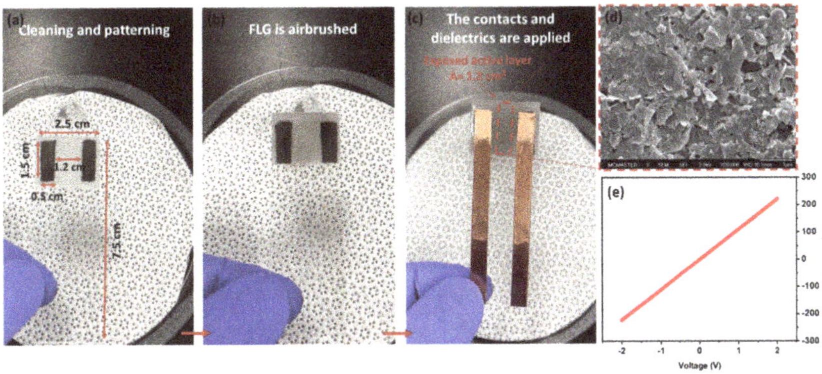

Figure 1. Sensor fabrication steps: (**a**) Cleaning and prepatterning with pencil-drawn contacts, (**b**) airbrushing FLG, and (**c**) attachment of Cu tape and dielectrics. (**d**) SEM image of the FLG airbrushed on the surface, (**e**) the I-V curve of the fabricated chemiresistive sensor indicating the ohmic device (sensor resistance was ∼10 kΩ).

Annealing: The airbrushed samples were placed into a furnace under the flow of N_2/H_2 (95%/5%) and heated to 500 °C and 350 °C for FLG and GO, respectively. A temperature of 350 °C has been shown as a safe temperature to anneal GO without considerable thermal decompositions or mass loss [43]. The samples are left to cool down gradually overnight in an N_2 atmosphere.

Pyrene derivatives functionalization: Pyrene carboxylic acid (Py-COOH), 1-amino Pyrene (Py-NH_2), and 1-hydroxypyrene (Py-OH) were purchased from Sigma Aldrich and used without further purification. To dope the samples, the fabricated sensors were placed into solutions of the respective pyrene in acetonitrile overnight (~16 h). At the end, the functionalized sensors were rinsed with pure acetonitrile to remove any excess pyrene.

pH measurement experiment: In preparation for the experiments, the devices were initially placed into a 3.42 mM NaCl solution and left overnight to equilibrate with the environment. An initial ionic conductivity above 0.33 mS/cm (equivalent to 3.42 mM NaCl) minimizes possible interferences due to changes in ionic strength when adjusting pH using NaOH (99%-Caledon Laboratories Ltd., ACS reagent) or HCl (37.2%-Caledon Laboratories Ltd., Ahmedabad, india. ACS reagent). The pH of the solution was adjusted by drop-wise addition of NaOH (0.1 M) or HCl (0.1 M) into the system until the desired pH (pH 3 to 9) was achieved. The devices were kept at each pH for 30 min while the current was recorded at intervals of 1 data point every two seconds. The last 60 data points (two minutes) from each step were used for further analysis.

2.3. *GO Preparation and SENSOR fabrication*

The GO powder was purchased from Zentek Ltd (Canada). Initially, 40 mg of GO was dispersed in 15 mL of ultrapure water and sonicated for 1h. The products were airbrushed on a preheated substrate to 200 °C. If annealing was required, the samples were placed at 350 °C for 24 h under N_2/H_2 (95%/5%) reducing environment. Then, the copper tapes were attached and subsequently covered by parafilm as dielectric (as discussed in Section 2.2).

2.4. *Characterization*

A Renishaw inVia Raman spectrometer was used to characterize the defect level of the graphene film. A Renishaw 633 nm HeNe laser with 17 mW power output was focused through the 50× objective lens with a spot size of about 1.5 µm. The laser power used for few-layer graphene analysis was 50% to minimize the noise and 5% power for graphene oxide to avoid film damage. The range of the Raman region was from 500 to 3500 cm^{-1} with a spectral resolution of 2 cm^{-1}. Spectra were recorded in at least two different spots for each sample to ensure reproducibility.

The XPS analyses were carried out with a Kratos AXIS Supra X-ray photoelectron spectrometer using a monochromatic Al K(alpha) source (15 mA, 15 kV). A charge neutralizer was used on all specimens. Survey scans were collected from an area of 300 × 700 μm^2 using a pass energy of 160 eV. High-resolution scans used a pass energy of 20 eV. All spectra were charge corrected to the mainline of C 1s (graphitic carbon, 284.5 eV). Spectra were analyzed using CasaXPS software (version 2.3.14).

3. Results and Discussions

3.1. *pH Response of Bare Graphene*

The response of bare graphene to pH was chosen as a starting point for understanding the impact of surface defectivity. Considering that most liquid-phase exfoliation methods to produce graphene derivatives, including ultrasonication, generate a high degree of defectivity (oxygen content between 10–15%), a defect-free (1–2%) CVD-grown SLG was used as a reference sample to obtain an insight into the response of pure graphene to pH. The SLG's low defectivity was confirmed by the XPS results, as shown in Figure 2. According to the high-resolution O 1s spectra, shown in Figure 2a, the peak at 533 eV is associated with the oxygen doubly bound to C, while the peak at 534.19 eV corresponds to oxygen in the SiO_2 substrate. The nature of oxygen–carbon bonds can be further analyzed using C 1s high-resolution spectra, as shown in Figure 2b. As seen, C-OH/C-O-C, and C=O peaks located at 286.50 and 287.90 eV are the dominant oxygen-based functional groups. The C=O is often interpreted as a result of graphene's ketone, aldehyde, and carboxyl groups [44]. The presence of C-OH indicates the formation of a primary or secondary alcohol and carboxyl groups, and the C-O-C implies the appearance of ether and epoxy sites on the surface [45]. These results confirm that oxygen-based functional groups are inevitably formed on graphene during the synthesis or transfer process. In this case, the acetone treatment to eliminate the PMMA and subsequent multiple rounds of bathing in DI water could cause this oxidation. Moreover, the possibility of oxygen-containing contaminants (e.g., carboxylic acids, alcohols, aldehydes, etc.) from the environment cannot be neglected. Nevertheless, the O/C atomic ratio of the transferred SLG is calculated at ~0.05, at the lower end of defect density. The atomic percentage of oxygen and carbon in SLG can be found in the XPS survey spectrum shown in Figure S1a.

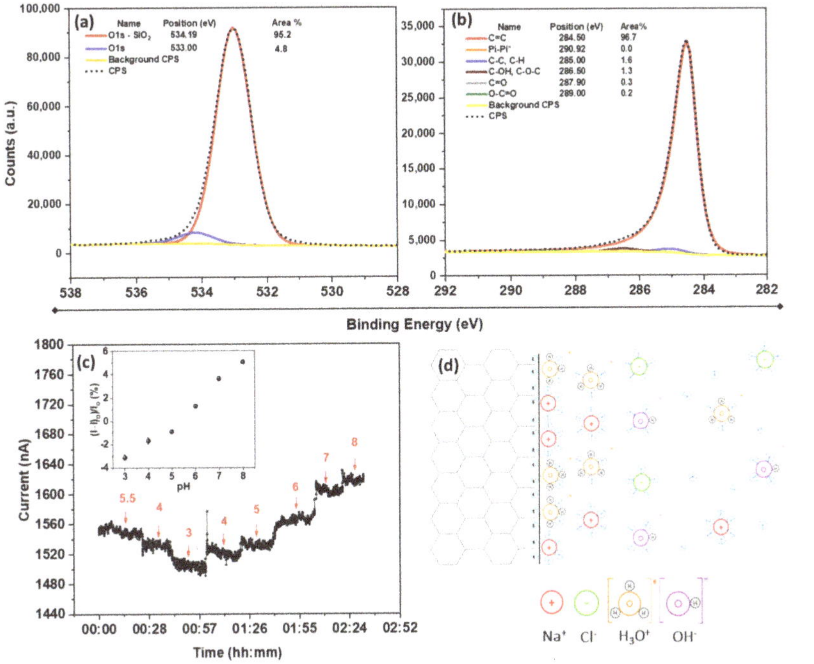

Figure 2. XPS high-resolution spectra of (**a**) O 1s and (**b**) C 1s of SLG, (**c**) pH response of SLG between pH 5.5–8 (the inset shows the corresponding calibration curve) with I_o = 1524 nA, and (**d**) schematics illustration of the formation of EDL on graphene in acidic solution and its electrostatic gating charging.

The pH response of an SLG chemiresistive device is shown in Figure 2c. The starting pH (5.5) was determined by an equilibrium of the aqueous solution with ambient air, established overnight. Due to the partition of CO_2 from the ambient air into water, the pH of deionized water gradually equilibrates to 5.6, assuming 400 ppm of CO_2 in the air. According to Figure 2c, the current through the SLG is observed to decrease when the pH is reduced to 3, and conversely increased when the pH was raised. The low pH response can be explained by the electrostatic gating effect of $H_3O^+{}_{aq}$, which n-dopes the surface. Since the holes are the majority carrier in graphene due to the presence of electron-withdrawing oxygen atoms, n-doping the surface reduces the charge carriers in the chemiresistor and makes it more resistive. In contrast, electrostatic p-doping of $OH^-{}_{aq}$ ions accumulated in the Stern layer increases the current at a high pH [26]. The schematic illustrations of the formation of the electrochemical double layer (EDL) in acidic solutions are shown in Figure 2d. It should be noted that the charge transfers through the formation of the EDL by H_3O^+ and OH^- are considered fully nonfaradaic. Therefore, the charges are electrostatically gated to the graphene surface. This mechanism is well-defined for SLG and is often deemed the typical graphene response to pH. This phenomenon has been studied in other devices such as FETs [46,47] or Schottky diodes [25], demonstrating the decrease (increase) in Fermi energy upon exposure to a high (low) pH. However, the Fermi energy of

SLG is prone to cross the Dirac point upon severe electrostatic doping [48,49]; therefore, it may not be a reliable system to further investigate the role of EDL and defects. Accordingly, the following parts of this research deploy FLG-based chemiresistive devices to explore the role of defects in pH sensitivity.

The transition from SLG to FLG requires a careful surface analysis, considering that a higher number of surface defects and functional groups are formed during the liquid-phase exfoliation method. The XPS survey spectrum of the synthesized FLG (Figure S1b) shows 13.9 at% of O and 79.5 at% of C. The O 1s high-resolution spectra (Figure 3a) show a peak at 532.4 eV attributed to trapped water and organic oxygen groups. The larger area of the peaks at 289.0 eV (O-C=O), 287.0 eV (C = O), and 286.5 eV (C-OH, C-O-C) in the C 1s high-resolution spectrum of the FLG compared to the SLG (Figure 3b) exhibit a higher oxygen content in the FLG lattice. Accordingly, the O/C ratio was calculated to be 0.24 by considering the areas of the C=C (284.5 eV) and O 1s (532.4 eV) features in the high-resolution spectra multiplied by their corresponding atomic percentages derived from the survey spectra. A summary of the oxygen-based functional groups in the SLG and FLG can be found in Table 1, demonstrating the greater defectivity of the FLG compared to SLG.

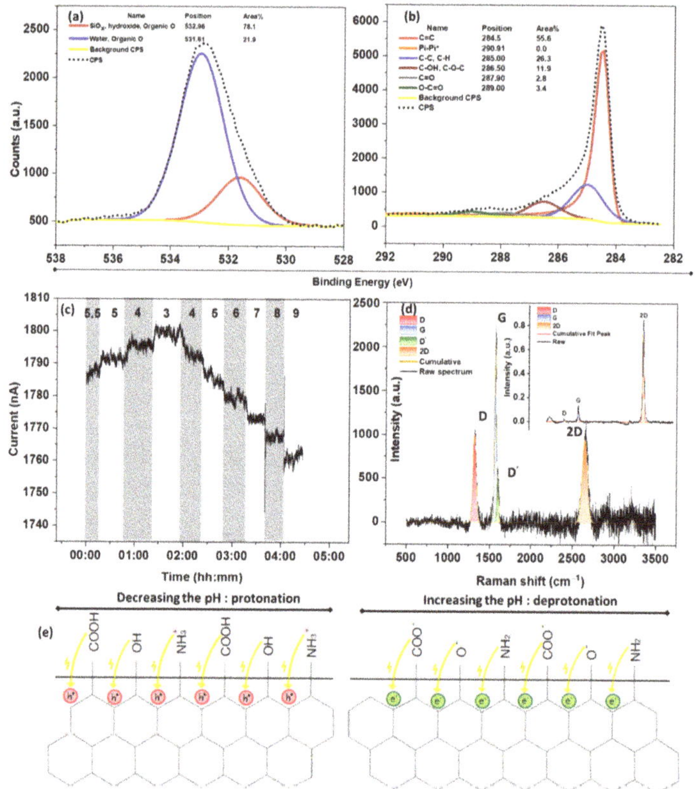

Figure 3. XPS high-resolution spectra of (**a**) O 1s and (**b**) C 1s of FLG, (**c**) pH response of FLG to pH 5.5–9 (I_o = 1788 nA), (**d**) Raman of FLG deconvoluted to the main graphene characteristics of D, G, D′, and 2D (inset shows the deconvoluted Raman spectrum of SLG), and (**e**) schematic illustration of defect-induced pH response of FLG through protonation/deprotonation of carboxyl, hydroxyl, and amine groups.

Table 1. Summary of the oxygen-containing groups obtained from XPS spectra of SLG and FLG.

	SLG			FLG		
	Area %	**at% ***	**Ratio to C=C at%**	**Area %**	**at% ***	**Ratio to C=C at%**
C=O	0.3	0.1119	0.003074	2.8	2.22	0.05
O-C=O	0.2	0.0746	0.002049	3.4	2.70	0.061
C-OH/O-C-O	1.3	0.4849	0.01332	11.9	9.46	0.214

* Atomic percentage is obtained by Area × total atomic percentage.

To explore the impact of the enhanced defectivity on the pH response, the chemiresistive response of the FLG is shown in Figure 3c. As seen, the current response to the variation in pH is entirely reversed compared to the SLG (Figure 2c), indicating that the dominant pH response mechanism is different. This means that decreasing the pH reduces the current, while increasing the pH toward the basic solution increases n-doping. We have recently shown that the defect-induced pH response of graphene originates from the protonation/deprotonation of carboxyl and amine at a low pH and hydroxyl groups at a high pH [26]. For example, in carboxyl groups, upon decreasing the pH to 3 (below the $pK_a = 3.1$), the $-COO^-$ is protonated to -COOH, and the surface becomes p-doped. The same concept can be applied to the protonation of $-NH_2$ to $-NH_3^+$ and the protonation of $-O^-$ to -OH at a pH around 3.7 and 8.2, respectively. Since this charge is transferred directly to the surface, the protonation causes p-doping, while deprotonation results in n-doping, giving an exact opposite behavior to the EDL-induced response. Apart from the defect-induced response, H_3O^+ ions have also been shown to enter the gaps between FLG flakes at a low pH. This proton injection in FLG can be confirmed by the blue shift of the 2D peak in the Raman spectra of FLG after the pH exposure (Figure S2). This result could be explained by the Faradaic charge transfer upon proton injection and p-doping of the graphene by lowering its Fermi energy [32,50]. This phenomenon becomes dominant when a porous structure is present [51] (e.g., porous graphitic electrodes in supercapacitors) [50]. Therefore, it is expected to observe a rise in conductance at low pH where H_3O^+ is present, mainly close to the surface. Accordingly, the main 2D peak of the FLG spectra (shown in Figure S2) shifts due to the lattice parameter modification by the stiffening/softening of the phonons–charge carrier interactions [52].

The defectivity level can also be estimated by Raman spectroscopy. The FLG (Figure 3d) and SLG (Figure 3d-inset) both exhibit the three main Raman characteristics of graphene, namely D, G, and 2D bands [53]. The D band stands for the presence of sp^3-hybridized environments, generated mainly by defects. The G band, however, represents the sp^2 hybridization of the graphene lattice. Thus, the ratio of intensity (or area) of D to G qualitatively characterizes the defectivity level of the structure. Accordingly, the I_D/I_G of the SLG is calculated at 0.11, while the FLG shows 0.43, supporting the XPS results. These values can also be well-fitted to previously published reports that low defect density graphene (less than a cross-over point of $I_D/I_G \sim 0.35$) demonstrates positive pH sensitivity (Figure 2c, inset), while the higher I_D/I_G ratio results in inverted pH sensitivity. It should be mentioned that the positive and negative sensitivities are defined relative to the variation of the current with the pH. This cross-over is where the graphene will become pH insensitive. The schematic illustration of such a protonation/deprotonation mechanism can be seen in Figure 3e.

3.2. *Selective Functionalization of Graphene*

Knowing that the type and density of the defects in graphene derivatives determine their pH response, the development of a pH-sensitive device can be achieved by selective functionalization. For this purpose, noncovalently attached pyrene derivatives with various pH-sensitive functional groups were employed as a model system to resemble the graphene surface terminated with pH-sensitive groups [26]. The charge transfer upon protonation/deprotonation of the functional groups is directly transduced to the FLG [46] via the interactions of the π-electron system of the pyrene ring with the FLG surface [54]. The maximum concentration of each pyrene derivative was chosen to give more than 90% surface coverage [55]. Moreover, the low pyrene solubility in water aids in the stability of the functionalization. However, to minimize the impact of the pre-existing defects of the FLG on pyrene functionalization, it is necessary to anneal the FLG samples under a reducing environment to eliminate the functionalities. The Raman spectra of the FLG before (Figure 3d) and after (Figure S3) annealing demonstrate the effective surface defect reduction, decreasing the I_D/I_G from 0.43 to 0.2. This surface defect reduction can be further confirmed by the pH response of the annealed samples (Figure S4). It is observed that the pH–current relationship is inverted compared to the bare (unannealed) FLG, rather more like the SLG (see Figure S5 for the calibration curve). This phenomenon demonstrates that the FLG has been annealed to a low defect state where the EDL response is dominant.

Figure 4 displays how the variation of carboxyl group concentration on graphene affects the pH response of a chemiresistive device. The sample exposed to 0.3 M Py-COOH (Figure 4a) demonstrates a selective response to the pH of around 3, giving a current change of ~55% (−21.58%/pH). This response is considered significant compared to the other graphene devices, where the maximum response barely exceeds 20% (see Figures 2c and 3c). The low pH sensitivity of the device at a high pH could be due to two simultaneous factors: (i) the annealing before pyrene functionalization has successfully eliminated the responsive functional groups, or (ii) the pyrene molecules have passivated the existing groups and defects. Notably, the pH sensitivity of the device to a pH range of 3–4 decreases upon reduction of the pyrene concentration to 0.15 mM (−12.31%/pH) and 0.1 mM (−2.11%/pH), as shown in Figure 4b,c, respectively. With doping concentrations around 0.05 mM (Figure 4d), a pH-insensitive platform (−0.2%/pH, Figure 4e, inset) is obtained, indicating the response due to carboxyl groups is balanced out by the response due to electrostatic gating by the EDL. The calibration curves representing the pH sensitivities are shown in Figure 4e. Although it is not easy to accurately measure the defect density of the -COOH group added to the surface, an estimate can be arrived at by making two assumptions: (i) the FLG film is flat with minimum surface roughness; (ii) a monolayer of molecules forms during functionalization. The last assumption has already been validated experimentally and can be considered realistic based on the selected concentrations in Table S1 [55]. Using these assumptions, and considering that each Py-COOH carries one carboxyl group, the approximate surface density of -COOH defects can be estimated:

$$Surface\ density = \frac{SC_{Rel}}{A_{Pyrene}}$$

where SC_{Rel} is the relative surface coverage of pyrene derivatives obtained from the literature [54,55], as shown in Table S1, and A_{Pyrene} is the area occupied by a single molecule of the respective pyrene derivative. The plot of the maximum pH response of the FLG as a function of the carboxyl group is shown in Figure 5a. Notably, the linear trend is not observed, and a carboxyl group density ~5.36 × 10^{13} cm^{-2} is estimated to be where the EDL becomes dominant. The nonlinear dependence of the maximum response to the carboxyl group surface density confirms that the pH response is not exclusively determined by the functional group mechanism. It should also be noted that any sensing data above the maximum concentration of Py-COOH (0.3 mM) may not be valid due to exposure of the FLG to a concentrated solution that may contain dimers or stacked molecules, leading to an invalidation of the above-noted assumption that only a monolayer of molecules is formed [53,54]. The detailed information of the data shown in Figures 4e and 5 have been provided in Tables S2 and S3, respectively.

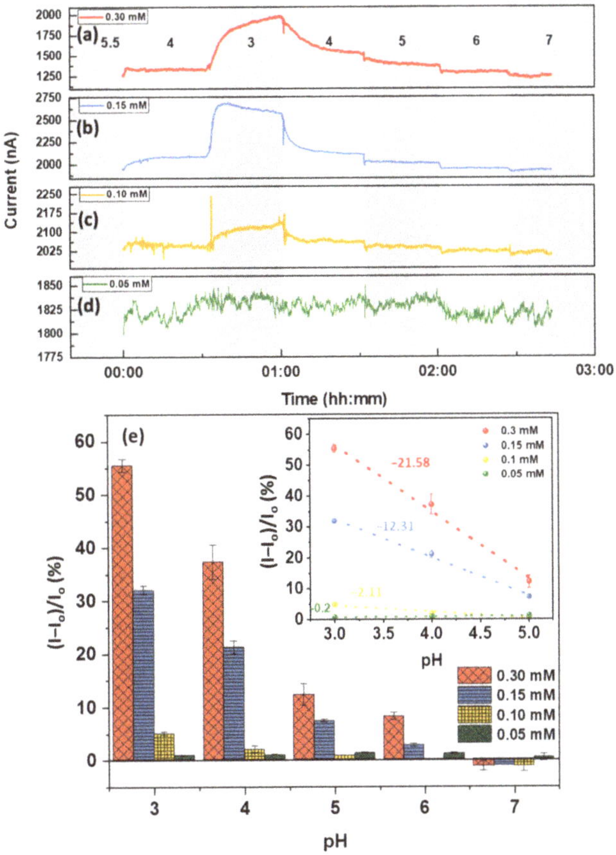

Figure 4. Sensing performance of 8 h annealed FLG functionalized by (**a**) 0.3 mM (I_o = 1274 nA), (**b**) 0.15 mM (I_o = 1980 nA), (**c**) 0.1 mM (I_o = 2052 nA), and (**d**) 0.05 mM (I_o = 1812 nA) of Py-COOH; (**e**) the calibration bar graph of the sensors demonstrating the highest pH response at around -COOH pK_a (3.1). The error bars represent average $\pm$ standard deviation of the last two minutes of the chemiresistive response (3 samples each).

Similar results can be obtained by using Py-NH_2, resembling amine groups of the FLG (see Figure 5b). The current increase at a pH of around 3–4 can be interpreted as the protonation of -NH_2 to -NH_3^+ and the p-doping of the surface. Remarkably, the highest concentration of Py-NH_2 (1.4 mM) does not lead to the maximum pH response of the surface (Figure 6a–d); this occurs at concentrations of around 0.7 mM, giving a maximum pH response of 6.8% (Figure 6e). Moreover, Figures 5b and 6e both reveal two phenomena: (i) 0.1 mM Py-NH_2, equivalent to a surface density of $\sim 5.35 \times 10^{13}$ cm^{-2} for -NH_2 groups, is not sufficient to overcome the EDL response of the surface, so that a maximum negative response of $\sim -4.5\%$ is obtained; (ii) at 0.35 mM (equivalent to 1.5×10^{14} cm^{-2}), a response of the device to both pH 3 and 4 is considerable, demonstrating that the lower surface coverage of Py-NH_2 may expose the leftover carboxyl group from the annealing process. It should be noted that the maximum response to pH drops from 3.1% to 2.7%, corresponding to 1.4 mM and 0.7 mM, respectively. Therefore, as expected, lower Py-NH_2 exposure leads to a lower response to a pH of 3. However, this trend is violated below 0.7 mM, equivalent to the surface density of 2.68×10^{14} cm^{-2}. One possible mechanism for the lower pH response of the high amine concentration could be an amide formation reaction between -NH_2 and -COOH of the surface. Upon these reactions, C in R-C̲OOH is reduced to R-

$CONH_2$, n-doping the surface. This n-doping counteracts the p-doping of the protonation and leads to a lower response.

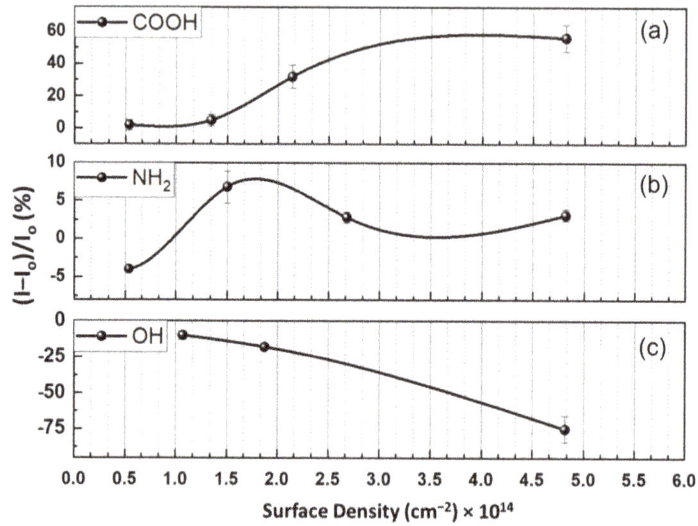

Figure 5. Variation of FLG pH sensitivity as a function of (**a**) carboxyl, (**b**) amine, and (**c**) hydroxyl defect densities. The error bars represent average ± standard deviation of the last two minutes of the chemiresistive response (3 samples each).

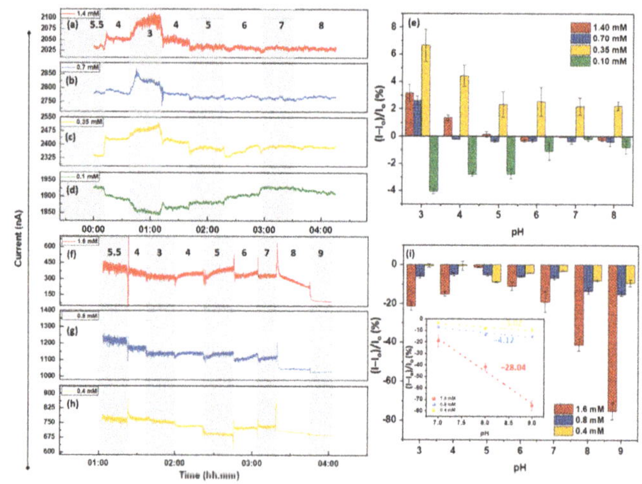

Figure 6. Sensing performance of 8 h annealed FLG functionalized by (**a**) 1.4 mM (I_o = 2035 nA), (**b**) 0.7 mM (I_o = 2775 nA), (**c**) 0.35 mM (I_o = 2330 nA), and (**d**) 0.1 mM (I_o = 1927 nA) of Py-NH_2. (**e**) Graph represents the calibration curve of the Py-NH_2-functionalized sensors. The sensing performance of 8 h annealed FLG functionalized with (**f**) 1.6 mM (I_o = 385 nA), (**g**) 0.8 mM (I_o = 1220 nA), and (**h**) 0.4 mM (I_o = 764 nA) of Py-OH. (**i**) The calibration bar graph of the sensors demonstrating the maximum pH response at pH around -OH pKa (8.7). The error bars represent average ± standard deviation of the last two minutes of the chemiresistive response (3 samples each).

The pH responses of the Py-OH-functionalized FLG at three different concentrations of 1.6, 0.8, and 0.4 mM are presented in Figure 6. The device pH response is dominated by the protonation/deprotonation of the -O^-/-OH groups at ahigh pH (pK_a = 8.7). Accordingly, the current drop at a pH of 9 of Figure 6f–h is due to this phenomenon. The corresponding calibration curves demonstrate that the device response to a pH of 7–9 is proportional to the -OH concentration and the density, so that a pH of 7–9 and sensitivity of 1.6, 0.8, and 0.4 mM can be calculated for −28.04, −4.12, and −3.02%/pH, respectively (Figure 6i, inset) Furthermore, the sensing behavior of the OH-functionalized FLG can be further analyzed: from a pH of 5.5 down to 3 and back up to 7, the device resembles the annealed FLG, while at a pH of 8 or above, it responds with OH groups. This manifestation reveals that even one sensor can respond differently at different pH ranges, depending on the dominant pH-sensing mechanisms. This result can be further confirmed by the significant impact of the high -OH concentration on the pH response, leading to an almost −75% response to a pH of 9 when the surface defect density is 1.07×10^{14} cm^{-2} (Figure 5c). The detailed information of Figure 6e,i can be found in Tables S4 and S5, respectively.

3.3. *pH Response of GO and Its Application towards the Development of GO-Based pH Sensors*

Based on the discussion above, enhancing graphene defectivity leads to a more defect-induced response and higher sensitivity. Therefore, highly defective graphene derivatives such as GO, carrying various surface/edge functional groups, are expected to be a worthwhile platform to study. The Raman spectrum of the GO with overlapping D and G bands is shown in Figure S6, demonstrating a Raman spectrum consistent with the literature. The Lorentzian deconvolution of the spectrum reveals the presence of multiple subpeaks under the curves, demonstrating an I_D/I_G ratio of 1.74. The significant overlap of the D' and G bands are also due to the enhanced intervalley scattering in the high defect region. Even though the presence of D** at 1479 cm^{-1} is often indicative of disordered carbons (amorphous), it could be due to the cumulative scattering of C=C stretching in sp^2 regions and the C-H wagging modes in a nanocrystalline diamond [56,57]. The higher defect density can enhance the GO pH sensitivity for aqueous solution applications. However, the stability of GO in water becomes an issue upon its exposure to aqueous solutions. The higher degree of local surface charges caused by functional groups decreases the GO water contact angle, increasing the GO dispersibility. The instability of GO-based chemiresistive device upon exposure to water can be seen in Figure 7a(1–3). Figure 7a(1) shows the airbrushed GO on the glass slide, and (b and c) display the same sensor after exposure to an aqueous solution after 30 and 60 min of exposure, respectively. This means that the delamination of the active layer of GO-based chemiresistors (Figure 7a(3)) limits their stability to less than 30 min. Moreover, the semiconductive nature of GO causes a high film resistance, which is impractical as a conductive active layer. Therefore, despite the remarkable properties of GO, it cannot be used as the active layer in a sensor in its pristine form.

To prevent disintegration, the GO samples were annealed at 350 °C under a reducing environment [43]. Exposure of GO to temperatures above 450 °C should either be done in vacuum systems or under a high flow of inert gas to avoid thermal decomposition [58]. This annealing treatment results in a visible color change from brown (Figure 7a(3)) to gray (Figure 7a(5–7)), indicating the successful reduction of the GO.

A comparison of the I_D/I_G ratio of pristine GO (Figure S6) with that of the 24 h annealed GO (Figure 7b) confirms an increase in surface sp^2 hybridization. However, due to the surface-insensitive nature of the Raman measurement, the obtained I_D/I_G = 1.3 (Figure 7b) indicates that a considerable number of defects still remain in the bulk. The low-temperature annealing of GO assists in the formation of a stable conductive GO film, while the intrinsic characteristics are preserved. It should be noted that the thermal annealing (reducing) of GO results in a more uniform and well-controlled product as compared to the chemically reduced GO [59]. Moreover, the I_D/I_G = 1.3 obtained from the 24 h-GO reveals that only a surface reduction has occurred, and the term reduced GO is not applied here.

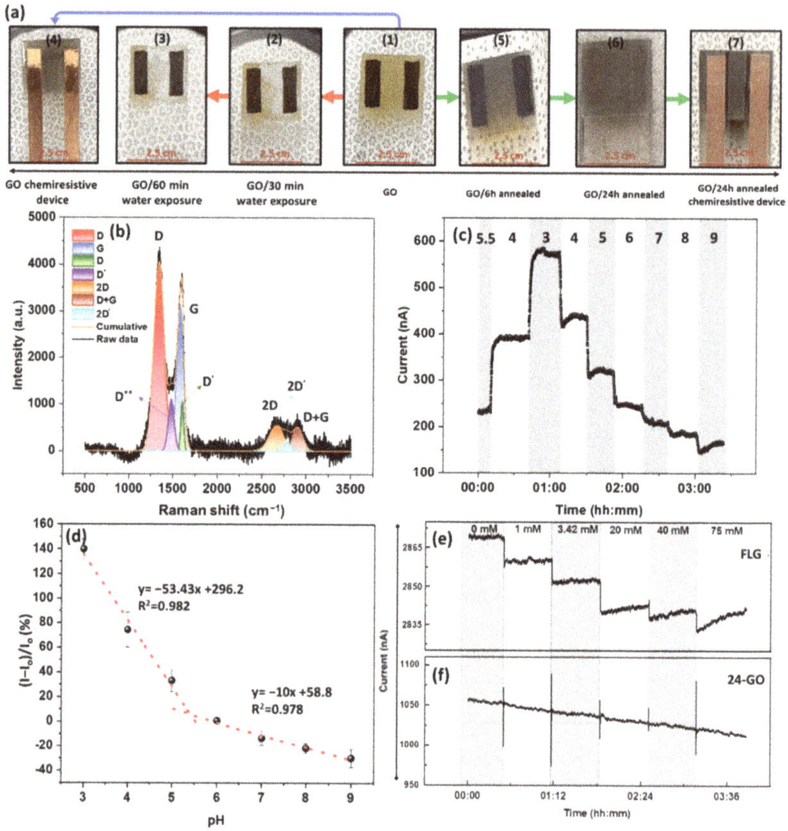

Figure 7. (**a**) The GO sensor fabrication: (**1**) Bare GO; (**2**) Bare GO exposed to the aqueous solution for 30 min; (**3**) Bare GO exposed to the aqueous solution for 6 min; (**4**) tTe GO chemiresistor without water exposure; (**5**) GO annealed for 6 h and (**6**) 26 h; (**7**) The 24 h annealed GO-based chemiresistor. (**b**) Deconvoluted Raman spectrum of 24 h-GO represents the presence of (from left to right): D, D*, G, D′, 2D, 2D′, D + G, and an I_D/I_G ratio of 1.3. (**c**) The pH response and (**d**) calibration curve of 24 h-GO (I_o = 242 nA). The solution conductivity response of (**e**) FLG (I_o = 2870 nA) and (**f**) 24 h-GO-based devices (I_o = 1050 nA).

The chemiresistive pH response of the 24 h annealed GO (24 h-GO) is shown in Figure 7c. Notably, the stepwise variation of the current in the 24 h-GO resembles the FLG response; however, the magnitude of the response is much higher. According to the calibration curve shown in Figure 7d, a more than 140% change in the current is obtained by changing the pH from 5.5 to 3, leading to a total response of 175% from a pH from 3 to 9. In order to allow for a calculation of sensitivity despite the overall nonlinear response, the working performance of the device is divided into low (3–5) and high (6–9) pH ranges. The selection of these two regimes is not arbitrary. The low range is chosen based on the pH response of -COOH and -NH_2, while the latter is based on the pH response of -OH groups. Accordingly, the low range offers a sensitivity of −53.4%/pH. In contrast, the high range sensitivity is calculated as −10%/pH. This difference in sensitivity is due to the highly favorable formation of -COOH during the GO synthesis and the dominant response of the carboxyl groups. Notably, the initial formation of -OH upon the oxidation of graphene to GO could subsequently produce -COOH through the two-step oxidations of hydroxyl → aldehyde → carboxyl [60].

As one of the possible interferences for the pH measurement is the ionic strength of the solution, the performances of the FLG and the 24 h-GO were measured against the changing the solution conductivity. For this purpose, NaCl was used to adjust the solution conductivity, giving up to 7.25 mS/cm for 75 mM. As seen in Figure 7e, the addition of NaCl gives rise to a small stepwise reduction of the current in the FLG. The mechanism justifying this behavior originates from the formation of an EDL on the SLG or FLG due to the long-range arrangement of Na^+ ions in the Stern layer. Therefore, the electrostatic gating charge screening is the primary mechanism, and the addition of Na^+_{aq} induces negative charges in the FLG. In contrast, the addition of NaCl does not affect the current through the GO film, and a prolonged drift in current is observed (Figure 7f). The cause of this drift is unclear at this point but could be due to a number of reported effects, including interactions of sodium ions with the oxygen sites over time [61] or variations in the concentrations of dissolved gases (such as CO_2 or O_2) as a function of the ionic strength [62]. After drift correction, however, there is no discernible response of the devices to ionic strength. In fact, the EDL response is negligible compared to the high density of defects present in GO-based devices; therefore, they exhibit an inherent potential for the selective detection of pH.

4. Conclusions

We have reported the development of a highly pH-sensitive platform based on thermally annealed GO for next-generation sensing devices. To understand the sensing principle of the proposed platform, we established the pH detection mechanisms of the two most commonly used graphene derivatives (i.e., single-layer and few-layer graphene). The contrast in pH responses of the former as a defect-free model (I_D/I_G~0.1, O/C ratio = 0.05), with the latter having a degree of defectivity (I_D/I_G~0.43, O/C = 0.24), elucidates the importance of defects in pH-sensing graphene. Therefore, the selective functionalization of graphene using various pH-sensitive functional groups demonstrates graphene's defect density dependence on pH response. An approximate surface density of 4.82×10^{14} cm^{-2} of the carboxyl group on graphene leads to a 55% response with an −21.58%/pH sensitivity for a pH of 3–5. The exact same surface density of the -OH groups, however, results in a −75% change in the current, leading to a sensitivity of −28.04%/pH for a pH of 9–7. To develop a GO-based pH-sensitive platform, we demonstrated the importance of the surface reduction treatment at a relatively low temperature (350 °C) to enhance its durability for long-term operation while retaining its high defectivity. As a result, a pH-sensitive device with a maximum current change of 175% (from a pH of 3–9) was reported, giving the sensitivity of −53.43 and −10%/pH pertaining to the pH range of 3–5 and 6–9, respectively. The proposed platform demonstrates minimum interference with ionic conductivity due to the dominance of the defects and offers a reagent-free pH-sensitive substrate for future pH devices.

Supplementary Materials: The following supporting information can be downloaded at: https://www.mdpi.com/article/10.3390/nano12111801/s1, Figure S1: XPS survey spectra of Single layer graphene and few layer graphene; Figure S2: Raman spectra of few layer graphene before and after exposure to pH; Figure S3: Raman spectrum of 8h annealed few layer graphene; Figure S4: pH response of 8h annealed few layer graphene; Figure S5: calibration curve of the pH response of 8h annealed few layer graphene; Figure S6: the Lorentzian deconvolution of GO Raman spectrum; Table S1: Summary of information on Pyrene concentrations and their corresponding relative surface coverage obtained from literature; Table S2: average ± Standard deviation of Py-COOH functionalized sensors to pH range 3-8 (3 sensors each); Table S3: average ± standard deviation of maximum response of pyrene derivative functionalized sensors as a functional surface density (3 sensors each); Table S4: average ± standard deviation of Py-NH2 functionalized sensors to pH range 3-8 (3 sensors each); Table S5: average ± standard deviation of Py-OH functionalized sensors to pH range 3-9 (3 sensors each) [63–65].

Author Contributions: S.A. wrote and prepared the manuscript for submission and was involved in planning the experiments; X.H. performed the experiments and measurements; L.H. performed the experiments and contributed to the discussions of the contents; M.A.A. contributed to the discussions of the contents; P.R.S. contributed to the conceptualization, writing, and review; P.K. supervised the manuscript and contributed to conceptualization, methodology, writing, review, and editing. All authors have read and agreed to the published version of the manuscript.

Funding: The authors acknowledge financial support from the Natural Sciences and Engineering Research Council of Canada (NSERC) through the Discovery Grant Program (RGPIN06145-18).

Acknowledgments: The authors thank Doris Stevanovic from the Centre for Emerging Device Technologies (CEDT), McMaster University, for granting access to the furnace; Alex Adronov (McMaster) for giving access to a Raman spectrometer; Mark Biesinger (Surface Science Western) for the XPS operation and data analysis.

Conflicts of Interest: The authors declare no conflict of interest.

References

1. Casey, J.R.; Grinstein, S.; Orlowski, J. Sensors and Regulators of Intracellular PH. *Nat. Rev. Mol. Cell Biol.* **2009**, *11*, 50–61. [CrossRef] [PubMed]
2. Posadas, E.; del Morales, M.M.; Gomez, C.; Acién, F.G.; Muñoz, R. Influence of PH and CO_2 Source on the Performance of Microalgae-Based Secondary Domestic Wastewater Treatment in Outdoors Pilot Raceways. *Chem. Eng. J.* **2015**, *265*, 239–248. [CrossRef]
3. Kasi, V.; Sedaghat, S.; M Alcaraz, A.; Kannan Maruthamuthu, M.; Heredia-Rivera, U.; Nejati, S.; Nguyen, J.; Rahimi, R. Low-Cost Flexible Glass-Based PH Sensor via Cold Atmospheric Plasma Deposition. *ACS Appl. Mater. Amp Interfaces* **2022**, *14*, 9697–9710. [CrossRef] [PubMed]
4. Wiora, A.; Wiora, J. Over One-Year Long-Term Laboratory Tests of PH Electrodes in Terms of Industrial Applications Checking Stabilities of Their Parameters and Their Influence on Uncertainties of Measurements. *Sensors* **2018**, *18*, 4102. [CrossRef] [PubMed]
5. Spijkman, M.-J.; Brondijk, J.J.; Geuns, T.C.T.; Smits, E.C.P.; Cramer, T.; Zerbetto, F.; Stoliar, P.; Biscarini, F.; Blom, P.W.M.; de Leeuw, D.M. Dual-Gate Organic Field-Effect Transistors as Potentiometric Sensors in Aqueous Solution. *Adv. Funct. Mater.* **2010**, *20*, 898–905. [CrossRef]
6. Bartic, C.; Palan, B.; Campitelli, A.; Borghs, G. Monitoring PH with Organic-Based Field-Effect Transistors. *Sens. Actuators B Chem.* **2002**, *83*, 115–122. [CrossRef]
7. Ghoneim, M.T.; Nguyen, A.; Dereje, N.; Huang, J.; Moore, G.C.; Murzynowski, P.J.; Dagdeviren, C. Recent Progress in Electrochemical PH-Sensing Materials and Configurations for Biomedical Applications. *Chem. Rev.* **2019**, *119*, 5248–5297. [CrossRef]
8. Chen, Y.; Mun, S.C.; Kim, J. A Wide Range Conductometric PH Sensor Made With Titanium Dioxide/Multiwall Carbon Nanotube/Cellulose Hybrid Nanocomposite. *IEEE Sens. J.* **2013**, *13*, 4157–4162. [CrossRef]
9. Avolio, R.; Grozdanov, A.; Avella, M.; Barton, J.; Cocca, M.; de Falco, F.; Dimitrov, A.T.; Errico, M.E.; Fanjul-Bolado, P.; Gentile, G.; et al. Review of PH Sensing Materials from Macro- to Nano-Scale: Recent Developments and Examples of Seawater Applications. *Crit. Rev. Environ. Sci. Technol.* **2022**, *52*, 979–1021. [CrossRef]
10. Magnusson, E.B.; Halldorsson, S.; Fleming, R.M.T.; Leosson, K. Real-Time Optical PH Measurement in a Standard Microfluidic Cell Culture System. *Biomed. Opt. Express* **2013**, *4*, 1749. [CrossRef]
11. Wencel, D.; Kaworek, A.; Abel, T.; Efremov, V.; Bradford, A.; Carthy, D.; Coady, G.; McMorrow, R.C.N.; McDonagh, C. Optical Sensor for Real-Time PH Monitoring in Human Tissue. *Small* **2018**, *14*, 1803627. [CrossRef] [PubMed]
12. Goldcamp, M.J.; Conklin, A.; Nelson, K.; Marchetti, J.; Brashear, R.; Epure, E. Inexpensive and Disposable PH Electrodes. *J. Chem. Educ.* **2010**, *87*, 1262–1264. [CrossRef]
13. Takeshita, Y.; Martz, T.R.; Johnson, K.S.; Dickson, A.G. Characterization of an Ion Sensitive Field Effect Transistor and Chloride Ion Selective Electrodes for PH Measurements in Seawater. *Anal. Chem.* **2014**, *86*, 11189–11195. [CrossRef]
14. Shao, Y.; Ying, Y.; Ping, J. Recent Advances in Solid-Contact Ion-Selective Electrodes: Functional Materials, Transduction Mechanisms, and Development Trends. *Chem. Soc. Rev.* **2020**, *49*, 4405–4465. [CrossRef] [PubMed]
15. Vonau, W.; Guth, U. PH Monitoring: A Review. *J. Solid State Electrochem.* **2006**, *10*, 746–752. [CrossRef]
16. Kim, D.-M.; Cho, S.J.; Cho, C.-H.; Kim, K.B.; Kim, M.-Y.; Shim, Y.-B. Disposable All-Solid-State PH and Glucose Sensors Based on Conductive Polymer Covered Hierarchical AuZn Oxide. *Biosens. Bioelectron.* **2016**, *79*, 165–172. [CrossRef]
17. Novoselov, K.S.; Geim, A.K.; Morozov, S.V.; Jiang, D.; Zhang, Y.; Dubonos, S.V.; Grigorieva, I.V.; Firsov, A.A. Electric Field Effect in Atomically Thin Carbon Films. *Science* **2004**, *306*, 666–669. [CrossRef]
18. Novoselov, K.S.; Mishchenko, A.; Carvalho, A.; Castro Neto, A.H. 2D Materials and van Der Waals Heterostructures. *Science* **2016**, *353*. [CrossRef]
19. Angizi, S.; Akbar, M.A.; Darestani-Farahani, M.; Kruse, P. Review—Two-Dimensional Boron Carbon Nitride: A Comprehensive Review. *ECS J. Solid State Sci. Technol.* **2020**, *9*, 083004. [CrossRef]
20. Hatamie, A.; Rahmati, R.; Rezvani, E.; Angizi, S.; Simchi, A. Yttrium Hexacyanoferrate Microflowers on Freestanding Three-Dimensional Graphene Substrates for Ascorbic Acid Detection. *ACS Appl. Nano Mater.* **2019**, *2*, 2212–2221. [CrossRef]

21. Rezvani, E.; Hatamie, A.; Berahman, M.; Simchi, M.; Angizi, S.; Rahmati, R.; Kennedy, J.; Simchi, A. Synthesis, First-Principle Simulation, and Application of Three-Dimensional Ceria Nanoparticles/Graphene Nanocomposite for Non-Enzymatic Hydrogen Peroxide Detection. *J. Electrochem. Soc.* **2019**, *166*, H3167–H3174. [CrossRef]
22. Wehling, T.O.; Novoselov, K.S.; Morozov, S.V.; Vdovin, E.E.; Katsnelson, M.I.; Geim, A.K.; Lichtenstein, A.I. Molecular Doping of Graphene. *Nano Lett.* **2008**, *8*, 173–177. [CrossRef] [PubMed]
23. Zubiarrain-Laserna, A.; Angizi, S.; Akbar, M.A.; Divigalpitiya, R.; Selvaganapathy, P.R.; Kruse, P. Detection of Free Chlorine in Water Using Graphene-like Carbon Based Chemiresistive Sensors. *RSC Adv.* **2022**, *12*, 2485–2496. [CrossRef]
24. Castro Neto, A.H.; Guinea, F.; Peres, N.M.R.; Novoselov, K.S.; Geim, A.K. The Electronic Properties of Graphene. *Rev. Mod. Phys.* **2009**, *81*, 109–162. [CrossRef]
25. Angizi, S.; Selvaganapathy, P.R.; Kruse, P. Graphene-Silicon Schottky Devices for Operation in Aqueous Environments: Device Performance and Sensing Application. *Carbon* **2022**, *194*, 140–153. [CrossRef]
26. Angizi, S.; Yu, E.Y.C.; Dalmieda, J.; Saha, D.; Selvaganapathy, P.R.; Kruse, P. Defect Engineering of Graphene to Modulate pH Response of Graphene Devices. *Langmuir* **2021**, *37*, 12163–12178. [CrossRef] [PubMed]
27. Leenaerts, O.; Partoens, B.; Peeters, F.M. Water on Graphene: Hydrophobicity and Dipole Moment Using Density Functional Theory. *Phys. Rev. B* **2009**, *79*, 235440. [CrossRef]
28. Taherian, F.; Marcon, V.; van der Vegt, N.F.A.; Leroy, F. What Is the Contact Angle of Water on Graphene? *Langmuir* **2013**, *29*, 1457–1465. [CrossRef]
29. Angizi, S.; Hatamie, A.; Ghanbari, H.; Simchi, A. Mechanochemical Green Synthesis of Exfoliated Edge-Functionalized Boron Nitride Quantum Dots: Application to Vitamin C Sensing through Hybridization with Gold Electrodes. *ACS Appl. Mater. Interfaces* **2018**, *10*, 28819–28827. [CrossRef]
30. Angizi, S.; Shayeganfar, F.; Azar, M.H.; Simchi, A. Surface/Edge Functionalized Boron Nitride Quantum Dots: Spectroscopic Fingerprint of Bandgap Modification by Chemical Functionalization. *Ceram. Int.* **2020**, *46*, 978–985. [CrossRef]
31. Tehrani, Z.; Whelan, S.P.; Mostert, A.B.; Paulin, J.V.; Ali, M.M.; Ahmadi, E.D.; Graeff, C.F.O.; Guy, O.J.; Gethin, D.T. Printable and Flexible Graphene PH Sensors Utilising Thin Film Melanin for Physiological Applications. *2D Mater.* **2020**, *7*, 024008. [CrossRef]
32. Jung, S.-H.; Seo, Y.-M.; Gu, T.; Jang, W.; Kang, S.-G.; Hyeon, Y.; Hyun, S.-H.; Lee, J.-H.; Whang, D. Super-Nernstian PH Sensor Based on Anomalous Charge Transfer Doping of Defect-Engineered Graphene. *Nano Lett.* **2021**, *21*, 34–42. [CrossRef] [PubMed]
33. Salvo, P.; Melai, B.; Calisi, N.; Paoletti, C.; Bellagambi, F.; Kirchhain, A.; Trivella, M.G.; Fuoco, R.; di Francesco, F. Graphene-Based Devices for Measuring PH. *Sens. Actuators B Chem.* **2018**, *256*, 976–991. [CrossRef]
34. Vivaldi, F.; Bonini, A.; Melai, B.; Poma, N.; Kirchhain, A.; Santalucia, D.; Salvo, P.; Francesco, F. di A Graphene-Based pH Sensor on Paper for Human Plasma and Seawater. In *Proceedings of the 2019 41st Annual International Conference of the IEEE Engineering in Medicine and Biology Society (EMBC), Berlin, Germany; 23–27 July 2019*; IEEE: New York, NY, USA; pp. 1563–1566.
35. Sha, R.; Komori, K.; Badhulika, S. Amperometric PH Sensor Based on Graphene–Polyaniline Composite. *IEEE Sens. J.* **2017**, *17*, 5038–5043. [CrossRef]
36. Konkena, B.; Vasudevan, S. Understanding Aqueous Dispersibility of Graphene Oxide and Reduced Graphene Oxide through p K_a Measurements. *J. Phys. Chem. Lett.* **2012**, *3*, 867–872. [CrossRef] [PubMed]
37. Ito, J.; Nakamura, J.; Natori, A. Semiconducting Nature of the Oxygen-Adsorbed Graphene Sheet. *J. Appl. Phys.* **2008**, *103*, 113712. [CrossRef]
38. Chinnathambi, S.; Euverink, G.J.W. Hydrothermally Reduced Graphene Oxide as a Sensing Material for Electrically Transduced PH Sensors. *J. Electroanal. Chem.* **2021**, *895*, 115530. [CrossRef]
39. Kim, T.; Hong, S.; Yang, S. A Solid-State Thin-Film Ag/AgCl Reference Electrode Coated with Graphene Oxide and Its Use in a PH Sensor. *Sensors* **2015**, *15*, 6469–6482. [CrossRef]
40. Neupane, S.; Subedi, V.; Thapa, K.K.; Yadav, R.J.; Nakarmi, K.B.; Gupta, D.K.; Yadav, A.P. An Alternative pH Sensor: Graphene Oxide-Based Electrochemical Sensor. *Emergent Mater.* **2021**, *5*, 509–517. [CrossRef]
41. Sohn, I.-Y.; Kim, D.-J.; Jung, J.-H.; Yoon, O.J.; Nguyen Thanh, T.; Tran Quang, T.; Lee, N.-E. PH Sensing Characteristics and Biosensing Application of Solution-Gated Reduced Graphene Oxide Field-Effect Transistors. *Biosens. Bioelectron.* **2013**, *45*, 70–76. [CrossRef]
42. Dalmieda, J.; Zubiarrain-Laserna, A.; Ganepola, D.; Selvaganapathy, P.R.; Kruse, P. Chemiresistive Detection of Silver Ions in Aqueous Media. *Sens. Actuators B Chem.* **2021**, *328*, 129023. [CrossRef]
43. Sengupta, I.; Sharat Kumar, S.S.S.; Pal, S.K.; Chakraborty, S. Characterization of Structural Transformation of Graphene Oxide to Reduced Graphene Oxide during Thermal Annealing. *J. Mater. Res.* **2020**, *35*, 1197–1204. [CrossRef]
44. Gupta, B.; Kumar, N.; Panda, K.; Kanan, V.; Joshi, S.; Visoly-Fisher, I. Role of Oxygen Functional Groups in Reduced Graphene Oxide for Lubrication. *Sci. Rep.* **2017**, *7*, 45030. [CrossRef] [PubMed]
45. Kwan, Y.C.G.; Ng, G.M.; Huan, C.H.A. Identification of Functional Groups and Determination of Carboxyl Formation Temperature in Graphene Oxide Using the XPS O 1s Spectrum. *Thin Solid Film.* **2015**, *590*, 40–48. [CrossRef]
46. Achtyl, J.L.; Vlassiouk, I.V.; Fulvio, P.F.; Mahurin, S.M.; Dai, S.; Geiger, F.M. Free Energy Relationships in the Electrical Double Layer over Single-Layer Graphene. *J. Am. Chem Soc.* **2013**, *135*, 979–981. [CrossRef]
47. Li, X.; Shi, J.; Pang, J.; Liu, W.; Liu, H.; Wang, X. Graphene Channel Liquid Container Field Effect Transistor as PH Sensor. *J. Nanomater.* **2014**, *2014*, 547139. [CrossRef]

48. He, H.; Kim, K.H.; Danilov, A.; Montemurro, D.; Yu, L.; Park, Y.W.; Lombardi, F.; Bauch, T.; Moth-Poulsen, K.; Iakimov, T.; et al. Uniform Doping of Graphene Close to the Dirac Point by Polymer-Assisted Assembly of Molecular Dopants. *Nat. Commun.* **2018**, *9*, 3956. [CrossRef]
49. Gierz, I.; Riedl, C.; Starke, U.; Ast, C.R.; Kern, K. Atomic Hole Doping of Graphene. *Nano Lett.* **2008**, *8*, 4603–4607. [CrossRef]
50. Kiani, M.J.; Ahmadi, M.T.; Karimi Feiz Abadi, H.; Rahmani, M.; Hashim, A.; Che harun, F.K. Analytical Modelling of Monolayer Graphene-Based Ion-Sensitive FET to PH Changes. *Nanoscale Res. Lett.* **2013**, *8*, 173. [CrossRef]
51. Zahed, M.A.; Barman, S.C.; Das, P.S.; Sharifuzzaman, M.; Yoon, H.S.; Yoon, S.H.; Park, J.Y. Highly Flexible and Conductive Poly (3, 4-Ethylene Dioxythiophene)-Poly (Styrene Sulfonate) Anchored 3-Dimensional Porous Graphene Network-Based Electrochemical Biosensor for Glucose and PH Detection in Human Perspiration. *Biosens. Bioelectron.* **2020**, *160*, 112220. [CrossRef]
52. Medina, H.; Lin, Y.-C.; Obergfell, D.; Chiu, P.-W. Tuning of Charge Densities in Graphene by Molecule Doping. *Adv. Funct. Mater.* **2011**, *21*, 2687–2692. [CrossRef]
53. Casiraghi, C.; Hartschuh, A.; Qian, H.; Piscanec, S.; Georgi, C.; Fasoli, A.; Novoselov, K.S.; Basko, D.M.; Ferrari, A.C. Raman Spectroscopy of Graphene Edges. *Nano Lett.* **2009**, *9*, 1433–1441. [CrossRef] [PubMed]
54. Osella, S.; Kiliszek, M.; Harputlu, E.; Unlu, C.G.; Ocakoglu, K.; Kargul, J.; Trzaskowski, B. Controlling the Charge Transfer Flow at the Graphene/Pyrene–Nitrilotriacetic Acid Interface. *J. Mater. Chem. C* **2018**, *6*, 5046–5054. [CrossRef]
55. Zhen, X.V.; Swanson, E.G.; Nelson, J.T.; Zhang, Y.; Su, Q.; Koester, S.J.; Bühlmann, P. Noncovalent Monolayer Modification of Graphene Using Pyrene and Cyclodextrin Receptors for Chemical Sensing. *ACS Appl. Nano Mater.* **2018**, *1*, 2718–2726. [CrossRef]
56. Tang, B.; Guoxin, H.; Gao, H. Raman Spectroscopic Characterization of Graphene. *Appl. Spectrosc. Rev.* **2010**, *45*, 369–407. [CrossRef]
57. Díez-Betriu, X.; Álvarez-García, S.; Botas, C.; Álvarez, P.; Sánchez-Marcos, J.; Prieto, C.; Menéndez, R.; de Andrés, A. Raman Spectroscopy for the Study of Reduction Mechanisms and Optimization of Conductivity in Graphene Oxide Thin Films. *J. Mater. Chem. C* **2013**, *1*, 6905. [CrossRef]
58. Sengupta, I.; Chakraborty, S.; Talukdar, M.; Pal, S.K.; Chakraborty, S. Thermal Reduction of Graphene Oxide: How Temperature Influences Purity. *J. Mater. Res.* **2018**, *33*, 4113–4122. [CrossRef]
59. Farah, S.; Farkas, A.; Madarász, J.; László, K. Comparison of Thermally and Chemically Reduced Graphene Oxides by Thermal Analysis and Raman Spectroscopy. *J. Therm. Anal. Calorim.* **2020**, *142*, 331–337. [CrossRef]
60. Vacchi, I.A.; Raya, J.; Bianco, A.; Ménard-Moyon, C. Controlled Derivatization of Hydroxyl Groups of Graphene Oxide in Mild Conditions. *2d Mater.* **2018**, *5*, 035037. [CrossRef]
61. Kwon, K.C.; Choi, K.S.; Kim, C.; Kim, S.Y. Role of Metal Cations in Alkali Metal Chloride Doped Graphene. *J. Phys. Chem. C* **2014**, *118*, 8187–8193. [CrossRef]
62. Sander, R. Compilation of Henry's Law Constants (Version 4.0) for Water as Solvent. *Atmos. Chem. Phys.* **2015**, *15*, 4399–4981. [CrossRef]
63. Zhen, X.V.; Swanson, E.G.; Nelson, J.T.; Zhang, Y.; Su, Q.; Koester, S.J.; Bühlmann, P. Noncovalent Monolayer Modification of Graphene Using Pyrene and Cyclodextrin Receptors for Chemical Sensing. *ACS Appl. Nano Mat.* **2018**, *1*, 2718–2726. [CrossRef]
64. An, X.; Simmons, T.; Shah, R.; Wolfe, C.; Lewis, K.M.; Washington, M.; Nayak, S.K.; Talapatra, S.; Kar, S. Stable Aqueous Dispersions of Noncovalently Functionalized Graphene from Graphite and Their Multifunctional High-Performance Applications. *Nano Lett.* **2010**, *10*, 4295–4301. [CrossRef]
65. Georgakilas, V.; Otyepka, M.; Bourlinos, A.B.; Chandra, V.; Kim, N.; Kemp, K.C.; Hobza, P.; Zboril, R.; Kim, K.S. Functionalization of Graphene: Covalent and Non-Covalent Approaches, Derivatives and Applications. *Chem. Rev.* **2012**, *112*, 6156–6214. [CrossRef]

nanomaterials

Article

A System of Rapidly Detecting Escherichia Coli in Food Based on a Nanoprobe and Improved ATP Bioluminescence Technology

Zhen Sun [1], Jia Guo [1], Wenbo Wan [2] and Chunxing Wang [1,*]

[1] College of Physics and Electronic Science, Shandong Normal University, Jinan 250358, China; szskyedu@163.com (Z.S.); 17862967173@163.com (J.G.)
[2] College of Information Science and Engineering, Shandong Normal University, Jinan 250358, China; wanwenbo@sdnu.edu.cn
* Correspondence: cxwang@sdnu.edu.cn

Abstract: Bacterial contamination is an important factor causing food security issues. Among the bacteria, *Escherichia coli* is one of the main pathogens of food-borne microorganisms. However, traditional bacterial detection approaches cannot meet the requirements of real-time and on-site detection. Thus, it is of great significance to develop a rapid and accurate detection of bacteria in food to ensure food safety and safeguard human health. The pathogen heat-treatment module was designed in this paper based on the techniques including nanoprobe, pathogen heat-treatment, graphene transparent electrode (GTE), and adenosine triphosphate (ATP) bioluminescence technology. The system mainly consists of two parts: one is the optical detection unit; the other is the data processing unit. And it can quickly and automatically detect the number of bacterial colonies in food such as milk etc. The system uses not only the probe to capture and enrich *E. coli* by antigen-antibody interaction but also the heat treatment to increase the amount of ATP released from bacterial cells within five minutes. To enhance the detecting accuracy and sensitivity, the electric field generated by GTE is adopted in the system to enrich ATP. Compared to the other conventional methods, the linear correlation coefficient of the system can be reached 0.975, and the system meets the design requirements. Under the optimal experimental conditions, the detection can be completed within 25 min, and the detectable concentration of bacteria is in the range of 3.1×10^1–10^6 CFU/mL. This system satisfies the demands of a fast and on-site inspection.

Keywords: nanoprobe; pathogen heat-treatment; adenosine triphosphate luminescence; graphene transparent electrode

Citation: Sun, Z.; Guo, J.; Wan, W.; Wang, C. A System of Rapidly Detecting Escherichia Coli in Food Based on a Nanoprobe and Improved ATP Bioluminescence Technology. *Nanomaterials* **2022**, *12*, 2417. https://doi.org/10.3390/nano12142417

Academic Editor: Thomas Pons

Received: 2 June 2022
Accepted: 12 July 2022
Published: 14 July 2022

1. Introduction

In recent years, with the development of economic and social progress, human living standards have gradually improved, increasing attention has been paid to food safety, and requirements for food safety technology that are more stringent have been established. According to news reports and data, bacterial infections cause about 40% of all diseases and pose a huge threat to public health and the social economy [1]. Today, common bacterial detection approaches include culture and colony counting methods [2], enzyme-linked immunosorbent assay (ELISA) [3], biosensor technology [4], and adenosine triphosphate (ATP) and bioluminescence assay [5].

Escherichia coli (*E. coli*) is a common conditionally pathogenic bacterium. Under certain conditions, a part of *E. coli* strains harboring virulence factors are considered to be pathogenic to humans, so they are often used as an indicator for monitoring the quality of food and drinking water. Currently, the approaches to detecting *E.coli* have been divided into traditional detection methods and new modern detection methods. The traditional detection methods include the plate dilution method [6], membrane filtration, fluorescence quenching [7], and the use of quartz-crystal microbalance-based sensors [8]. Although these methods have relatively high reliability and sensitivity, most of them still have some drawbacks. The plate dilution method, for example, was once regarded as a classic microbial detection method. However, this approach is comparatively complex because it requires the culturing and enriching of bacteria in the lab prior to testing. Thus, the detection cycle is longer, making real-time detection impossible. It is difficult to meet the current domestic and foreign requirements of bacteria detection in food. New modern detection methods include ATP bioluminescence technology [9], magnetic-activated cell sorting (MACS) [10], the biosensor detection method [11], PCR detection technology [12], the gene chip method [13] and genome sequencing [14]. Among these modern detection methods, such as immunoassays, which are being applied in industry, most modern methods cannot meet the demand for portable and on-site detection due to the high equipment cost and professional laboratory operation. At the same time, most of the traditional bacteria detection methods generally have disadvantages, such as being time-consuming, complicated operation, low signal-to-noise ratio and non-specific bacterial identification, and these shortcomings lead that not all traditional methods can meet industry needs in every context. Therefore, it is very important to develop a fast, portable and highly accurate bacterial detection system to ensure food safety and maintain public health and environmental sanitation.

The combination of immunomagnetic separation technology and ATP bioluminescence technology provides a way to detect bacteria. However, how to improve the accuracy and expand the detection limit is an important question we face.

Immunomagnetic separation technology is one of the research hotspots in the field of bacterial detection. Magnetic nanoprobes are an indispensable part of immunomagnetic separation technology, based on nanotechnology and immunology, using various magnetic nanoparticles coated with immunoactive substances for immunological or biological analysis. Fish et al. realized a method to detect the concentration of *Bacillus spores* rapidly using immunomagnetic separation technology incorporating a chromatography technique in 2009 [15]. Moreover, Kuang et al. used immunomagnetic separation combined with a fluorescent probe to detect *Salmonella* with a sensitivity of 500 CFU/mL [16]. In 2016, Mengxu et al. designed an electrochemical immunosensor to detect *E. coli* and *Salmonella typhimurium* in food based on immunomagnetic separation technology and screen-printed interdigitated microelectrodes [17]. In recent years, immunomagnetic separation technology has developed rapidly and become a common detection method.

ATP bioluminescence technology is one of the most common microbiological detection methods, which can determine the total number of microorganisms in food. ATP is an important energy substance in cells that can store and provide energy. It is immanent in all cells and usually measured by the fluorescein–luciferase luminescence system. The wide adhibition of ATP bioluminescence technology in food detection and the medical industry is due to the creation and application of portable ATP Detectors. Murphy et al. used ATP bioluminescence technology to detect bacteria in liquid milk [18]. Although this method was proven feasible by comparing the results from the standard plate count method, it is difficult to use on-site. In 2017, Zhang et al. utilized ATP bioluminescence technology to detect the number of *E. coli* bacteria in food. The quantitative results demonstrated that their method was useful in detecting the bacteria with a sensitivity of 3.0×10^2 CFU mL^{-1} [19]. Xu et al. combined transparent graphene electrodes with ATP bioluminescence technology to improve the accuracy of *E. coli* detection in food [20], but sensitivity remained low.

According to the research, temperature change has an obvious influence on the bacteria-detecting process. Heat-treatment pathogen technology is used to adopt heating, thermal insulation, and cooling to determine the properties or reaction states of pathogens under different temperatures, which can be better used in the field of bacterial detection or bacterial inactivation. Today, in addition to inactivating bacteria at high temperatures, the pathogen heat-treatment technology can be used to enhance the signal strength for bacterial detection. In 2017, Lee et al. found that the amount of ATP released from bacterial cells, such as *Salmonella enteritidis*, *E. coli O157:H7*, and *Bacillus cereus*, can be increased by subjecting the samples to heat treatment, which helps detect the number of bacteria colonies using ATP bioluminescence technology [21].

Because photoelectric conversion is applied in this system, graphene, a new bioassay material, has attracted our attention because of its excellent light transmission and electrical conductivity. In 2004, Novoselov et al. prepared graphene films by mechanical exfoliation and discovered their unique electronic properties [22]. Li et al. prepared graphene films by graphene chemical vapor deposition and developed a process to transfer graphene film to various substrates, making it possible to prepare transparent graphene electrodes [23]. In 2010, Bae et al. produced transparent graphene electrodes using a roll-to-roll production method and wet chemicals [24]. Afterwards, it was found that the optical transmittance of transparent graphene electrodes is 97.4% and that their conductance can be as low as 125 Ω^{-1}, which is better than the other conventional transparent electrodes, such as those fabricated from indium tin oxide. Moreover, graphene also has excellent thermal conductivity, of up to 5300 $W/m \cdot K$, higher than that of carbon nanotubes and diamonds.

Based on the above studies, our final goal is to further enhance the accuracy and sensitivity of the ATP bioluminescence detection system and expand its detection limit. Therefore, we designed a rapid detection system based on immunomagnetic separation technology, ATP bioluminescence technology, and pathogen heat-treatment technology. In addition, the system utilizes electric field force to enrich ATP. The system uses the biotinylated *E. coli* antibodies and streptavidin-modified magnetic nanoparticles to prepare the magnetic nanoprobes with an avidin-biotin link reaction. The detection tube is prepared by the chemical vapor deposition (CVD) of graphene [25]. The probes capture the pathogens, and then to enhance the amount of ATP the target pathogens release, heat treatment is carried out on the samples before the target pathogen is lysed. Among them, thin-film heaters and resistance temperature detectors (RTDs), the main components of heat treatment, have been widely applied in sensor chips, biological chips, and microfluid chips [26,27]. The released ATP can be enriched by an electric field applied through transparent graphene electrodes because ATP is negatively charged under weakly alkaline conditions. The experiment showed that the number of bacterial colonies is proportional to the luminescence intensity. As the intensity of ATP bioluminescence attenuates over time, we add another light source to reduce the error and use photomultiplier tubes (PMT) for luminescence acquisition and photoelectrical signal conversion. We then carry out a series of optimizations in the subsequent signal processing of the system to meet the requirements of measurement accuracy and achieve quantitative detection. Compared to the previous system, the heat treatment module is added, and the ATP bioluminescence reaction chamber is improved. Because temperature change has a significant impact on the bacteria-detecting process, the addition of heat treatment modules can eliminate the influence of ambient temperature change on ATP bioluminescence reaction, increase the relative luminescence unit (RLU), enhance the luminescence signal, and improve the accuracy and sensitivity of detection results. The results show that under the optimal experimental conditions, the linear correlation coefficient of the system could reach 0.975, and the detection concentration of bacteria was in the range of 1.7×10^1–10^6 CFU/mL. The combination of these technologies not only improves detection results and makes the system more convenient but also shows good performance in field detection.

2. Detection Principle

Adenine nucleoside triphosphate, abbreviated as adenosine triphosphate (ATP), is an energy-supplying cell substance. It exists in the cells of all kinds of organisms [28]. ATP consists of an adenine molecule, a ribose molecule and nucleotides formed with three linked phosphate groups [29]. ATP releases a large amount of energy when it hydrolyzes, the most direct energy source in living organisms. Figure 1 shows the molecular constitution of adenosine triphosphate.

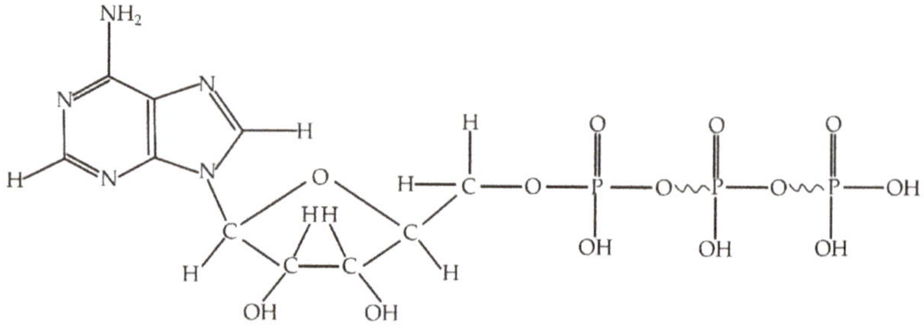

Figure 1. Structure of Adenosine triphosphate.

In general, the quantity of ATP is roughly the same in each bacterium, at approximately 1–2 fg per cell [30]. Moreover, the mechanism, luciferin reacting with oxygen under the action of luciferase to produce bioluminescence, was discovered in 1947, and ATP can provide the energy for this reaction. According to our previous experiments, the experimental reagent's luminescence intensity has a strong linear relationship with the concentration of ATP in a certain concentration range. The above findings provide theoretical bases for measuring the amount of bacteria utilizing ATP bioluminescence technology. In the experiments, we need to use a reagent group including an erase reagent, lysis reagent, luciferase, and fluorescein. The function of the erase reagent is to remove the effects of somatic cells and free ATP. The function of the lysis reagent is to lyse *E. coli* and release ATP. Luciferase and fluorescein provide raw materials for the bioluminescence process and react in the presence of Mg^{2+} and O_2. In addition, we subject the captured *E. coli* pathogen to heating during detection, which can increase the amount of ATP the pathogen releases and enhance detection sensitivity. In our previous study, we found that the reagent's concentration affected the luminescence intensity. Therefore, we subjected the reagent concentration to 30–40 mg/L to stabilize the peak luminescence intensity. When the ATP concentration is less than 10^{-7} mol/L, we observe a strong linear relationship between the ATP content and light intensity when the other reactants are sufficient in the reaction process. Because the solution's pH affects luciferase activity, it will directly cause the emitting light to deviate from its wavelength peak. The results of experimental data show that a stable peak wavelength of light is acquired when the pH is between 7.5 and 7.8. In this process, almost all of the energy from ATP is converted into light energy. The amount of *E. coli* cells can be measured indirectly using Equation (1), where hν represents luminous energy.

$$\mathrm{ATP} + \mathrm{O_2} + \mathrm{Luciferin} \xleftrightarrow{\mathrm{Mg^{2+}} + \mathrm{Luciferase}} \mathrm{Oxyluciferin} + \mathrm{AMP} + \mathrm{CO_2} + \mathrm{ppi} + h\nu \quad (1)$$

A nanoprobe is a nanoscale biosensor that can detect single living cells, which have the characteristics of nanoscale size and real-time monitoring and causes little damage to cells [31]. The immunomagnetic beads (IMB) separation technique is a new immunological technique that combines the high specificity of immunological response with the unique magnetic responsiveness of magnetic beads. It is a kind of immunological detection method with strong specificity and high sensitivity. IBMs are magnetic beads wrapped with monoclonal antibodies, which have the strengths of high speed, high efficiency and simple operation. Therefore, they have a wide range of applications in the enrichment and separation of bacteria [32]. Compared to the traditional methods, IBMs do not have the disadvantages of time-consuming techniques and a pre-enrichment requirement. Magnetic nanoparticles can be collected by a magnetic field, and this property can be used to combine bacteria with magnetic nanoparticles for specific separation and to capture target bacteria.

In our previous study, we designed medium-size IBMs (120–200 nm) to capture *E. coli* [19]. First, we utilized streptavidin to modify IMBs to obtain magnetic beads whose surface is modified with streptavidin. Then we mixed the *E. coli*-specific antibodies and biotin to get biotinylated *E. coli* antibodies. We prepared immunomagnetic nanoprobes by combining magnetic beads modified with streptavidin with biotinylated *E. coli* antibodies by enzyme-linked immunosorbent assay [33,34]. We used immunomagnetic nanoprobes to capture *E. coli* by antigen-antibody reaction, and we prepared the IMB/antibody–*Escherichia coli*-immune compounds. An antibody is an immunoglobulin produced by plasma cells differentiated from B cells in response to the stimulation of antigenic substances, and antibodies can specifically bind to the corresponding antigen. In our experiments, we studied *E. coli O157:H7* and used *E. coli O157:H7* monoclonal antibodies to capture them. These antibodies bound specifically to the somatic antigen (O antigen) and the flagellar antigen (H antigen) of *E. coli O157:H7*, respectively, to achieve the effect of capturing the bacterium. Figure 2 shows the preparation and capture processes. To enrich and isolate the magnetic immune compounds, we designed a magnetic field at the two polar ends of the samples.

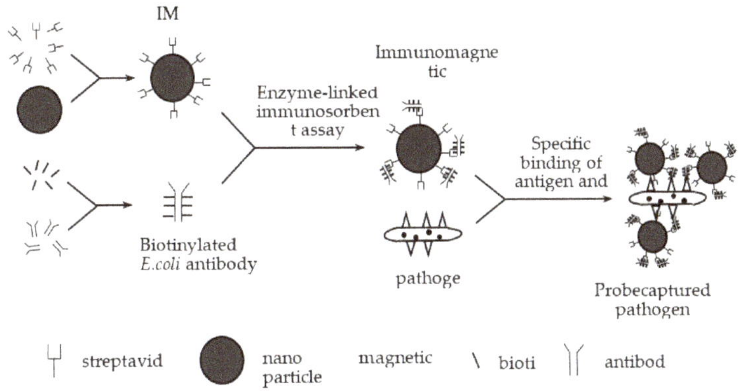

Figure 2. Magnetic nanoprobe and capture of *E. coli*.

Graphene is a material with a mono-layered two-dimensional hexagonal lattice structure [35]. Due to graphene's excellent properties, such as good electrical conductivity [36], light transmission, excellent mechanical properties, and electron mobility at room temperature, graphene and its chemically modified derivatives can be used as graphene transparent electrodes. Moreover, graphene has high thermal conductivity, which allows us to apply it for heat and cooling.

In previous studies, we found that temperature has a significant impact on the amount of ATP bacterial cells release. Therefore, in this study, we found a method that improves the sensitivity by increasing the amount of ATP that *E. coli* releases by subjecting the samples to heat treatment [21]. Here, we designed a temperature control module accurately. A heat treatment unit comprises three parts: a thermal cycler, reaction chambers, and a temperature detection and signal feedback web. The thermal cycler comprises a heating component, a cooling component and a thermal signal controller. However, the great difficulties of designing the thermal cycle structure lie in the accurate control of temperature and excellent thermal uniformity around the reaction chamber.

To solve those problems, we designed a thin-film heater. The thin-film heater, which is formed by depositing polycrystalline or metallic materials on the substrate surfaces [37,38], is very popular because of its advantages, such as low power consumption, high operation speed, and strong temperature control performance, and it is crucial for the portability of bacterial detection equipment. To meet space and environmental needs, we designed a flexible polyimide thin-film heater that mainly comprises external insulation material and internal heating wire. We chose polyimide as external insulation material because it has high-temperature resistance, high insulation and excellent mechanical tractility properties. Commonly, the materials of the heating wire used in thin-film heaters are divided into

Fe-Cr-Al alloy and nickel-chromium alloy. Here, we selected nickel-chromium alloy as the material for the thin-film heater because it has excellent thermoelectric properties. It is difficult to deform at high temperatures, and it has the advantageous qualities of having a long service life, being non-magnetic, having strong corrosion resistance, etc.

A temperature detector is an essential part of the temperature detection and signal feedback unit. It is used not only to detect the temperature of the heater and the reaction chamber but also to feed the temperature signal to the MCU. The heating or cooling element is run by the thermal signal controller to maintain a stable temperature. We used the resistance temperature detector (RTD) in our system because of its high accuracy, high stability, and wide temperature-control range. In general, metallic materials are used for RTDs; temperature monitoring relies on the relationship between the resistance and the temperature [39–41]. Platinum, copper and nickel are commonly used to manufacture RTDs, for they have the advantages of a high-temperature coefficient and quick response to temperature changes. Moreover, they are easily manufactured into refined coils. Among them, platinum, with its high stability, does not react physically or chemically in a certain temperature range. Platinum-resistance temperature detectors have a wide measuring range. In addition, they not only have better linearity than thermocouples and thermistors, but they are also more accurate and stable temperature sensors than any we have obtained up to now.

3. Testing Prototype Design

The system primarily includes two parts. One is the optical detection unit, and the other is the photoelectric signal processing and display unit, as Figure 3 shows.

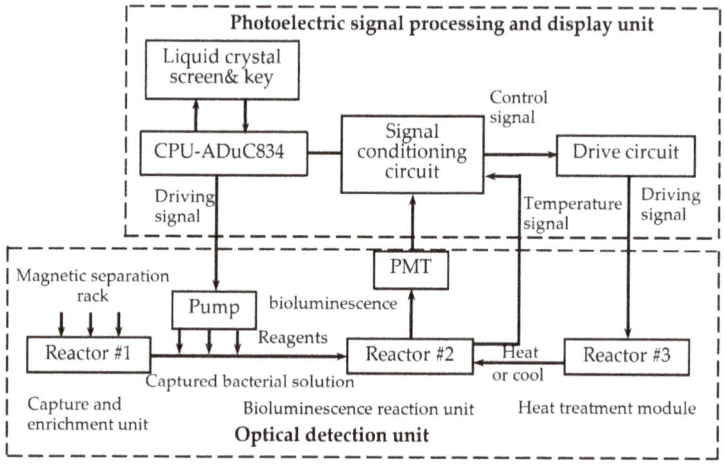

Figure 3. Frame diagram of the system. Reactor #1 is the capture and enrichment unit, which is intended to capture *E. coli* pathogens by our designed immunomagnetic nanoprobes and to enrich bacteria by a magnetic separation rack. Reactor #2 is the bioluminescence reaction unit, and reactor #3 is the heat treatment module, both of which cooperate to process the captured *E. coli* solution by pathogens heat-treating and adding reagents by a peristaltic pump. The bioluminescence signal is collected by the PMT and sent to the CPU through the signal processing circuit. The control signals of the peristaltic pump and the heat treatment module are sent by the CPU. The liquid crystal screen displays the results, and the device can be operated by key.

The optical detection unit comprises four components. Reactor #1 is a capture and enrichment unit, reactor #2 is a bioluminescence reaction unit, and reactor #3 is a heat treatment module. Additionally, the unit contains an automatic filling-up unit.

Reactor #1 is the capture and enrichment unit, which is intended to use our designed immunomagnetic nanoprobes to capture *E. coli* pathogens and enrich bacteria by magnetic separation rack, as shown in Figure 3. The details of *E. coli O157: H7* cultured are as follows. *E. coli O157: H7* standard strains were added to the modified *E. coli* medium (37 °C, 24 h) and then homogenized in 300 μL sterile saline. After centrifugation at 5000 rpm, the precipitated supernatant was discarded and placed in saline suspension again. Finally, the original bacterial solution was obtained at a concentration of 10^7–10^8 CFU/mL. Multiple gradient concentrations of the samples were obtained by diluting the original bacterial solution. The samples were divided into two parts, one was for the conventional culture method, and the other was for detection by our method. In the following, we will describe the technical details. First, 50 μL of the immunomagnetic nanoprobes solution is added to the *E. coli* solution and wait 10 min for the probe to capture the bacteria. Due to utilizing an enzyme-linked immunosorbent assay to bind *E. coli* antibodies to the immunomagnetic nanoprobes, the probe can capture *E. coli* through antigen-antibody reaction. Second, after the *E. coli* is captured, magnetic separation is carried out. And under the action of the magnetic separation rack, the IMB/antibody–Escherichia coli-immune compounds are adsorbed onto the bottom of the test tube. Then, the supernatant is discarded after magnetic separation. Last but not least, to obtain the captured *E. coli* solution, the enriched samples are washed three times with 150 mL of cleaning solution (10-mmol L-1 PBS, 0.05% Tween-20).

Reactor #2 is the bioluminescence reaction unit, which consists of a special detection tube and a photomultiplier (PMT). The detection tube is characterized by a graphite layer attached to the bottom. The heat treatment module (Reactor #3) is a temperature control device implemented using a heating component, a cooling component and a thermal signal controller. They cooperate to process the captured *E. coli* solution. First, the solution used for the ATP bioluminescence testing is subjected to heat treatment to increase the amount of ATP released from *E. coli*. We designed a kind of flexible thin-film heater as the heating component in the temperature control device, and its design was as shown in Figure 4. We used the polyimide film as the substrate material (thickness 0.15 mm). Nickel-chromium alloy foil was selected as the resistance circuit material (thickness 0.05 mm). Nickel-chromium alloy foil was laminated onto the substrate (polyimide film). This step was achieved using a thin thermosetting adhesive layer with excellent bonding properties when bonded to both materials. After lamination was stabilized, the substrate was drilled and positioned so that each layer within the heater remained aligned during a period of manufacturing. Next, the photoresist was smeared evenly on the panel (nickel-chromium alloy foil). After the photoresist was attached well, we placed the advanced photomask tool on the resist. The mask, which was generated based on a CAD design tool, was the final design template of the heating element, including the applicable element width required to produce the correct resistive heater. Then, the exposure was carried out. In this process, the resist was exposed to ultraviolet light to cure for use as a chemical etching resist. Next, the uncured resist was removed to expose the foil for etching and foil removal. At the same time, the cured resist can protect the pattern of the heater element. The panel was then chemically etched, stripped, and cleaned to chemically remove the foil not protected by the resistor and to retain the heating element pattern on the panel. The next step was to place a top overlay on the panel and laminate it. The top layer, the polyimide film (thickness 0.15 mm), provided access to the heating elements, including connecting wires and the openings for assemblies. Finally, after the heating film was removed from the panel, the wire was connected to the heating film, and the high-performance pressure-sensitive adhesive was coated on the back of the heating film. We adhered the heating film to the inner wall of the container that holds the test tube, and the heating film was wrapped tightly on the outer wall of the test tube for heating. Resistance temperature detector (RTD) PT100′s connector was fixated in the reserved position between the thin-film heater and the test tube for monitoring and feedback on the temperature of the test tube. Both the thin-film heater and RTD were controlled by the external drive circuit and the thermal signal controller. The thin-film heater was controlled using proportional-integral-derivative (PID) closed-loop controllers, which were implemented on CPU-ADUC834. CPU-ADUC834 is a microcontroller that has two independent ADCs [42]. In addition, the thin film heater was driven by a double MOS parallel active output circuit. A PT100 resistance temperature detector (RTD) was connected with a MAX31865 chip to convert the voltage signal corresponding to test tube temperature into a digital signal. The MAX31865 chip was then connected with an ADUC834 to control the drive circuit according to the collected temperature. Because the optimal temperature for the bioluminescence reaction is 24.5 °C, we heated the tube to 50 °C for 5 min and then let it cool to 24.5 °C for 20 s for the next bioluminescence reaction. The cooling component was implemented by a low-power fan, which was also controlled by the ADUC834.

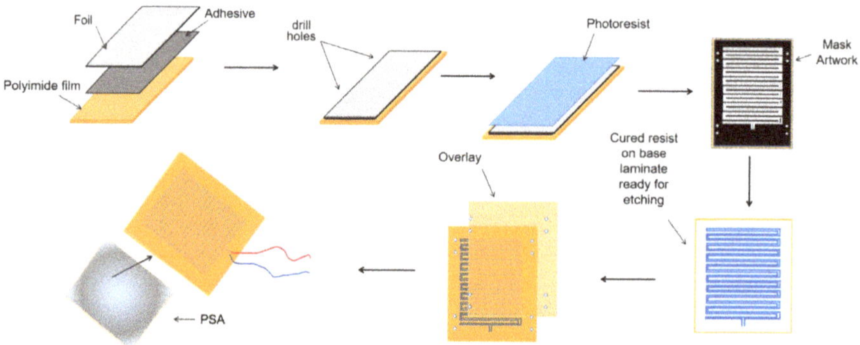

Figure 4. The production flow chart of the thin-film heater.

ATP bioluminescence reaction occurred in the detection tube in the fluorescent reaction unit (Reactor #2). We connected the graphene electrode to the upper electrode through a steady current source, creating a weak electric field in the detection tube. The studies show that graphene has good optical transmittance, so the PMT, which is used for light trapping and photoelectric conversion, can be placed on the side of the test tube. To avoid the interference of external light as much as possible to ensure the accuracy of detection, the entire bioluminescence reaction and light capture and detection by PMT were performed in a completely closed environment. The procedure for this part of the assay is as follows: 30 μL of enriched *E. coli* suspension was added to the special test tube with a graphene electrode attached to the bottom. The solution in the tube is heat-treated for 5 min using the flexible thin-film heater we designed and then cooled to the optimum ATP bioluminescence reaction temperature using the cooling fan. Then 270 μL of the detection reagent was accurately injected with a peristaltic pump. The lysis agent in the detection reagent can lyse *E. coli* and release ATP. Meanwhile, the graphene electrode is energized by a built-in current source to create an electric field outside the test tube. Moreover, the experiments indicate that the luminescence from the ATP bioluminescence reaction is stronger under weak alkaline conditions compared to other acid-base conditions, and ATP has a negative charge under weak alkaline conditions. ATP will be enriched at the bottom of the test tube under the action of the electric field. The electric field also can enhance the activity of ATPASE and accelerate the oxidation of luciferase, thus enhancing the stability and accuracy of the system. After adding the detection reagent, the energy for the bioluminescence reaction is provided by ATP, and the luminescence emitted from the ATP bioluminescence reaction is captured by the PMT. The system automatically reads the optical signal after 60 s. And an internal calibration and compensation unit was used to detect and compensate for light attenuation, which will be described in the following. It is essential to add a filtering circuit in the design to reduce external noise interference due to the unavoidable system noise and weak input signals.

For portability and automation, we added an automated filling-up unit to the system, which is controlled by a microcontroller and enables the system to automatically select the appropriate reagents and quickly fill the required dose of reagents as required, as shown in Figure 5. In this unit, a single reagent hose line is represented by a thin line, and multiple reagent line hoses are represented by thick lines. For the tubes used in each unit, we choose the hoses characterized by high elasticity, low adhesiveness and low permeability. The different types of reagents for system selection are stored in different containers. Erase reagent, lysis reagent (50 mmol/LTris-HCl, 150 mmol/L NaCl, 0.02% NaN_3, 100 μg/mL PMSF, 1 μg/mL Aprotinin, 1% Triton X-100) and bioluminescent reaction reagent (luciferase/luciferin reagent substrate, from Beijing YPH Biotechnology Co., Ltd., Beijing, China) is stored in three tubes a, b and c, respectively. Using a peristaltic pump in the automatic filling device, the required reagents are added to each unit via hoses to complete the test.

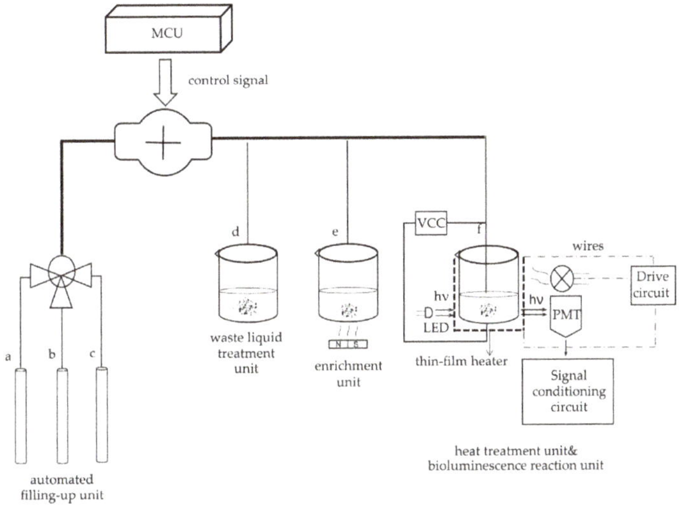

Figure 5. Automatic filling-up unit. *Escherichia coli* is captured using immunomagnetic nanoprobes and enriched in tube e. Pathogen heat treatment and bioluminescence reaction are performed in tube f. Tubes a, b and c contain erase reagent, lysis reagent and bioluminescence reagent, respectively. The peristaltic pump is controlled by MCU to fill the required reagents during the detection process. Tube d is used to store the waste liquid.

Although the test is carried out in a closed device, some external light still enters the reactor through cracks and transparent tube refraction, which forms scatting light. The light emitted by ATP luminescence is also absorbed by the solution because the ATP luminescence reaction takes place in the solution. These two points may affect the testing results, so the optical calibration method is adopted for compensation, and its schematic diagram is shown in Figure 6. Figure 6a is the schematic diagram without using the optical calibration method. The sample and detection reagent react to release bioluminescence in an optical reaction cell, which is directly detected by photomultiplier tubes. Therefore, it is inevitable to be affected by the absorption of luminescence and scatting light. To solve this problem, a light-emitting diode (LED) is installed on the opposite side of the photomultiplier tube, as shown in Figure 6b. The luminescence wavelength of the LED is required to be consistent with ATP bioluminescence, whose peak wavelength is 562 nm. And under the control of a constant current source, make its luminescence as stable as possible. When no detection reagent is added, the photomultiplier tube first detects and records the original light intensity of the LED and then turns it off. The approaches to close include power supply and shutter. After adding the detection reagent and bioluminescence detection, the LED is turned on again, and its luminescence is detected by a photomultiplier tube after passing through the optical reaction cell. Because the absorption and scattering of the light attenuate it, the detected bioluminescence signal can be corrected by the degree of attenuation to achieve the calibration function.

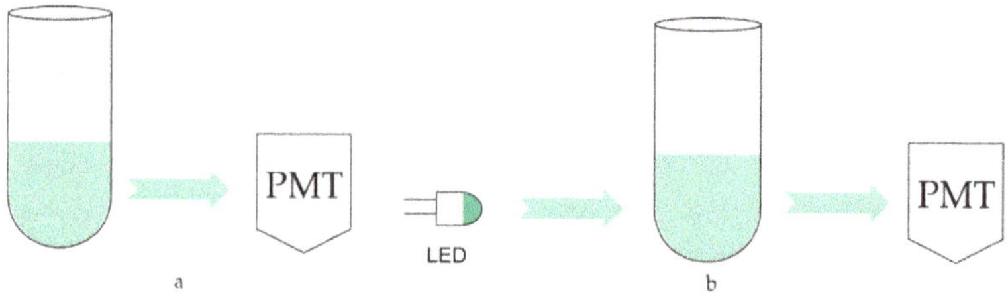

Figure 6. The schematic diagram of the optical calibration method. (**a**) PMT luminescence collection without optical calibration; (**b**) PMT luminescence collection with optical calibration.

After processing, the bioluminescence reaction is carried out in a special detector tube in reactor #2, with graphene electrodes attached to the bottom. The tube is chosen to use a high-transmittance glass test tube to improve the light transmission. The PMT is responsible for collecting the light signal, obtaining the relative luminous intensity, and completing the photoelectric conversion. As shown in Figure 7, current signals derived from photoelectric conversion are turned into analog voltage signals by an integrated operational amplifier circuit and become digital signals after analog-to-digital conversion. Then they are sent to a microcontroller for processing. A complete inspection process is controlled in 30 min. Moreover, to automate and improve the system's efficiency, we designed an automatic cleaning function, which can be automatically cleaned between two inspection processes. The waste liquid obtained after cleaning is delivered to the waste liquid pool. This function can reduce the interference between two consecutive assays.

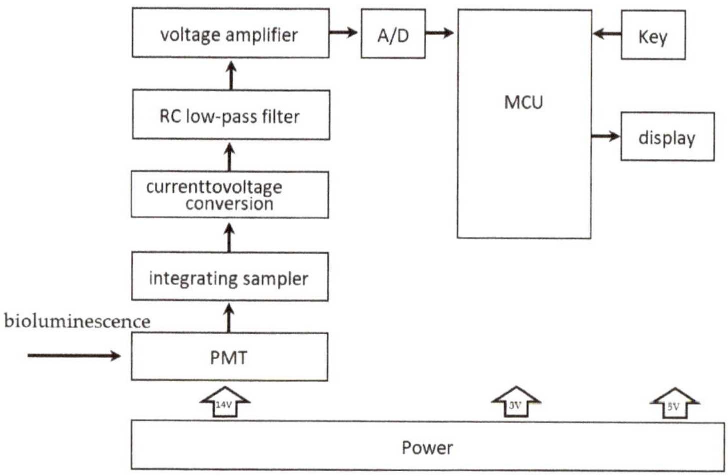

Figure 7. Signal transmission unit. The bioluminescence signal is converted into the current signal by the PMT. The current signal is sampled, and current-to-voltage is converted and then sent to the RC low-pass filter for filtering. The filtered analog voltage signal is amplified by the voltage amplifier circuit, which becomes a digital signal after analog-to-digital conversion and is transmitted to the MCU.

4. Results and Discussion

4.1. System Noise and Signal Level

During the testing process, the accuracy of the measurement results is improved using a filtering circuit in signal processing due to the interference from system noise and weak input signals are two important factors affecting the measurement accuracy. The specific filtering process is as follows. Firstly, the signal from PMT acquisition and photoelectric conversion is sampled and integrated. The sampling integral processing consists of two processes: sampling and integration. The sampling interval is determined according to the accuracy of the signal to be recovered. Then the sampling is integrated, and the integration process is implemented with an analog circuit. Both sampling and integration can be implemented using an integrating sampler, whose purpose is to extract the weak input signal from the noise. Later, the obtained samples are accumulated synchronously using an RC low-pass filter to suppress the noise and extract the desired signal, which takes advantage of the non-correlation between the noise and the signal. Figure 8 shows the system noise and the effective signal levels before and after filtering. It is clear from the line chart in Figure 8 that the effective signal levels before filtering ranges from 13 mV to 17.5 mV, and the effective signal levels after filtering ranges from 14.5 mV to 16 mV, which has a significant difference between the two sets of results. The system noise level ranges from −0.2 mV to 0.15 mV. Thus, it can be observed that the output signal level is smoother, and the output signal results are more stable than before filtering.

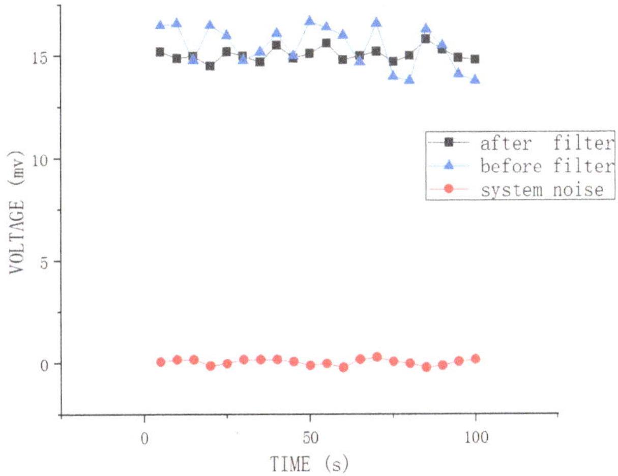

Figure 8. System noise and output signal level before and after filtering.

4.2. The Efficiency of Binding *E. coli* Antibodies to Immunomagnetic Beads

For the immunomagnetic nanoprobes to capture *E. coli*, the IMBs are required to bind with a sufficient amount of *E. coli* antibodies. In this study, the binding efficiency refers to the binding efficiency of the IMBs to the antibodies. Many instruments are available to measure the binding efficiency, and we chose the UV spectrophotometer. The specific operation is as follows: The absorbance of the *E. coli* antibody solution before and after adding IMBs is measured separately with the UV spectrophotometer. When IMBs are not yet bound to the antibody, the absorbance of the solution before the addition of IMBs can be recorded as A0. When the binding of IMB and antibody is completed, the absorbance of the solution after the addition of IMBs can be recorded as A1. The binding efficiency (R_C) can be calculated using Equation (2).

$$\text{Rc} = \left[\frac{\text{A0} - \text{A1}}{\text{A0}}\right] \times 100\% \tag{2}$$

The results obtained from the preliminary data analysis in Table 1 show the average binding efficiency of 86.3%, with a maximum fluctuation of 1.4%. The above results indicate that the binding efficiency of IMB and *E. coli* antibodies is satisfactory, which meets the requirements of the binding efficiency of the assay. Moreover, according to the results of the subsequent experiments, the efficiency of immunomagnetic nanoprobes for capturing *E. coli* reaches 99.5%, which can meet the detection requirements for capturing accuracy.

Table 1. The efficiency of binding *E. coli* antibodies to immunomagnetic beads.

A0	A1	R_C
		%
34.2	4.6	86.5
35.8	5.4	84.9
33.9	4.8	85.8
37.4	5.1	86.4
38.2	4.9	87.2
36.5	4.7	87.1

4.3. *ATP Bioluminescence Detection*

During the detection process, the intensity of the luminescence signal gradually decayed over time due to the consumption of the substrate. To calculate the concentration of *E. coli* in solution using the luminescence intensity at the initial moment, we need to record the luminescence intensity at the initial moment. Therefore, we should perform ATP bioluminescence reaction kinetic fitting experiments. The experimental method is as follows. We choose standard ATP solutions with five different concentrations of 10^{-15}, 10^{-14}, 10^{-13}, 10^{-12} and 10^{-11} mol/L. Under the same circumstances, different concentrations of standard ATP solutions and sufficient amounts of other reactants are reacted, which is repeated five times. The average of the five testing results is taken as the final experimental data, and the results are shown in Figure 10. It is apparent from Figure 10 that the luminescence intensity of ATP is negatively correlated with time, and the higher the concentration of ATP at the same experimental moment, the stronger the luminescence intensity of the solution. The experimental data of the samples with a concentration of 10^{-13} mol/L can be compared with the other data in Figure 10, which shows that the initial luminescence intensity produced by the samples with the concentration of 10^{-15} and 10^{-14} mol/L are too low and easily disturbed by noise. The luminescence intensity produced by the standard ATP solutions of 10^{-12} and 10^{-11} mol/L is not detected in the first 100 s because its intensity is too high beyond the detection range. Therefore, to improve the accuracy of the detection results, we choose samples with a concentration of 10^{-13} mol/L to perform the kinetic fitting of the ATP bioluminescence reaction.

Figure 9 indicates the bioluminescence reaction kinetic fitted curve of the sample with a concentration of 10^{-13} mol/L. A closer inspection of Figure 9 shows that the output voltage obtained from the photoelectric conversion decays from an initial value of 2187.63 mV to 1013.72 mV within the first 100 s. The data in this figure are consistent with the kinetic Equation (3) [43], which will be used in subsequent experiments to calculate the initial luminous intensity:

$$y = \mathrm{A1} \times \mathrm{e}^{\mathrm{Kt}} + \mathrm{y0} \tag{3}$$

where y is the output voltage corresponding to the luminous intensity at time t, A1 is the output voltage corresponding to the luminous intensity at the initial moment, K is the attenuation constant, and y0 is the error. The calculated correlation coefficient of the fitted curve is 0.9943.

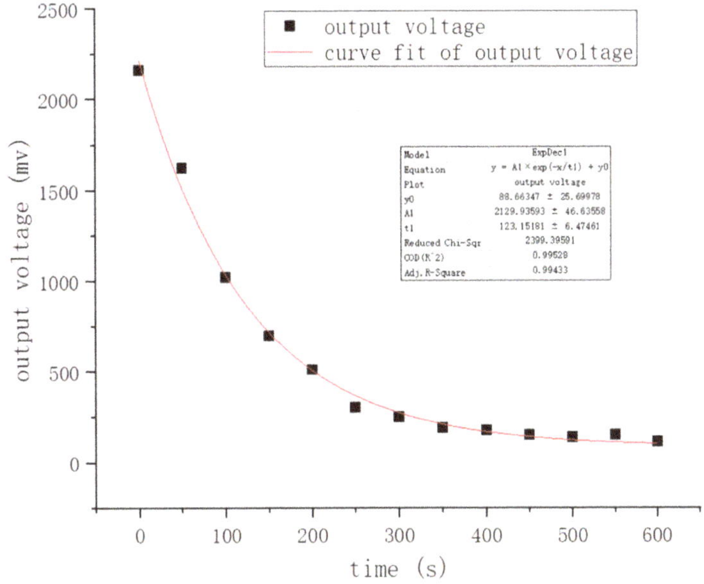

Figure 9. Kinetic fitted curve of the sample of concentration 10^{-13} mol/L.

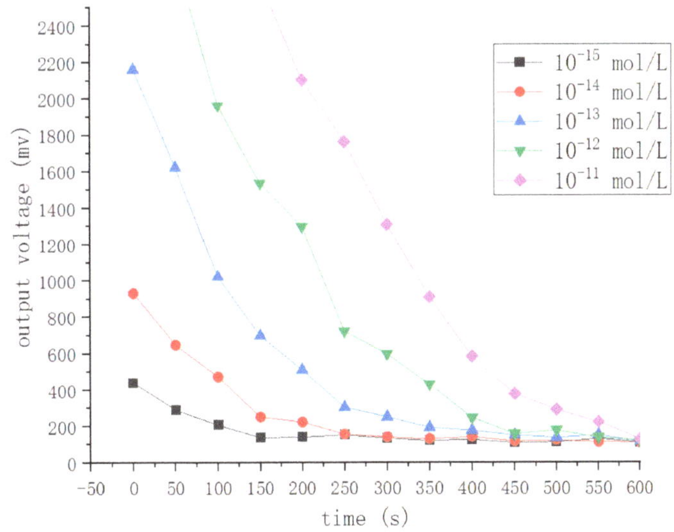

Figure 10. Kinetic fitting curves for the samples of five different ATP concentrations.

4.4. Effect of pH on Test Results

Luciferase is indispensable for ATP bioluminescence reaction, whose activity is easily affected by pH value. The experiments indicate that the fluorescence from ATP bioluminescence reaction is stronger under weak alkaline conditions than in other acid-base conditions. To achieve the desired effect of the bioluminescence reaction, we need to gain the pH interval where the luciferase activity reaches its peak. To obtain this pH interval, ATP solutions with a concentration of 10^{-13} mol/L were used for the test. The luminescence intensity of the solution, which reflects the luciferase activity, could be measured by a fluorescence detector. The details of the experiment are as follows. Multiple experiments were performed with the same concentration of ATP solution, and the pH value of the solution was adjusted with HCL and NAOH. Experiments showed that the more suitable temperature for ATP bioluminescence reaction is 24.5 °C [19], so the temperature of this experiment was controlled within a range of 24 ± 1 °C. The experimental results are shown in Figure 11. It is apparent from Figure 11 that the luminescence intensity is almost 0 when the pH value is too large (greater than 10) or too small (less than 4), which may be that the luciferase is inactivated and the reaction cannot be carried out. The luminescence intensity is significantly better when the pH value is between 7.5 and 7.9, and the luminescence intensity reaches its peak when the pH value is 7.7, which is 31.6 × 10^4 RLU. Thus, the pH value of the solution in detection should be adjusted to a weak alkaline condition. ATP has a negative charge under weak alkaline conditions, which is conducive to the enrichment of ATP by the electric field force.

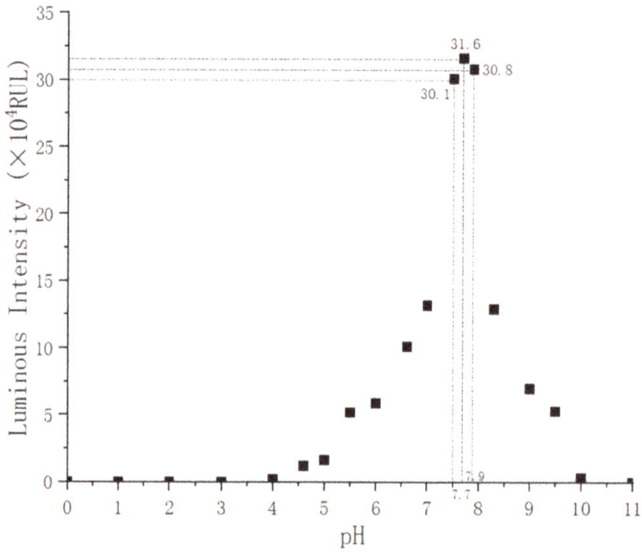

Figure 11. Effect of pH on luminescence intensity.

4.5. System Precision

System precision is a prerequisite to ensuring the system's accuracy, which can be expressed by the reproducibility of the measurement results. Although high precision cannot guarantee high accuracy, low precision must not have good accuracy. Precision refers to the degree of dispersion between the measured data in the measurement of multiple identical samples. Coefficient of variation (*CV*) is generally used to determine the level of precision. *CV* can be calculated by Formula (4):

$$CV = \frac{SD}{Mean} \times 100\% \tag{4}$$

where *SD* is standard deviation; *Mean* is the average value. A low coefficient of variation demonstrates good testing precision, while a high variation coefficient indicates poor testing precision. We used 10 independent bacterial sample solutions with a concentration of 3 × 10^5 CFU/mL to analyze the system precision in the experiment, and the results are shown in Figure 12. A simple analysis of the data in Figure 12 showed that the coefficient of variation was 3.96%.

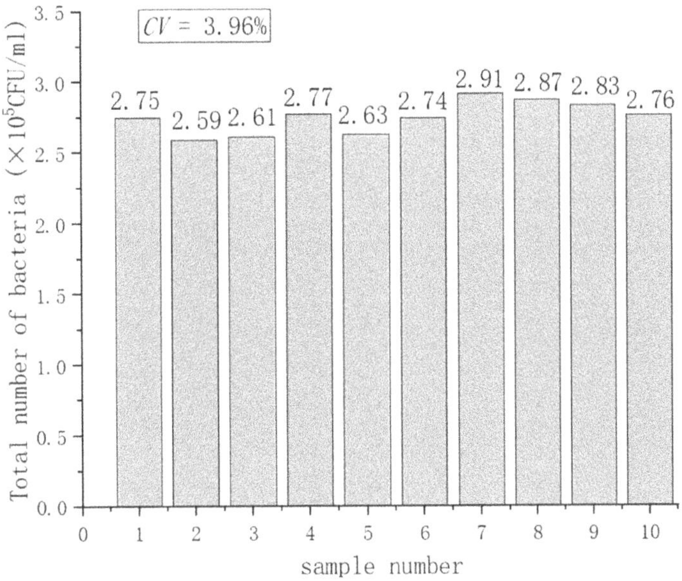

Figure 12. System precision testing.

4.6. Linear Relationship

In general, the lower limit of detection (LLD) is defined as the point where the signal-to-noise ratio equals three [44]. In this study, we applied the standard deviation of luminescence intensity as the system noise and calculated the lower limit of detection for luminescence intensity as 600 RLU. We selected several standard *E. coli* solutions ranging from 10^1 to 10^7 CFU/mL as sample solutions to test the relationship between luminescence intensity and bacterial concentration. The tests were divided into two groups: solutions without heat treatment and solutions after heat treatment. Figure 13 demonstrated that the luminous intensity was linearly related to the *E. coli* concentration. The linear correlation coefficient was 0.972 before and 0.975 after heat treatment. For the solutions without heat treatment, the luminescence intensity was below the lower detection limit when the concentration of *E. coli* was below 10^2 CFU/mL. For the solutions after heat treatment, the system cannot detect when *E. coli* concentration is below 30 CFU/mL. As Figure 13 shows, there was a significant difference between the two groups of fitted curves. Comparing the fitted curves before and after heat treatment, it can be seen that the RLU of the heat-treated solution increased several times in the solution containing the same concentration of *E. coli*. Thus, the limit of detection (LOD) of *E. coli* detection using the ATP bioluminescence technique was increased by about one magnitude because of the enhanced bioluminescence signal. Our results suggest that the heat treatment step of the pathogen is useful for improving the sensitivity of the ATP bioluminescence technique for the detection of *E. coli*. As Table 2 shows, our results suggest that adding a pathogen heat-treatment step is useful to enhance the sensitivity for *E. coli* detection by the ATP bioluminescence technique.

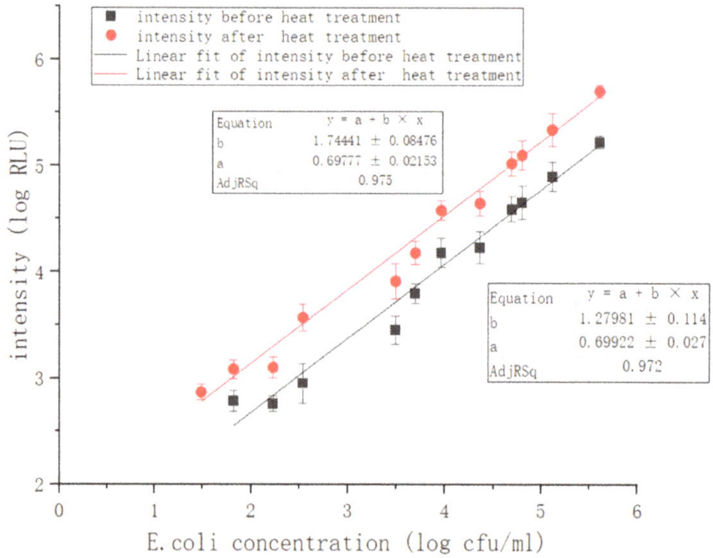

Figure 13. Correlation between luminescence intensity and *E. coli* concentration.

Table 2. Results compared with other test methods.

Method	Bacteria	Detection Range	Detection Time	Linear Coefficient	Reference
Electrochemical method	*E. coli K1:H7*	22 CFU/mL	30 min	0.841	[45]
Resistive method	*E. coli Rosetta 2pLyss*	10^7 CFU/mL	2 h	0.875	[46]
Magnetic silica nanotubes method	*S. Typhimurium*	10^3–10^7 CFU/mL	30 min	0.901	[47]
Magnetic nanoprobe-ATP method	*E. coli O157:H7*	10^2–10^8 CFU/mL	<20 min	0.964	[19]
Nanoprobe transparent graphene electrode-ATP method	*E. coli O157:H7*	10^2–10^6 CFU/mL	<20 min	0.972	[20]
Nanoprobe- improve ATP method	*E. coli O157:H7*	31–10^6 CFU/mL	25–30 min	0.975	Present work

4.7. Detection Accuracy

Absolute error is often regarded as a classical parameter to measure detection accuracy, which can be described as:

$$E = ABS[\log(PM) - \log(TCM)] \quad (5)$$

where E is the absolute error, PM is the detection results of the present method, and TCM is the detection results of the traditional culture method. Generally speaking, we regard the traditional culture method results as actual values and the present method results as measured values. When E < 1, the present method results and the traditional culture method results can be considered almost identical [48]. The detection system tested various foods and beverages, and the results were compared with those of the traditional culture method to calculate absolute error, represented in Figure 14b. Samples included four categories: drinks, meats, grain and quick-frozen food, Examples of items included milk, juice, beef, and frozen fish from supermarkets and farmers' markets. To prepare the testing samples, we used the dilution method and the mixture ratio method [20]. In processing the experiment, the experimental conditions were adopted as follows: ATP bioluminescence reaction was preceded by heat treatment (50 °C, 5 min), and then the sample solution was cooled to the optimum temperature for bioluminescence reaction (24.5 °C, pH 7.4). During the assay, for solid and semi-solid samples, such as beef, we mixed 25 g samples with 225 mL normal saline in a sterile homogenization cup and homogenized it at 5000 rpm for 3 min to produce a 1:10 dilution solution. Liquid samples, such as milk, were collected with a sterile pipette. 25 mL sample was mixed with 225 mL normal saline in a sterile conical flask, and then it was shaken well at 200 rpm for 3 min to produce a 1:10 homogenized sample. Figure 14b shows the testing results before the addition of the heat treatment module, and Figure 14d shows the testing results after the addition of the heat treatment module. The comparison shows that the detection accuracy was more than 94% after adding the heat treatment module, especially in beverage and grain samples. The detection accuracy of quick-frozen food was also significantly improved.

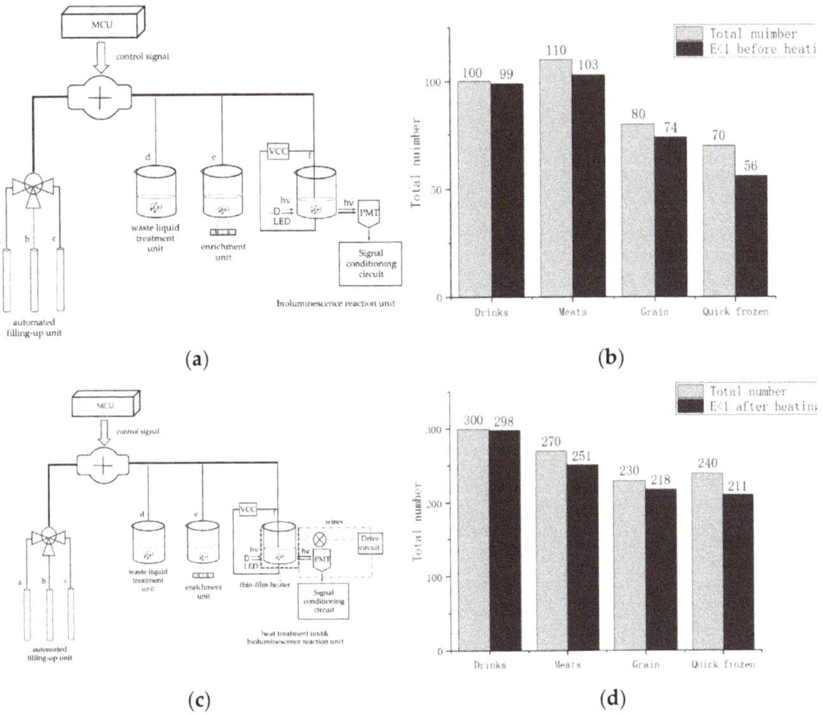

Figure 14. Comparison of the system accuracy. (**a**) detection model before adding the heat treatment module; (**b**) testing results before adding the heat treatment module; (**c**) detection model after adding the heat treatment module; (**d**) testing results after adding the heat treatment module.

5. Conclusions

Escherichia coli is a conditionally pathogenic bacterium, and some of its strains carrying virulence factors are pathogenic to humans. *E. coli* detection research has been a hot issue in the food hygiene and health field. This study is undertaken to design a real-time, rapid and accurate portable *E. coli* detection system and evaluate the detection effectiveness of the system. The system is innovative in combining three technologies: immunomagnetic separation technology, improved ATP bioluminescence technology, and graphene transparent electrodes. Among these, the improved ATP bioluminescence technology is a combination of heat-treatment pathogen technology and ATP bioluminescence detection to further improve detection range and detection accuracy. Based on the analysis of the testing results, the system can be used to detect *E. coli* in food and beverages with a detection accuracy of more than 94%. The detection time was within 30 min, and the range of detection colonies was 3.1×10^1 to 10^6 CFU/mL. The coefficient of variation was 3.96%, indicating that the system has high reliability and reproducibility. In addition, the system's results showed a good linear correlation with those obtained by the traditional detection method, with a linearity coefficient of 0.975. Compared with the unimproved system [18], the detection accuracy was higher, the detection range was larger, the coefficient of variation was lower, and the linear correlation was stronger. A limitation of this study is that the system can only detect *E. coli* and cannot detect multiple pathogens simultaneously. In the future, further studies will improve on this limitation.

Author Contributions: Conceptualization, Z.S. and C.W.; methodology, Z.S.; software, Z.S.; validation, Z.S., J.G. and C.W.; formal analysis, Z.S.; investigation, Z.S.; resources, C.W.; data curation, C.W.; writing—original draft preparation, Z.S.; writing—review and editing, Z.S., C.W. and W.W.; visualization, Z.S.; supervision, C.W.; project administration, C.W. and W.W.; funding acquisition, C.W. and W.W. All authors have read and agreed to the published version of the manuscript.

Funding: This research received no external funding.

Conflicts of Interest: The authors declare no conflict of interest.

References

1. Scharff, R.L. Economic Burden from Health Losses Due to Foodborne Illness in the United States. *J. Food Prot.* **2012**, *75*, 123–131. [CrossRef]
2. Leoni, E.; Legnani, P.P. Comparison of selective procedures for isolation and enumeration of Legionella species from hot water systems. *J. Appl. Microbiol.* **2001**, *90*, 27–33. [CrossRef] [PubMed]
3. Crowther, J.R. ELISA: Theory and practice. *Methods Mol. Biol.* **1995**, *42*, 1–218. [PubMed]
4. Ahmed, A.; Rushworth, J.V.; Hirst, N.A. Biosensors for Whole-Cell Bacterial Detection. *Clin. Microbiol. Rev.* **2014**, *27*, 631–646. [CrossRef] [PubMed]
5. Molin, O.; Nilsson, L.; Ansehn, S. Rapid detection of bacterial growth in blood cultures by bioluminescent assay of bacterial ATP. *J. Clin. Microbiol.* **1983**, *18*, 521–525. [CrossRef]
6. Haltalin, K.C.; Markley, A.H.; Woodman, E. Agar plate dilution method for routine antibiotic susceptibility testing in a hospital laboratory. *Am. J. Clin. Pathol.* **1973**, *60*, 384–394. [CrossRef]
7. Gauthier, T.D.; Shane, E.C.; Guerin, W.F.; Seitz, W.R.; Grant, C.L. Fluorescence quenching method for determining equilibrium constants for polycyclic aromatic hydrocarbons binding to dissolved humic materials. *Environ. Sci. Technol.* **1986**, *20*, 1162–1166. [CrossRef]
8. Mao, X.L.; Yang, L.J.; Su, X.L.; Li, Y.B. A nanoparticle amplification based quartz crystal microbalance DNA sensor for detection of Escherichia coli O157: H7. *Biosens. Bioelectron.* **2006**, *21*, 1178–1185. [CrossRef]
9. Luo, J.; Liu, X.; Tian, Q.; Yue, W.; Zeng, J.; Chen, G.; Cai, X. Disposable bioluminescence-based biosensor for detection of bacterial count in food. *Anal. Biochem.* **2009**, *394*, 1–6. [CrossRef]
10. Geens, M.; Van de Velde, H.; De Block, G.; Goossens, E.; Van Steirteghem, A.; Tournaye, H. The efficiency of magnetic-activated cell sorting and fluorescence-activated cell sorting in the decontamination of testicular cell suspensions in cancer patients. *Hum. Reprod.* **2007**, *22*, 733–742. [CrossRef]
11. Velusamy, V.; Arshak, K.; Korostynska, O.; Oliwa, K.; Adley, C. An overview of foodborne pathogen detection: In the perspective of biosensors. *Biotechnol. Adv.* **2010**, *28*, 232–254. [CrossRef] [PubMed]
12. Schmidt, H.; Knop, C.; Franke, S.; Aleksic, S.; Heesemann, J.; Karch, H. Development of PCR for screening of enteroaggregative Escherichia coli. *J. Clin. Microbiol.* **1995**, *33*, 701–705. [CrossRef] [PubMed]
13. Güldener, U.; Seong, K.Y.; Boddu, J.; Cho, S.; Trail, F.; Xu, J.R.; Adam, G.; Mewes, H.W.; Muehlbauer, G.J.; Kistler, H.C. Development of a Fusarium graminearum Affymetrix GeneChip for profiling fungal gene expression in vitro and in planta—ScienceDirect. *Fungal Genet. Biol.* **2006**, *43*, 316–325. [CrossRef]
14. Fournier, P.E.; Drancourt, M.; Raoult, D. Bacterial genome sequencing and its use in infectious diseases. *Lancet Infect. Dis.* **2007**, *7*, 711–723. [CrossRef]
15. Fisher, M.; Atiya-Nasagi, Y.; Simon, I.; Gordin, M.; Mechaly, A.; Yitzhaki, S. A combined immunomagnetic separation and lateral flow method for a sensitive on-site detection of Bacillus anthracis spores—Assessment in water and dairy products. *Lett. Appl. Microbiol.* **2009**, *48*, 413–418. [CrossRef]
16. Kuang, H.; Cui, G.; Chen, X.; Yin, H.; Yong, Q.; Xu, L.; Peng, C.; Wang, L.; Xu, C. A One-Step Homogeneous Sandwich Immunosensor for Salmonella Detection Based on Magnetic Nanoparticles (MNPs) and Quantum Dots (QDs). *Int. J. Mol. Sci.* **2013**, *14*, 8603–8610. [CrossRef] [PubMed]

17. Xu, M.; Wang, R.; Li, Y. Rapid detection of Escherichia coli O157:H7 and Salmonella Typhimurium in foods using an electrochemical immunosensor based on screen-printed interdigitated microelectrode and immunomagnetic separation. *Talanta* **2016**, *148*, 200–208. [CrossRef] [PubMed]
18. Murphy, S.C.; Kozlowski, S.M.; Bandler, D.K.; Boor, K.J. Evaluation of adenosine triphosphate bioluminescence hygiene monitoring for trouble-shooting fluid milk shelf-life problems. *J. Dairy Sci.* **1998**, *81*, 817–820. [CrossRef]
19. Zhang, Z.; Wang, C.; Zhang, L.; Meng, Q.; Zhang, Y.; Sun, F.; Xu, Y. Fast detection of Escherichia coli in food using nanoprobe and ATP bioluminescence technology. *Anal. Methods* **2017**, *9*, 5378–5387. [CrossRef]
20. Xu, Y.; Zhang, L.; Wang, C. A Rapid Detection System Design for Escherichia Coli in Food Based on a Nanoprobe and Graphite Electrode Coupled With ATP Bioluminescence Technology. *IEEE Access* **2019**, *7*, 106882–106889. [CrossRef]
21. Lee, J.; Park, C.; Kim, Y.; Park, S. Signal Enhancement in ATP Bioluminescence to Detect Bacterial Pathogens via Heat Treatment. *Biochip J.* **2017**, *11*, 287–293. [CrossRef]
22. Novoselov, K.S.; Geim, A.K.; Morozov, S.V.; Jiang, D.; Zhang, Y.; Dubonos, S.V.; Grigorieva, I.V.; Firsov, A.A. Electric field effect in atomically thin carbon films. *Science* **2004**, *306*, 666–669. [CrossRef] [PubMed]
23. Li, X.; Cai, W.; An, J.; Kim, S.; Nah, J.; Yang, D.; Piner, R.; Velamakanni, A.; Jung, I. Large-Area Synthesis of High-Quality and Uniform Graphene Films on Copper Foils. *Science* **2009**, *324*, 1312–1314. [CrossRef]
24. Bae, S.; Kim, H.; Lee, Y.; Xu, X.; Park, J.S.; Zheng, Y.; Balakrishnan, J.; Lei, T.; Kim, H.R.; Song, Y.I. Roll-to-roll production of 30-inch graphene films for transparent electrodes. *Nat. Nanotechnol.* **2010**, *5*, 574–578. [CrossRef] [PubMed]
25. Kim, E.; Jain, N.; Jacobs-Gedrim, R.; Xu, Y.; Yu, B. Exploring carrier transport phenomena in a CVD-assembled graphene FET on hexagonal boron nitride. *Nanotechnology* **2012**, *23*, 125706. [CrossRef]
26. Johnson, C.L.; Wise, K.D.; Schwank, J.W. A thin-film gas detector for semiconductor process gases. In Proceedings of the International Electron Devices Meeting, San Francisco, CA, USA, 11–14 December 1988; IEEE: Piscataway, NJ, USA, 1988.
27. Zhao, Z.; Cui, Z.; Cui, D.F.; Xia, S.H. Monolithically integrated PCR biochip for DNA amplification. *Sens. Actuators A* **2003**, *108*, 162–167. [CrossRef]
28. Hoffmann-Berling, H. Adenosinetriphosphate as the energy substance for cell movement. *Biochim. Biophys. Acta* **1954**, *14*, 182–194.
29. Kennard, O.; Isaacs, N.W.; Coppola, J.C.; Kirby, A.J.; Warren, S.; Motherwell, W.D.; Watson, D.G.; Wampler, D.L.; Chenery, D.H.; Larson, A.C.; et al. Three dimensional structure of adenosine triphosphate. *Nature* **1970**, *225*, 333–336. [CrossRef]
30. Hawronskyj, J.-M.; Holah, J. ATP: A universal hygiene monitor. *Trends Food Sci. Technol.* **1997**, *8*, 79–84. [CrossRef]
31. Veigas, B.; Jacob, J.M.; Costa, M.N.; Santos, D.S.; Viveiros, M.; Inacio, J.; Martins, R.; Barquinha, P.; Fortunato, E.; Baptista, P.V. Gold on paper-paper platform for Au-nanoprobe TB detection. *Lab Chip* **2012**, *12*, 4802–4808. [CrossRef]
32. Li, L.; Yin, D.; Xu, K.; Liu, Y.; Song, D.; Wang, J.; Zhao, C.; Song, X.; Li, J. A sandwich immunoassay for brucellosis diagnosis based on immune magnetic beads and quantum dots. *J. Pharm. Biomed. Anal.* **2017**, *141*, 79–86. [CrossRef] [PubMed]
33. Zhao, Z.J.; Liu, X.M. Preparation of monoclonal antibody and development of enzyme-linked immunosorbent assay specific for Escherichia coli O157 in foods. *Biomed. Environ. Sci.* **2005**, *18*, 254–259. [PubMed]
34. Valdivieso-Garcia, A.; Riche, E.; Abubakar, O.; Waddell, T.E.; Brooks, B.W. A double antibody sandwich enzyme-linked immunosorbent assay for the detection of Salmonella using biotinylated monoclonal antibodies. *J. Food Prot.* **2001**, *64*, 1166–1171. [CrossRef] [PubMed]
35. Blake, P.; Brimicombe, P.D.; Nair, R.R.; Booth, T.J.; Jiang, D.; Schedin, F.; Ponomarenko, L.A.; Morozov, S.V.; Gleeson, H.F.; Hill, E.W.; et al. Graphene-based liquid crystal device. *Nano Lett.* **2008**, *8*, 1704–1708. [CrossRef] [PubMed]
36. Li, W.; Tan, C.; Lowe, M.A.; Abruna, H.D.; Ralph, D.C. Electrochemistry of Individual Monolayer Graphene Sheets. *Acs Nano* **2011**, *5*, 2264–2270. [CrossRef] [PubMed]
37. Dai, C.-L. A capacitive humidity sensor integrated with micro heater and ring oscillator circuit fabricated by CMOS-MEMS technique. *Sens. Actuators B* **2007**, *122*, 375–380. [CrossRef]
38. Resnik, D.; Vrtacnik, D.; Mozek, M.; Pecar, B.; Amon, S. Experimental study of heat-treated thin film Ti/Pt heater and temperature sensor properties on a Si microfluidic platform. *J. Micromech. Microeng.* **2011**, *21*, 025025. [CrossRef]
39. Yan, W.; Li, H.; Liu, J.; Guo, J. EPMA and XRD study on nickel metal thin film for temperature sensor. *Sens. Actuators A* **2007**, *136*, 212–215. [CrossRef]
40. Yi, C.; Lee, J.-H.; Kwak, B.S.; Lin, M.X.; Kim, H.O.; Jung, H.-I. Diagnosis of diabetes mellitus using sialic acid expression of erythrocyte and a microfluidic resistive temperature detector (micro-RTD). *Sens. Actuators B* **2014**, *191*, 305–312. [CrossRef]
41. Han, J.; Cheng, P.; Wang, H.; Zhang, C.; Zhang, J.; Wang, Y.; Duan, L.; Ding, G. MEMS-based Pt film temperature sensor on an alumina substrate. *Mater. Lett.* **2014**, *125*, 224–226. [CrossRef]
42. Purtov, K.V.; Petushkov, V.N.; Baranov, M.S.; Mineev, K.S.; Rodionova, N.S.; Kaskova, Z.M.; Tsarkova, A.S.; Petunin, A.I.; Bondar, V.S.; Rodicheva, E.K.; et al. The Chemical Basis of Fungal Bioluminescence. *Angew. Chem. Int. Ed.* **2015**, *54*, 8124–8128. [CrossRef] [PubMed]
43. Hunter, D.M.; Lim, D.V. Rapid Detection and Identification of Bacterial Pathogens by Using an ATP Bioluminescence Immunoassay. *J. Food Prot.* **2010**, *73*, 739–746. [CrossRef] [PubMed]
44. Pu, Q.S.; Sun, Q.Y.; Hu, Z.; Su, Z.X. Application of 2-mercaptobenzothiatole-modified silica gel to on-line preconcentration and separation of silver for its atomic absorption spectrometric determination. *Analyst* **1998**, *123*, 239–243. [CrossRef]
45. West, J.A.; Ramamurthy, P. Methods of Minimizing Temperature Cross-Sensitivity in Vapor Sensors and Compositions Therefor. U.S. Patent 7,708,947, 4 May 2010.

46. Mallya, A.N.; Sowmya, P.; Ramamurthy, P.C. Organic nanocomposite sensor for detection of Escherichia coli. In Proceedings of the IEEE International Conference on Emerging Electronics, Bengaluru, India, 3–6 December 2014; IEEE: Piscataway, NJ, USA, 2014.
47. Phuong Diem, N.; Trong Binh, T.; Dung Thi Xuan, N.; Min, J. Magnetic silica nanotube-assisted impedimetric immunosensor for the separation and label-free detection of Salmonella typhimurium. *Sens. Actuators B* **2014**, *197*, 314–320.
48. Wang, C.; Zhang, B.; Zhuang, X. A biochemical system of rapidly detecting bacteria based on ATP bioluminescence technology. *Eur. Food Res. Technol.* **2013**, *236*, 41–46. [CrossRef]

nanomaterials

Article

Comprehensive Study of the Chemistry behind the Stability of Carboxylic SWCNT Dispersions in the Development of a Transparent Electrode

Jovana Stanojev [1,*], Stevan Armaković [2], Sara Joksović [1], Branimir Bajac [1], Jovan Matović [1] and Vladimir V. Srdić [3]

1 BioSense Institute, University of Novi Sad, Dr Zorana Djindjica 1, 21000 Novi Sad, Serbia; sara.joksovic@biosense.rs (S.J.); branimir.bajac@biosense.rs (B.B.); jovan.matovic@biosense.rs (J.M.)
2 Department of Physics, Faculty of Sciences, University of Novi Sad, Trg Dositeja Obradovica 4, 21000 Novi Sad, Serbia; stevan.armakovic@df.uns.ac.rs
3 Faculty of Technology Novi Sad, University of Novi Sad, Bulevar Cara Lazara 1, 21000 Novi Sad, Serbia; srdicvv@uns.ac.rs
* Correspondence: jovana.stanojev@biosense.rs; Tel.: +381-214852137

Abstract: Single-walled carbon nanotubes (SWCNTs) are well-known for their excellent electrical conductivity. One promising application for SWCNT-based thin films is as transparent electrodes for uncooled mid-IR detectors (MIR). In this paper, a combination of computational and experimental studies were performed to understand the chemistry behind the stability of carboxylic SWCNTs (SWCNTs-COOH) dispersions in different solvents. A computational study based on the density functional tight-binding (DFTB) method was applied to understand the interactions of COOH-functionalized carbon nanotubes with selected solvents. Attention was focused on understanding how the protonation of COOH groups influences the binding energies between SWCNTs and different solvents. Thin film electrodes were prepared by alternately depositing PEI and SWCNT-COOH on soda lime glass substrates. To prepare a stable SWCNT dispersion, different solvents were tested, such as deionized (DI) water, ethanol and acetone. The SWCNT-COOH dispersion stability was tested in different solvents. Samples were prepared to study the relationship between the number of depositions, transparency in the MIR range (2.5–5 μm) and conductivity, looking for the optimal thickness that would satisfy the application. The MIR transparency of the electrode was reduced by 20% for the thickest SWCNT layers, whereas sheet resistance values were reduced to 150–200 kΩ/sq.

Keywords: SWCNT; layer-by-layer; transparent electrode; DFTB; DFT; binding energies

Citation: Stanojev, J.; Armaković, S.; Joksović, S.; Bajac, B.; Matović, J.; Srdić, V.V. Comprehensive Study of the Chemistry behind the Stability of Carboxylic SWCNT Dispersions in the Development of a Transparent Electrode. *Nanomaterials* **2022**, *12*, 1901. https://doi.org/10.3390/nano12111901

Academic Editors: Dong Liu and Baiqing Yuan

Received: 27 April 2022
Accepted: 30 May 2022
Published: 1 June 2022

1. Introduction

Optics and optoelectronics are a field of electronics and physics that has been attracting attention in recent times; it includes devices such as lasers, LEDs, solar cells, touch screens, photo diodes, detectors, etc. [1–3]. A substantial number of optoelectronic devices are focused on the development of infrared components that work in near-infrared (NIR) and mid-infrared (MIR) spectral ranges. The MIR spectral range (2.5–5 μm) is particularly interesting for organic molecule detection and functional group identification. With respect to this, the development of a new generation of MIR detectors is more than justified and will be beneficial for future technologies and devices. Transparent electrodes are the vital components of any optoelectronic device, including detectors. Until now, indium tin oxide (ITO) has been the most commonly used material for transparent electrodes [4]. However, ITO production is limited by the shortage of indium resources in the world and its high cost [5]. Furthermore, the fabrication process of transparent ITO films requires high processing temperatures of over 300 °C, which is not appropriate for many applications [6]. Contrary to this, carbon nanotubes—both single-walled (SWCNT) and multi-walled (MWCNT)—have the potential for application as ITO replacements. However, a lot of issues need to be addressed and solved.

Over the last 20 years, single-walled carbon nanotubes (SWCNTs) have gained significant attention in materials science due to their exceptional mechanical, electrical, optical, chemical and thermal properties. The molecular structure of a SWCNT is constructed by sp^2 hybridized carbon in the form of rolled-up graphene sheet. Depending on the diameter and chiral vector, SWCNTs can have metallic, semimetallic or semiconducting properties [5,7,8]. Their high value of mobility (~10,000 cm^2/Vs^{-1}) [7], low resistivity [7,8], high current-carrying capacities (~109 A/cm^2) [7,9,10], high thermal conductivity (~3500 W/mK), ballistic transport and high point of stress fracture (~50 GPa) are some of the reasons why SWCNTs have a very wide field of application [7,8].

Progress in science and technology today inevitably requires the application of computational methods aimed at investigating the properties of molecules and materials [11–13]. Quantum-mechanical methods, fundamentally based on the density functional theory (DFT), are among the most frequently utilized levels of theory and are beneficial for calculating the binding strength between molecules [14–17]. Although DFT methods compromise accuracy and computational cost, some molecular systems are too complex for this level of theory. Fortunately, semi-empirical techniques have also been developed, and they offer reasonable accuracy at a fraction of the computation resources required for regular DFT calculations. One of the best-known semi-empirical methods is the density-functional tight-binding method (DFTB) [18–22]. This method enables calculations orders of magnitude faster than DFT while retaining the explicit description of the electronic structure [23,24]. Due to the size of the studied systems, which consisted of more than 300 atoms each, the DFTB method was applied in this study to obtain the geometries of the studied molecular systems. Later, to understand the interactions of COOH-functionalized SWCNTs with different dispersants, DFT calculations were applied. The application of the DFTB level of theory ensured the obtaining of reliable geometries for these huge systems at a reasonable computational cost [25–30], while single point energy calculations via the DFT method ensured the obtaining of reliable information on noncovalent interactions [31–33]. Particular attention was focused on understanding how the protonation of COOH groups influenced the binding energies between the SWCNTs and selected dispersants.

In this paper, similar to our previous work [34], we developed an easily processed low-cost mid-IR transparent electrode fabricated from SWCNTs, based on a well-established layer-by-layer (LbL) deposition technology [34]. Hence, this work is more oriented toward the computational and experimental study of dispersibility of 80% pure SWCNTs in different dispersants, since dispersion stability has a crucial impact on LbL deposition quality. The literature has shown [35,36] that SWCNTs that lack purity require additional purification and dispersing steps to obtain dispersion with long term stability. It has been shown that surface modification with molecular groups with covalent bonding helps with the dispersibility and long-term stability of SWCNTs [37]. It was also found that the addition of HCl contributed to the dispersion stability in different ways, by removing metal impurities and/or the protonation of SWCNT walls [35,36]. Little or no attention has been paid to what happens with -COOH side groups, and how such groups may interact with solvents. Here, we try to understand how the protonation of -COOH functional groups influences the later dispersion of SWNCTs in water, acetone and ethanol. The experimental study was supported by the computational part of this research paper.

2. Experimental Part

2.1. Experimental Procedure

The experimental procedure for the successful deposition of high quality, conductive, uniform and mid-IR transparent films using the LbL deposition process requires the stable dispersion of the carboxylic SWCNTs (SWCNT-COOH). Meeting these requirements was demanding, as SWCNT-COOH with lower purity levels (below 80% purity) is prone to agglomeration, and may contain certain impurities—thus making dispersions hard to keep stable for even a short period. The direct dispersion of SWCNT-COOH was not possible, because immediately after the dispersion precipitates were formed, and deposited films had near-infinite resistivity with no homogenous or visible layers. Pretreatment of the 80% pure SWCNT-COOH with dilute HCl resulted in the formation of stable dispersions susceptible to LbL deposition. The explanations behind such behavior are supported by DFT calculations, pointing out the effects of various solvents used for dispersion, followed by the characterization of transparency and electrical properties.

The LbL technique is a simple, low-cost method used for the alternating deposition of polyethyleneimine (PEI) and carboxylic functionalized carbon nanotube monolayers [38]. In this research, we used SWCNT-COOH (purity < 80%) with an average diameter of 2 nm, and an average length of a few μm, purchased from Nanocyl S.A., Belgium. As a positively charged layer, polyethyleneimine (Sigma Aldich, St. Louis, MO, USA) was used, due to its -NH_3^+ and -NH_2^+—protonated groups, and carboxylic SWCNTs were used as a negatively charged layer because of their -COO^- groups.

In the first step, the commercial SWCNT-COOH was pretreated with HCl to ensure the preparation of stable dispersions. SWCNTs were dispersed in deionized (DI) water and sonicated for 10 min. The sonication process was a crucial step for achieving high dispersibility in the SWCNT-COOH. Therefore, sonication was performed using a Bandelin sonopuls HD 70 sonicator, with 60% of the 60 W RF power and a working frequency of 20 kHz. The microtip used for sonication was 2 mm in diameter. Afterwards, the pH value was adjusted to 3.5 for SWCNT-COOH using 0.1 M HCl, followed by sonication for 30 min. The prepared dispersion was centrifuged (1100 rpm for 15 min), and washed with deionized water several times, until slightly acidic dispersion was obtained (pH~5.5). After the treatment with HCl, the SWCNT-COOH was dispersed in 15 mL of water, ethanol and acetone as different dispersants and sonicated for about 30 min to achieve stable SWCNT-COOH dispersion. After this process, the only stable dispersion was obtained in acetone.

For the deposition of films, soda lime glass substrates were cleaned according to standard procedures described in our previous work [34]. Polyethyleneimine and polyacrylic acid (PAA) were diluted in DI water. The prepared substrates were soaked in 1% aqueous solutions of PEI and polyacrylic acid (PAA—Sigma Aldich) in order to achieve better adhesion. Substrates were dipped in PEI and PAA solutions for 10 and 15 min, respectively, followed by a washing step with DI water inbetween. The substrates were soaked in the PEI solution for 10 min and then in the SWCNT-COOH dispersion for 60 min, which represents one cycle. After each deposited layer, the substrates were washed with 10 MΩ deionized water (for 15 s) to remove excess material bound by the weak van der Waals forces and to form an uniform monolayer. Each deposited SWCNT-COOH monolayer was dried at 120 °C for 10 min. The prepared SWCNT multilayer structure is built of uniform bilayer units, consisting of one PEI and one SWCNT-COOH layer (Figure 1). The cycle was repeated 2, 4, 6, 8 and 10 times to study the correlation between the number of deposited bilayers and important functional properties such as transparency and electrical conductivity in the mid IR range.

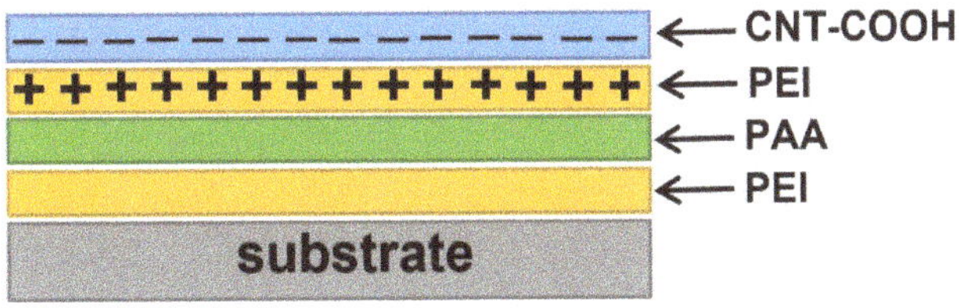

Figure 1. Schematic representation of the PEI+SWCNT-COOH structure.

Fabricated samples were examined by Raman spectroscopy. Raman spectra were measured using a Thermo Scientific (Waltham, MA, USA) DXR Raman microscope with a green laser (λ = 532 nm) and a red laser (780 nm) at a power of 8 mW. The expositions during the registration of the spectrum consisted of 10 expositions per 20 s, with 10× magnification. Spectroscopic analysis (UV-Vis (Jasco V-750) and FTIR (Thermo Scientific NicoletiS20) spectrophotometer) was used to investigate transparency in the UV-Vis and mid-IR spectra range. FTIR measurements were carried out in the wavelength range of 2500–3500 nm, with a resolution of 4 cm^{-1} and 32 scans. The sheet resistance of the films was measured by a sheet resistance multimeter (Hewlett Packard, Palo Alto, CA, USA, 3457A).

2.2. Computational Details

The computational analysis of interactions between SWCNT-COOH and solvent molecules was performed by applying the density functional theory (DFT) and density functional tight-binding (DFTB) methods. The DFTB method was used to geometrically optimize very large systems (consisting of more than 300 atoms) and calculate binding energies. In the case of the DFTB calculations, the Hamiltonian based on the extended tight-binding (xTB) model combined with the GFN1-xTB parametrization was applied. This type of approach, developed by Grimme and coworkers [39], offers a wide coverage of elements compared to other model Hamiltonians. Binding energies were based on DFTB calculations according to the following equation:

$$E_b = E_{tot}(SWCNT - COOH + solvent\ molecule) - E(SWCNT - COOH) \\ -E(solvent\ molecule), \quad (1)$$

where $E_{tot}(SWCNT - COOH + solvent\ molecule)$ denotes the total energy of the optimized complex consisting of the SWCNT-COOH and solvent molecule, *E(SWCNT - COOH)* denotes the total energy of the optimized carbon nanotube functionalized with COOH, while $E(solvent\ molecule)$ denotes the total energy of the solvent molecule. A dispersion corrected variant of the B3LYP exchange-correlation functional [40] (the B3LYP-D3) was applied for DFT calculations [39,41], combined with the 6-31G(d,p) basis set [42–44]. To study noncovalent interactions between selected systems, ground state geometries were exported and subjected to DFT calculations. DFTB calculations were performed with the DFTB engine of the Amsterdam Modeling Suite 2021.1, by Software for Chemistry and Materials (SCM) [45]. DFT calculations were carried out with the Jaguar [46,47] program, as implemented in the Schrödinger Materials Science Suite, version 2022-1.

3. Results and Discussion

3.1. Computational Results

The aim of the computational study was to explore the interactions between different solvents and -COOH groups in SWCNTs, and the suitability of each one for the dispersion of SWCNT-COOH. Attention was paid to the nature of interactions that may promote the formation of stronger noncovalent bonds between the SWCNT and the dispersant. In this section, we shed light on the effects that HCl may have on 80% purity SWCNT-COOH in dispersions.

For the purposes of the computational study, a total number of ten systems containing more than 300 atoms were subjected to DFTB calculations to obtain their ground state geometries. Due to the size of the considered systems, the application of DFT or some other non-semiempirical method would not be feasible. The experimentally obtained results indicated that acetone had the best dispersive properties among all the studied solvents, especially after the treatment with HCl. To understand why acetone had the most substantial impact after the treatment with HCl, it was reasonable to suggest that the nature of the noncovalent interactions between the solvents and the SWCNT material governed the dispersion stability in this case. We have already pointed out the variety of effects that HCl may have on the stability of 80% SWCNT dispersions. Simple purification with HCl (removal of metal traces) seemed not to have a major impact on the stability, since the dispersion was not stable with all solvents—except with acetone; thus, we focused on the influence of the interactions between the acetone and SWCNT-COOH. Specifically, we suggest that the important factors for stability may be found in the noncovalent molecule interactions and molecule orientations that may occur in the protonated -COOH group state. To explore these complex effects, we performed a detailed computational analysis that involved the consideration of molecule structure and orientation, binding energies and noncovalent surface interactions.

From a computational standpoint, the first task was to investigate the structural properties of SWCNT-COOH, and the protonated form of SWCNT-COOH (pSWCNT-COOH), interacting with different solvent molecules. Specific structural features, such a thes intermolecular distances between SWCNT-COOH/pSWCNT-COOH and the solvent molecules, might give us initial assumptions about how protonation influences the interactions with solvent molecules. Geometrically optimized systems at the DFTB level of theory are presented in Figure 2.

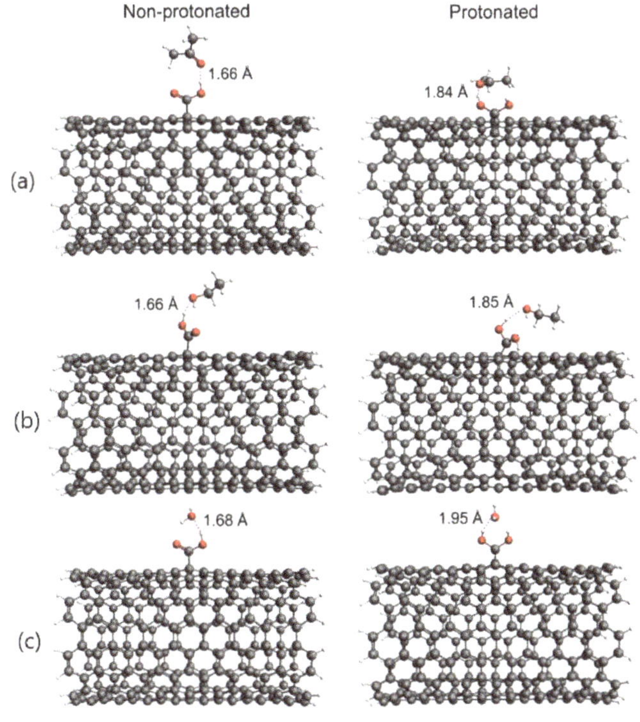

Figure 2. DFTB/GFN-xTB optimized systems with intermolecular distances between nanotubes and (**a**) acetone, (**b**) ethanol and (**c**) H_2O.

As shown in Figure 2, the protonation of COOH groups leads to increased values for the shortest intermolecular distances and the quite different orientation of the solvent molecules in relation to the SWCNT-COOH. Regarding the non-protonated form, the intermolecular distances in the SWCNT-COOH were around 1.66 Å in acetone, ethanol and water. The intermolecular distances between pSWCNT-COOH and acetone and between pSWCNT-COOH and ethanol were around 1.85 Å and were 1.95 Å for water. Our next computational task was to study the binding energies between carbon nanotubes and solvent molecules. These results are summarized in Table 1.

Table 1. Binding energies [kcal/mol] between carbon nanotubes and solvent molecules.

	Binding Energies [kcal/mol]	
	Non-Protonated	**Protonated**
Carbon nanotube + acetone	−11.60	−13.29
Carbon nanotube + ethanol	−11.75	−4.44
Carbon nanotube + H_2O	−11.49	−7.66
Carbon nanotube + HCl	−8.02	−5.42

At first sight, it might seem that the results presented in Table 1 and Figure 2 are not in agreement. Namely, the analysis of the specific intermolecular distances presented in Figure 2 shows that protonation increased the distances between the nanotubes and solvent molecules. For this reason, it is expected that the binding energies would decrease as a consequence of this increase in distance. The distance between the pSWCNT-COOH and acetone was significantly higher than the distance between the SWCNT-COOH and acetone. According to this, the binding energy is expected to decrease. This was indeed reflected in the stability of the dispersed SWCNT in water—with a value of pH ~3.5—as the SWCNTs were even more susceptible to agglomeration, forming a precipitate immediately after sonication. Contrary to this, the binding energy between the pSWCNT-COOH and acetone increased by almost 2 kcal/mol. In all other cases of protonation, the binding energy decreased.

The results obtained so far impose the necessity of investigating the structural and charge distribution properties in more detail. For this purpose, we analyzed the effects of protonation on the structure near the carboxyl group attached to the nanotube. We also investigated the number and intensity of noncovalent interactions between acetone and SWCNT-COOH/pSWCNT-COOH. We first refer to the effects of protonation on the structure. Figure 3 contains the extracted structures of the SWCNT-COOH and pSWCNT-COOH in close proximity to the carboxyl group, for easier visualization.

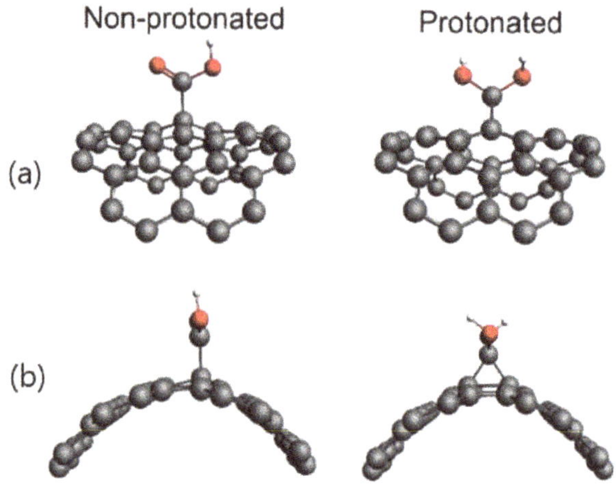

Figure 3. Close proximity of carboxyl groups in the SWCNT-COOH and pSWCNT-COOH (**a**) top views and (**b**) side views, as obtained by DFT/GFN-xTB optimization.

The side view of the SWCNT-COOH and pSWCNT-COOH indicated the clear structural difference between these two structures. Namely, the protonation of the carboxyl group led to the bending of the hydrogen atoms towards the nanotube, and an additional bond was formed between the carbon atom of a carboxyl group and a nanotube.

The next step to explain these results regarding the binding energies is to identify and quantify the noncovalent interactions formed between the nanotubes and solvent molecules. According to the literature data [48,49], the analysis of noncovalent interactions has been performed previously by analyzing the electron density between all the atoms.

The identification and quantification of noncovalent interactions were performed by DFT calculations. Since the systems contained more than 300 atoms, DFT calculations would not be possible at a reasonably accurate level of theory. Therefore, the relevant region presented in Figure 3 was used for DFT calculations. Hydrogen bonds were added to edge the atoms to take care of any dangling bonds. To properly optimize bonds to added hydrogens, all bonds except those with hydrogen atoms were fixed—after which, the optimization was performed at the DFTB level of theory. Finally, the simplified systems of nearly 70 atoms were used for DFT calculations at the B3LYP-D3/6-31G(d,p) level of theory with the Jaguar program. The noncovalent interactions are presented in Figure 4.

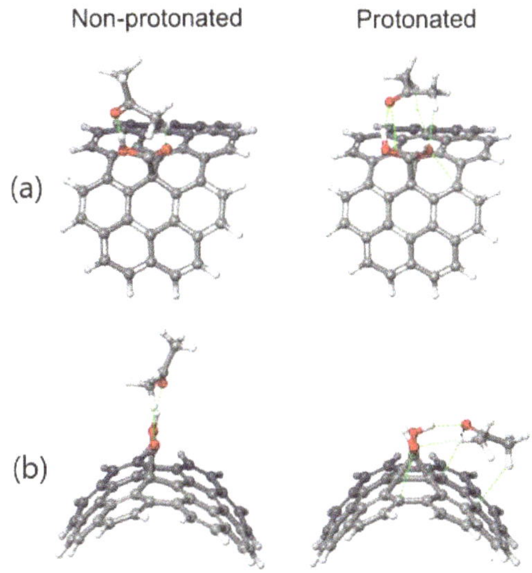

Figure 4. Noncovalent interactions (green dotted line) between SWCNT-COOH/pSWCNT-COOH and acetone (**a**) top and (**b**) side views, as obtained with the B3LYP-D3/6-31G(d,p) level of theory.

The results in Figure 4 explain why the binding energy between pSWCNT-COOH and acetone increased compared to the binding energy between SWCNT-COOH and acetone, despite the distance between pSWCNT-COOH and acetone being significantly higher than the distance between SWCNT-COOH and acetone. Namely, as shown in Figure 4, in the case of the non-protonated form of the SWCNT-COOH, acetone molecules were more likely to interact with -COOH, rather than the nanotube itself; therefore, only two noncovalent interactions were formed. As a result of protonation, the carboxyl group leaned sideways, while the acetone still interacted with the carboxyl group via two noncovalent interactions. As a consequence, this modeling has shown that protonated -COOH/acetone species are likely to bend toward the nanotube and form two additional noncovalent bonds. This effect, which we propose here, mainly contributed to the overall stability of the dispersion by reducing and minimizing interactions between nanotubes. Therefore, in the protonated form, the number of noncovalent interactions was roughly doubled, and consequently, stronger binding was achieved. This explains the results presented in Table 1, and why acetone had a more substantial influence after the treatment of nanotubes with HCl. With the performed set of experiments being in agreement with the computational study presented above, in the following section we present the properties of the obtained SWCNT-based electrodes.

3.2. Film Characterization

Figure 5 shows the Raman spectra of the 80% pure and acid-treated SWCNT-COOH powder. The peak position at ~1592 cm^{-1} originated from a G-band due to the in-plane stretching of sp^2-hybridized carbon atoms, which arose along the axis of a nanotube [50,51]. The vibration observed at ~1346 cm^{-1} is called a D-band and it is related to the breaking symmetry in the structure. The intensity of the D-band is not negligible and indicates disorders and defects such as vacancies, amorphous carbon, dangling sp^2 bonds, etc. [50–52]. The Raman spectra show that there was no significant difference between the 80% pure and acid-treated nanotubes spectra, which indicates that there was no change in the structure after acid treatment [37,53], which is important from the standpoint of the properties of these materials. What is more, Raman analysis was also carried out using a red laser to detect the RBM modes of the SWCNTs (Figure S1—Supplementary Material). The peak positions at 157 cm^{-1} and 260 cm^{-1} were from the RBM mode, indicating that the SWCNT-COOH has a difference in diameter [54]. The results of the Raman spectrometry analysis were also in agreement with the computational studies, which showed that only noncovalent interactions were formed between the -COOH and acetone molecules. Carboxylic groups tended to only lean sideways, without permanently affecting the chemical changes in the nanotubes. Hence, the Raman results and computational calculations added up to one conclusion, supporting our claim that HCl had not affected the SWCNT-COOH structure.

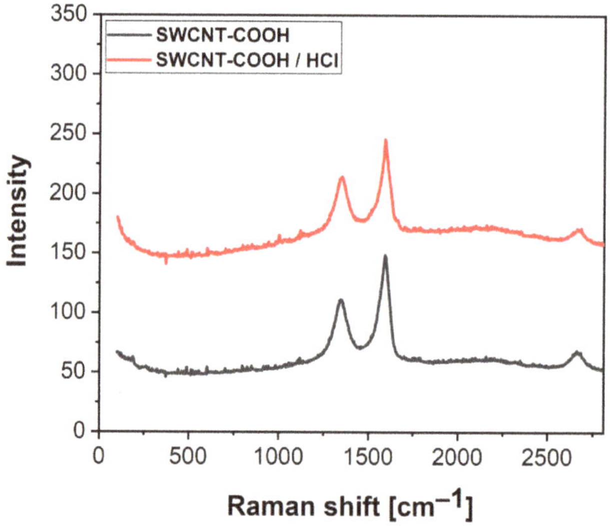

Figure 5. Raman spectra of SWCNT-COOH before and after the HCl treatment.

A crucial property of the fabricated films, for application in uncooled MIR detectors, is a transparency of above 50% in the range of between 2.5 and 3.5 μm. However, the conductivity of the film is dependent on the film thickness (the number of deposited layers). Therefore, this research was directed towards finding the balance between these two functional properties. We also analyzed the transparency in the visible range of the spectrum. This may also be important for the consideration of such electrodes in a wider range of applications. For up to 10 layers, the film showed a transparency of above 50% in general, with a mild increase towards higher wavelengths (Figure 6). A gradual reduction in film transparency was simply a direct repercussion of the deposited layer number.

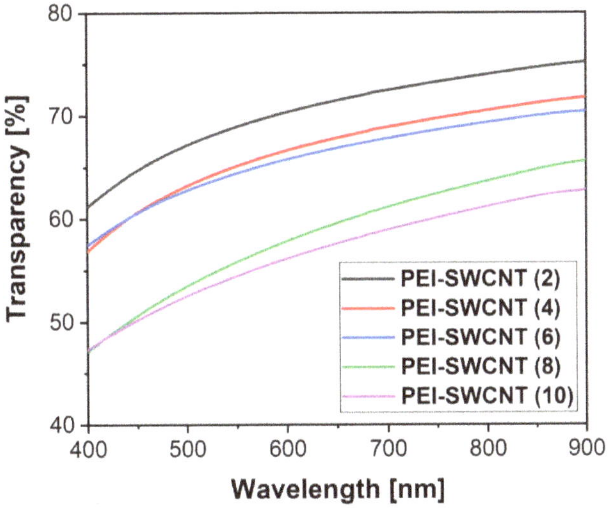

Figure 6. The transmittance of PEI+SWCNT with up to 10 bilayers in the visible spectral range.

The FTIR spectra of the samples showed a gradual decrease in transparency in the IR detector working range with the number of deposited layers. Given that the transparency of the glass substrate was around 70%, the film with six deposited bilayers transmitted over 50% of the IR waves. We consider that further deposition of the material would not be beneficial for applications of the film; thus, we conclude that between four and six bilayers is an optimal number regarding optical properties. FTIR spectra for all samples are presented in Figure 7.

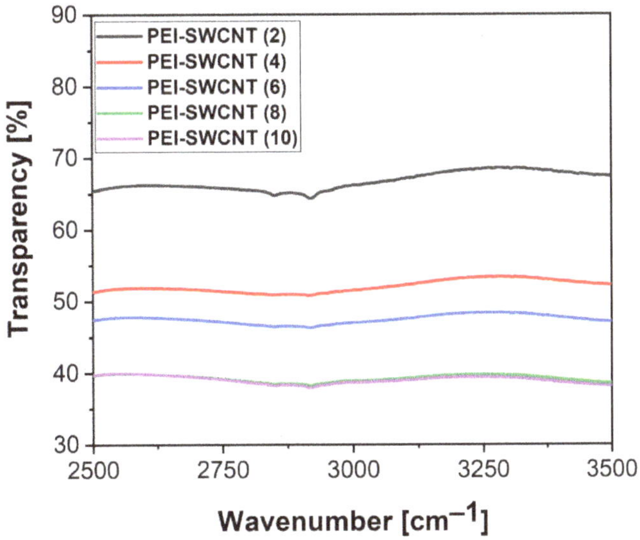

Figure 7. FTIR spectra of PEI+SWCNT with up to 10 bilayers in the IR spectral range.

The electrical conductivity of the films was measured by four-point probe method, repeating the procedure over at least three positions across the samples to confirm homogeneity. Even though the sheet resistance was not as low as expected, a decreasing trend was recorded with the number of bilayers deposited with low-purity SWCNT-COOH. The values for the sheet resistance were not as low as expected, most probably because the conductivity of the 80% pure nanotubes had a major impact on the electrical properties. Although the pretreatment procedure had a positive impact on the SWCNT-COOH dispersibility, film uniformity and transparency, solving several important obstacles towards the application of low-cost SWCNTs, it still did not improve the conductivity to desired levels. In comparison to these results, the thin film electrodes based on the higher purity MWCNT-COOH from our previous studies [34] had better sheet resistance values, in the range 10–20 kΩ. The sample with eight bilayers had a sheet resistance of around 200 kΩ/sq. The sheet resistance values are presented in Table 2. It is assumed that by depositing more layers, more SWCNT-SWCNT interconnections are formed—thus increasing the conductivity. Such high values for the sheet resistance of the two bilayers were probably recorded because even in stable dispersions, some agglomerate formation may occur due to impurities present—where even long sonication and the proposed pretreatment do not completely break them apart. Thus, more depositions are needed to reach the appropriate conductivity, balanced with optical transparency. Nevertheless, here we report the successful deposition of the glass substrate and the good electrical properties of a much cheaper, low-purity SWCNT-COOH. The further deposition of bilayers saturated the conductivity not far from six to eight bilayer samples, additionally confirming that this is the optimal number with the given material and processing procedure.

Table 2. Sheet resistance of PEI+SWCNT with up to 10 bilayers.

Sample Name	Sheet Resistance (kΩ/sq)
PEI + SWCNT (2)	9500
PEI + SWCNT (4)	1000–2000
PEI + SWCNT (6)	300–400
PEI + SWCNT (8)	200–250
PEI + SWCNT (10)	200–220

4. Conclusions

This research was focused on the systematical understanding of the influence of HCl and different solvents (water, ethanol, acetone) on SWCNT-COOH dispersion stability through theoretical and experimental approaches, as well as the preparation of high-quality thin films with foreseen applications in mid-IR transparent thin film electrodes. Carboxylic SWCNTs of 80% purity have shown great potential as an alternative to expensive high-purity SWCNTs for the development of thin films transparent in the UV/Vis and MIR range. The nanotubes were treated successfully with HCl and dispersed in acetone. The dispersion showed long-term stability.

The computational results proposed that even though the intermolecular distances between the solvents and -COOH groups increased after protonation, the binding energy increased when acetone was tested as a solvent. Concerning such behavior, we propose that in the protonated state, -COOH groups are more likely to bend towards the SWCNT wall and form additional noncovalent bonds. This effect, combined with partial purification with HCl, contributed to the overall stability of the dispersions.

The Raman spectroscopy showed that treatment with HCl had no significant impact on SWCNT-COOH regarding its chemical structure. The optical transparency of the films in the visible range was between 55 and 70%, while the transparency in the mid-IR was between 40 and 70%, depending on the number of deposited bilayers. The sheet resistance also decreased with the number of bilayers; for the PEI+SWCNT(2) film it was 9500 kΩ/sq and for the PEI+SWCNT(8) film it was 200 kΩ/sq. The sheet resistance value decreased with increases in the number of deposited bilayers, as expected. The values of the sheet resistance were not sufficiently low for the envisioned application until now; therefore, the pretreatment procedure was highly beneficial for the dispersion stability and film uniformity only, reflecting the scientific contribution of this research. We estimate that the optimal film thickness regarding the transparency and the sheet resistance values was recorded in the sample with six bilayers. Even though we have demonstrated that stable dispersions can be successfully prepared with 80% pure SWCNT-COOH—and as such, can be used for the LbL deposition of transparent films with improved conductivity—future research should be directed towards the more detailed study of SWCNT crosslinking, considering different processing conditions or the introduction of compatible carbon-based conductive materials.

Supplementary Materials: The following supporting information can be downloaded at: https://www.mdpi.com/article/10.3390/nano12111901/s1, Figure S1: Raman spectra of SWCNT-COOH after the HCl treatment, analyzed with the red laser (780 nm).

Author Contributions: Conceptualization, J.S., S.A., S.J. and B.B.; methodology, J.S., S.A. and S.J.; software, S.A.; validation, J.S. and S.J.; formal analysis, J.S., S.A. and B.B.; investigation, J.S., S.A., S.J. and B.B.; resources, J.S., S.A., S.J. and B.B.; data curation, S.A. and S.J.; writing—original draft preparation, J.S.; writing—review and editing, J.S., S.A., J.M., V.V.S. and B.B.; visualization, S.J. and B.B.; supervision, J.M. and V.V.S.; project administration, J.S. All authors have read and agreed to the published version of the manuscript.

Funding: This research was funded by The Ministry of Education, Science and Technological Development of the Republic of Serbia (Grant No. 451-03-68/2022-14/200358 and Grant No. 451-03-68/2022-14/200125).

Conflicts of Interest: The authors declare no conflict of interest.

References

1. Morag, A.; Jelinek, R. "Bottom-up" transparent electrodes. *J. Colloid Interface Sci.* **2016**, *482*, 267–289. [CrossRef] [PubMed]
2. Wang, X.; Zhi, L.; Tsao, N.; Tomović, Ž.; Li, J.; Müllen, K. Transparent carbon films as electrodes in organic solar cells. *Angew. Chem. Int. Ed.* **2008**, *47*, 2990–2992. [CrossRef] [PubMed]
3. Spanggaard, H.; Krebs, F.C. A brief history of the development of organic and polymeric photovoltaics. *Sol. Energy Mater. Sol. Cells* **2004**, *83*, 125–146. [CrossRef]
4. Armstrong, N.R.; Veneman, P.A.; Ratcliff, E.; Placencia, D.; Brumbach, M. Oxide contacts in organic photovoltaics: Characterization and control of near-surface composition in indium-tin oxide (ITO) electrodes. *Acc. Chem. Res.* **2009**, *42*, 1748–1757. [CrossRef]
5. Wang, T.; Lu, K.; Xu, Z.; Lin, Z.; Ning, H.; Qiu, T.; Yang, Z.; Zheng, H.; Yao, R.; Peng, J. Recent developments in flexible transparent electrode. *Crystals* **2021**, *11*, 511. [CrossRef]
6. Zhu, Y.; Deng, Y.; Yi, P.; Peng, L.; Lai, X.; Lin, Z. Flexible Transparent Electrodes Based on Silver Nanowires: Material Synthesis, Fabrication, Performance, and Applications. *Adv. Mater. Technol.* **2019**, *4*, 1900413. [CrossRef]
7. Cao, Q.; Rogers, J.A. Ultrathin films of single-walled carbon nanotubes for electronics and sensors: A review of fundamental and applied aspects. *Adv. Mater.* **2009**, *21*, 29–53. [CrossRef]
8. Gupta, N.; Gupta, S.M.; Sharma, S.K. Carbon nanotubes: Synthesis, properties and engineering applications. *Carbon Lett.* **2019**, *29*, 419–447. [CrossRef]
9. Hecht, D.S.; Hu, L.; Irvin, G. Emerging transparent electrodes based on thin films of carbon nanotubes, graphene, and metallic nanostructures. *Adv. Mater.* **2011**, *23*, 1482–1513. [CrossRef]
10. Hu, L.; Hecht, D.S.; Gruner, G. Percolation in transparent and conducting carbon nanotube networks. *Nano Lett.* **2004**, *4*, 2513–2517. [CrossRef]
11. Paularokiadoss, F.; Jeyakumar, T.C.; Thomas, R.; Sekar, A.; Bhakiaraj, D. Group13 Monohalides [AX (A= B, Al, Ga, In; X= Halogens)] as Alternative Ligands for Carbonyl in Organometallics: Electronic Structure and Bonding Analysis. *Comput. Theor. Chem.* **2022**, *1209*, 113587. [CrossRef]
12. Al-Zaqri, N.; Pooventhiran, T.; Alsalme, A.; Rao, D.J.; Rao, S.S.; Sankar, A.; Thomas, R. First-Principle Studies of Istradefylline with Emphasis on the Stability, Reactivity, Interactions and Wavefunction-Dependent Properties. *Polycycl. Aromat. Compd.* **2020**, 1–15. [CrossRef]
13. Bielenica, A.; Beegum, S.; Mary, Y.S.; Mary, Y.S.; Thomas, R.; Armaković, S.; Armaković, S.J.; Madeddu, S.; Struga, M.; van Alsenoy, C. Experimental and computational analysis of 1-(4-chloro-3-nitrophenyl)-3-(3, 4-dichlorophenyl) thiourea. *J. Mol. Struct.* **2020**, *1205*, 127587. [CrossRef]
14. Mary, Y.S.; Kumar, V.S.; Mary, Y.S.; Resmi, K.S.; Thomas, R. Detailed Quantum Mechanical Studies on Three Bioactive Benzimidazole Derivatives and Their Raman Enhancement on Adsorption over Graphene Sheets. *Polycycl. Aromat. Compd.* **2020**, 1–10. [CrossRef]
15. Al-Otaibi, J.S.; Mary, Y.S.; Mary, Y.S.; Thomas, R. Evidence of cluster formation of pyrrole with mixed silver metal clusters, Agx-My (x = 4,5, y = 2/1 and M = Au/Ni/Cu) using DFT/SERS analysis. *Comput. Theor. Chem.* **2022**, *1208*, 113569. [CrossRef]

16. Al-Otaibi, J.S.; Mary, Y.S.; Mary, Y.S. Adsorption of a thione bioactive derivative over different silver/gold clusters-DFT investigations. *Comput. Theor. Chem.* **2022**, *1207*, 113497. [CrossRef]
17. Al-Otaibi, J.S.; Mary, Y.S.; Mary, Y.S.; Ullah, Z.; Kwon, H.W. Adsorption behavior and solvent effects of an adamantane-triazole derivative on metal clusters—DFT simulation studies. *J. Mol. Liq.* **2022**, *345*, 118242. [CrossRef]
18. Kohler, C.; Seifert, G.; Frauenheim, T. Density functional based calculations for Fen (n 32). *Chem. Phys.* **2005**, *309*, 23–31. [CrossRef]
19. Elstner, M.; Porezag, D.; Jungnickel, G.; Elsner, J.; Haugk, M.; Frauenheim, T.; Suhai, S.; Seifert, G. Self-consistent-charge density-functional tight-binding method for simulations of complex materials properties. *Phys. Rev. B* **1998**, *58*, 7260. [CrossRef]
20. Elstner, M. The SCC-DFTB method and its application to biological systems. *Theor. Chem. Acc.* **2006**, *116*, 316–325. [CrossRef]
21. Seifert, G.; Porezag, D.; Frauenheim, T. Calculations of molecules, clusters, and solids with a simplified LCAO-DFT-LDA scheme. *Int. J. Quantum Chem.* **1996**, *58*, 185–192. [CrossRef]
22. Porezag, D.; Frauenheim, T.; Köhler, T.; Seifert, G.; Kaschner, R. Construction of tight-binding-like potentials on the basis of density-functional theory: Application to carbon. *Phys. Rev. B* **1995**, *51*, 12947. [CrossRef] [PubMed]
23. Van den Bossche, M. DFTB-assisted global structure optimization of 13-and 55-atom late transition metal clusters. *J. Phys. Chem. A* **2019**, *123*, 3038–3045. [CrossRef] [PubMed]
24. Elstner, M.; Seifert, G. Density functional tight binding. *Philos. Trans. R. Soc. A Math. Phys. Eng. Sci.* **2014**, *372*, 20120483. [CrossRef] [PubMed]
25. Xu, Y.; Friedman, R.; Wu, W.; Su, P. Understanding intermolecular interactions of large systems in ground state and excited state by using density functional based tight binding methods. *J. Chem. Phys.* **2021**, *154*, 194106. [CrossRef]
26. Barone, V.; Carnimeo, I.; Scalmani, G. Computational spectroscopy of large systems in solution: The DFTB/PCM and TD-DFTB/PCM approach. *J. Chem. Theory Comput.* **2013**, *9*, 2052–2071. [CrossRef]
27. Kurban, H.; Dalkilic, M.; Temiz, S.; Kurban, M. Tailoring the structural properties and electronic structure of anatase, brookite and rutile phase TiO_2 nanoparticles: DFTB calculations. *Comput. Mater. Sci.* **2020**, *183*, 109843. [CrossRef]
28. Wu, L.; Han, Y.; Zhao, Q.; Zhang, L. Effects of chiral indices on the atomic arrangements and electronic properties of Si double-walled nanotubes (6, min)@(9, mout)(min= 0 to 6, mout= 0 to 9) by SCC-DFTB calculations. *Mater. Sci. Semicond. Process.* **2021**, *129*, 105775. [CrossRef]
29. Timsorn, K.; Wongchoosuk, C. Adsorption of NO_2, HCN, HCHO and CO on pristine and amine functionalized boron nitride nanotubes by self-consistent charge density functional tight-binding method. *Mater. Res. Express* **2020**, *7*, 55005. [CrossRef]
30. Xu, H.; Li, L.; Fan, G.; Chu, X. DFT study of nanotubes as the drug delivery vehicles of Efavirenz. *Comput. Theor. Chem.* **2018**, *1131*, 57–68. [CrossRef]
31. Laplaza, R.; Peccati, F.; Boto, R.A.; Quan, C.; Carbone, A.; Piquemal, J.; Maday, Y.; Contreras-García, J. NCIPLOT and the analysis of noncovalent interactions using the reduced density gradient. *Wiley Interdiscip. Rev. Comput. Mol. Sci.* **2021**, *11*, e1497. [CrossRef]
32. Li, W.; Miao, W.; Cui, J.; Fang, C.; Su, S.; Li, H.; Hu, L.; Lu, Y.; Chen, G. Efficient corrections for DFT noncovalent interactions based on ensemble learning models. *J. Chem. Inf. Model.* **2019**, *59*, 1849–1857. [CrossRef] [PubMed]
33. Shikhaliyev, N.Q.; Kuznetsov, M.L.; Maharramov, A.M.; Gurbanov, A.V.; Ahmadova, N.E.; Nenajdenko, V.G.; Mahmudov, K.T.; Pombeiro, A.J.L. Noncovalent interactions in the design of bis-azo dyes. *Cryst. Eng. Comm.* **2019**, *21*, 5032–5038. [CrossRef]
34. Stanojev, J.; Bajac, B.; Cvejic, Z.; Matovic, J.; Srdic, V.V. Development of MWCNT thin film electrode transparent in the mid-IR range. *Ceram. Int.* **2020**, *46*, 11340–11345. [CrossRef]
35. Wulan, P.P.D.K.; Ulwani, S.H.; Wulandari, H.; Purwanto, W.W.; Mulia, K. The effect of hydrochloric acid addition to increase carbon nanotubes dispersibility as drug delivery system by covalent functionalization. In *IOP Conference Series: Materials Science and Engineering*; Institute of Physics Publishing: Bristol, UK, 2018; Volume 316, pp. 1–8.
36. Wulan, P.P.D.K.; Permana, G.; Putri, W.A. The effect of optimization HCl addition to dispersibility in carbon nanotubes functionalization as drug delivery. In *AIP Conference Proceedings*; American Institute of Physics Inc.: College Park, MD, USA, 2020; Volume 2255, p. 060014.
37. Engtrakul, C.; Davis, M.F.; Gennett, T.; Dillon, A.C.; Jones, K.M.; Heben, M.J. Protonation of carbon single-walled nanotubes studied using ^{13}C and ^{1}H-^{13}C cross polarization nuclear magnetic resonance and Raman spectroscopies. *J. Am. Chem. Soc.* **2005**, *127*, 17548–17555. [CrossRef]
38. Mamedov, A.A.; Kotov, N.A.; Prato, M.; Guldi, D.M.; Wicksted, J.P.; Hirsch, A. Molecular design of strong single-wall carbon nanotube/polyelectrolyte multilayer composites. *Nat. Mater.* **2002**, *1*, 190–194. [CrossRef] [PubMed]
39. Grimme, S.; Bannwarth, C.; Shushkov, P. A robust and accurate tight-binding quantum chemical method for structures, vibrational frequencies, and noncovalent interactions of large molecular systems parametrized for all spd-block elements (Z = 1–86). *J. Chem. Theory Comput.* **2017**, *13*, 1989–2009. [CrossRef]
40. Becke, A.D. Density-functional thermochemistry. III. The role of exact exchange. *J. Chem. Phys.* **1993**, *98*, 5648–5652. [CrossRef]
41. Grimme, S.; Ehrlich, S.; Goerigk, L. Effect of the damping function in dispersion corrected density functional theory. *J. Comput. Chem.* **2011**, *32*, 1456–1465. [CrossRef]
42. Hariharan, P.C.; Pople, J.A. The influence of polarization functions on molecular orbital hydrogenation energies. *Theor. Chim. Acta* **1973**, *28*, 213–222. [CrossRef]
43. Hehre, W.J.; Ditchfield, R.; Pople, J.A. Self—Consistent molecular orbital methods. XII. Further extensions of Gaussian—Type basis sets for use in molecular orbital studies of organic molecules. *J. Chem. Phys.* **1972**, *56*, 2257–2261. [CrossRef]

44. Ditchfield, R.; Hehre, W.J.; Pople, J.A. Self-consistent molecular-orbital methods. IX. An extended Gaussian-type basis for molecular-orbital studies of organic molecules. *J. Chem. Phys.* **1971**, *54*, 724–728. [CrossRef]
45. Rüger, R.; Franchini, M.; Trnka, T.; Yakovlev, A.; van Lenthe, E.; Philipsen, P.; van Vuren, T.; Klumpers, B.; Soini, T. *AMS 2021.1 AMS 2021.1, SCM, Theoretical Chemistry*; Vrije Universiteit: Amsterdam, The Netherlands, 2021. Available online: https://www.scm.com (accessed on 29 May 2022).
46. Bochevarov, A.D.; Harder, E.; Hughes, T.F.; Greenwood, J.R.; Braden, D.A.; Philipp, D.M.; Rinaldo, D.; Halls, M.D.; Zhang, J.; Friesner, R.A. Jaguar: A high-performance quantum chemistry software program with strengths in life and materials sciences. *Int. J. Quantum Chem.* **2013**, *113*, 2110–2142. [CrossRef]
47. Jacobson, L.D.; Bochevarov, A.D.; Watson, M.A.; Hughes, T.F.; Rinaldo, D.; Ehrlich, S.; Steinbrecher, T.B.; Vaitheeswaran, S.; Philipp, D.M.; Halls, M.D. Automated transition state search and its application to diverse types of organic reactions. *J. Chem. Theory Comput.* **2017**, *13*, 5780–5797. [CrossRef] [PubMed]
48. Contreras-Garcia, J.; Johnson, E.R.; Keinan, S.; Chaudret, R.; Piquemal, J.-P.; Beratan, D.N.; Yang, W. NCIPLOT: A program for plotting noncovalent interaction regions. *J. Chem. Theory Comput.* **2011**, *7*, 625–632. [CrossRef] [PubMed]
49. Johnson, E.R.; Keinan, S.; Mori-Sánchez, P.; Contreras-García, J.; Cohen, A.J.; Yang, W. Revealing noncovalent interactions. *J. Am. Chem. Soc.* **2010**, *132*, 6498–6506. [CrossRef]
50. Yan, X.; Suzuki, T.; Kitahama, Y.; Sato, H.; Itoh, T.; Ozaki, Y. A study on the interaction of single-walled carbon nanotubes (SWCNTs) and polystyrene (PS) at the interface in SWCNT-PS nanocomposites using tip-enhanced Raman spectroscopy. *Phys. Chem. Chem. Phys.* **2013**, *15*, 20618–20624. [CrossRef]
51. Dasari, B.S.; Navaraj, W.; Taube, W.R.; Agarwal, P.B.; Rajput, M.; Kumar, A.; Akhtar, J. Room Temperature Single Walled Carbon Nanotubes (SWCNT) Chemiresistive Ammonia Gas Sensor Development of Surface Plasmon Resonance based Instrument View project 'Flexsensotronics'-Flexible Sensors and Electronics. *Sens. Transducers* **2015**, *190*, 24–30.
52. Hussain, S.; Jha, P.; Chouksey, A.; Raman, R.; Islam, S.S.; Islam, T.; Choudhary, P.K. Spectroscopic Investigation of Modified Single Wall Carbon Nanotube (SWCNT). *J. Mod. Phys.* **2011**, *2*, 538–543. [CrossRef]
53. Zhao, W.; Song, C.; Pehrsson, P.E. Water-soluble and optically pH-sensitive single-walled carbon nanotubes from surface modification. *J. Am. Chem. Soc.* **2002**, *124*, 12418–12419. [CrossRef]
54. Costa, S.; Borowiak-Palen, E.; Kruszynska, M.; Bachmatiuk, A.; Kalenczuk, R.J. Characterization of carbon nanotubes by Raman spectroscopy. *Mat. Sci.* **2008**, *26*, 433–441.

nanomaterials

Article

A Micro Electrochemical Sensor for Multi-Analyte Detection Based on Oxygenated Graphene Modified Screen-Printed Electrode

Baiqing Yuan *, Liju Gan, Gang Li, Chunying Xu and Gang Liu *

School of Chemistry and Materials Science, Ludong University, Yantai 264025, China; gan17737233930@163.com (L.G.); 1258331922@126.com (G.L.); chunyingxu@126.com (C.X.)
* Correspondence: bqyuan@ldu.edu.cn (B.Y.); shdliugang@163.com (G.L.)

Abstract: Electrode interfaces with both antibiofouling properties and electrocatalytic activity can promote the practical application of nonenzymatic electrochemical sensors in biological fluids. Compared with graphene, graphene oxide (GO) possesses unique properties such as superior solubility (hydrophilicity) in water, negative charge, and abundant oxygenated groups (oxo functionalities) in the plane and edge sites, which play an essential role in electrocatalysis and functionalization. In this work, a micro electrochemical sensor consisting of GO-modified screen-printed electrode and PDMS micro-cell was designed to achieve multi-analyte detection with excellent selectivity and anti-biofouling properties by electrochemically tuning the oxygen-containing functional species, hydrophilicity/hydrophobicity, and electrical conductivity. In particular, the presented electrodes demonstrated the potential in the analysis of biological samples in which electrodes often suffer from serious biofouling. The interaction of proteins with electrodes as well as uric acid was investigated and discussed.

Keywords: micro electrochemical sensor; multi-analyte detection; graphene oxide (GO); biofouling; oxo functionalities

Citation: Yuan, B.; Gan, L.; Li, G.; Xu, C.; Liu, G. A Micro Electrochemical Sensor for Multi-Analyte Detection Based on Oxygenated Graphene Modified Screen-Printed Electrode. *Nanomaterials* **2022**, *12*, 711. https://doi.org/10.3390/nano12040711

Academic Editor: Seung Hwan Ko

Received: 20 January 2022
Accepted: 17 February 2022
Published: 21 February 2022

1. Introduction

Electrochemical sensing represents a potential tool to perform clinical applications due to its low cost, simplicity and high selectivity and sensitivity. However, when the electrodes are exposed to biological fluids such as plasma or blood, biofouling often occurs due to the non-specific absorption of biological macromolecules, especially proteins, which hinders the electron transfer and thus results in a rapid loss of sensitivity [1,2]. This issue could be solved for the glucose meter combination with a semi-permeable membrane, in which small-molecule glucose could diffuse through the semi-permeable membrane and reach the electrode, but not for macromolecule proteins [3]. In addition to the enzyme and affinity-based electrochemical sensors, nonenzymatic electrochemical sensors coupled with antibiofouling strategies would be a promising commercial tool used in clinical analysis for small and electroactive biological molecules which are related to human health, such as uric acid (UA), ascorbic acid (AA), dopamine (DA), guanine (G) and adenine (A). In addition, voltammetry demonstrates high selectivity and thus could be used for multi-analyte measurements. Many chemical and physical strategies have been reported to alleviate the biofouling issues, including nanoengineered surfaces (i.e., nanoporous metals and nanocarbons), antifouling layers (PEG, zwitterionic polymers, and biopolymers), nanoporous membranes, and hydrogels [4]. For example, a nanoporous gold (NPG) electrode derived from the dealloying of a silver–gold alloy was recently employed as an antibiofouling electrode for the simultaneous determination of AA and UA [5]. However, the construction of these antibioflouling electrodes is often cumbersome and complex. It would be interesting if an electrode material possessed both excellent antibiofouling performance and high electrocatalytic activity toward these molecules.

Graphene oxide (GO) is a complex carbon-based macromolecule that has a two-dimensional (2D) structure and possesses abundant oxygen-containing functional species and hydrogen atoms on its basal plane and edges. Many models have been demonstrated to elucidate the structure of GO with dazzling confusion. Recently, Siaj et al. [6] overviewed all the aspects of GO structures and presented a precise structure clarifying the chemical nature of GO based on earlier and more modern models and new discoveries by other top researchers. The structural model comprises the presence of double bond (C=C), aromatic entity, epoxyl (C-O-C), hydroxy (C-OH), carboxylic acid (O=C-OH or salt), ketone (C=O), organic carbonate, phenol, quinone, lactol, ester carbonyl, carbon vacancies, sulfate ester, carbon radicals, implicit carbon-hydrogen bond (C-H), and allylic alcohol [6]. GO exhibits excellent water dispersity and hydrophilicity due to the ionizable edge O=C-OH species. In fact, carboxylic acid distributes at the edges while phenol hydroxyl and epoxide groups mainly present on the basal plane, endowing GO an amphiphile with hydrophilic edges and a more hydrophobic basal plane [7]. The versatile surface chemistry also offers GO a powerful platform for further construction with various chemical moieties (organic, inorganic, and nanocomposites) by covalent and noncovalent interactions [8].

Unfortunately, these oxo functionalities present on GO result in poor conductivity. In efforts tailoring GO with high conductivity for applications in electrochemistry and electronics, tremendous efforts have been directed toward the elimination of the oxidized functionalities from GO by means of microorganisms, chemicals, electrochemistry, heat, UV, microwave irradiation, ion bombardments or multiphase methods [9–11]. Although the removal of oxygen-containing functional species increases the conductivity of the 2D material, it was recently found that the residual oxo functionalities play an essential role in promoting the electrocatalytic reaction. For example, mildly reduced GO with high concentrations of epoxy or ring ether groups located either on their basal planes or at plane edges exhibit excellent activity, high selectivity and stability for electrochemical H_2O_2 production from oxygen [12,13]. Furthermore, carbon defects related to quinone/catechol groups or carboxylic acid edge sites play a more pivotal role in boosting 2 e^- ORR peroxide formation activity than other oxygen-containing functional species under alkaline conditions for nitrogen-doped reduced GO [14]. The residual oxygen-containing functional species present on graphene also enhanced the electrocatalytic oxidation of glutathione [15], dihydroxybenzene isomers, L-methionine [16], and ascorbic acid [17], as well as the reduction of polyphenol [18].

It has been proved that electrochemical reduction is an ideal method to tune both conductivity and electrocatalytic activity of GO under different degrees of reduction [19]. Here, electrochemically tuned GO was explored for the electrochemical sensing of multi-analytes, and the effect of oxygen-containing functional species on both electrocatalysis and antibiofouling ability was also investigated in detail. The results showed that the residual oxo functionalities were beneficial for the selectivity and electro-oxidation of UA, DA, AA, G and A. In addition, the oxo functionalities were responsible for the binding of proteins in serum, which forms a coating layer. The binding layer facilitated the mass transport and did not hinder the electron transfer of electroactive probe $Fe(CN)_6{}^{4-}/Fe(CN)_6{}^{3-}$.

2. Experimental Section

2.1. Chemicals and Solutions

Bovine Serum Albumin (BSA), UA, DA, AA, G, A, and Polydimethylsiloxanes (PDMS) were purchased from Sigma-Aldrich (St. Louis, MO, USA). GO was purchased from Nanjing XFNano Materials Tech Co., Ltd (Nanjing, Jiangsu, China). The screen-printed electrodes (DRP 110) were purchased from DROPSENS (Oviedo, Spain). All other chemicals were of analytical reagent grade and used without further purification. The aqueous solutions were prepared with doubly distilled water. An amount of 0.1 M pH = 7.2 PBS solution was used as the background electrolyte for measurements. The oxo functionalities present on GO were tuned by electrochemical treatment for 500 s at different potentials from −0.6 to −1.2 V using an I-T curve in acetic buffer (pH = 4). The electrochemical tuned

GO electrode was signed as $GO_{potential}$. For example, $GO_{-0.75V}$ refers to the GO electrode that was electrochemically treated at −0.75 V. Unless otherwise specified, $GO_{-0.75V}$ was used in the text.

2.2. Apparatus

The X-ray photoelectron spectroscopy (XPS) was recorded on a Thermo ESCALAB 250 Xi spectrometer fitted with a monochromatic Al Kα X-ray source. The morphologies were characterized by scanning electron microscopy (SEM) (Hitachi SU8010). All the electrochemical experiments were carried out with a CHI 750E electrochemical workstation with a conventional three-electrode system consisting of a modified GCE working electrode or screen-printed electrode, platinum coil auxiliary electrode, and Ag/AgCl (saturated KCl) reference electrode. Differential pulse voltammetry technology was used for electrochemical detection. Electrochemical impedance spectra (EIS) were carried out in 0.1 M KCl containing 5 mM $Fe(CN)_6{}^{3-/4-}$ in the frequency range of 1 MHz to 0.1 Hz at 0.24 V.

The screen-printed electrode modified with $GO_{-0.75V}$ was integrated with a PDMS-based micro electrochemical cell for the analysis of five biological molecules and serum (Schemes 1 and 2). In order to prevent the leaking of the solution, two powerful magnets were used to compress the cell together with the electrode. The micro PDMS cell was fabricated according to Scheme 2 in four steps, including mold construction, casting with PDMS mixture, baking, and mold removal. Two well-cut plastic tubes were concentrically fixed. The outer tube was sealed to a flat glass with instant adhesive. The inner tube was fixed on a screen-printed electrode template with adhesive, which was inserted through the sidewall of the outer tube and adhered to. The proper amount of degassed mixture of PDMS monomer and curing agent (in a 10:1 ratio) was poured into the ring part, followed by baking for 40 min at 100 °C. Finally, PDMS was peeled from the mold, and the cell body was obtained.

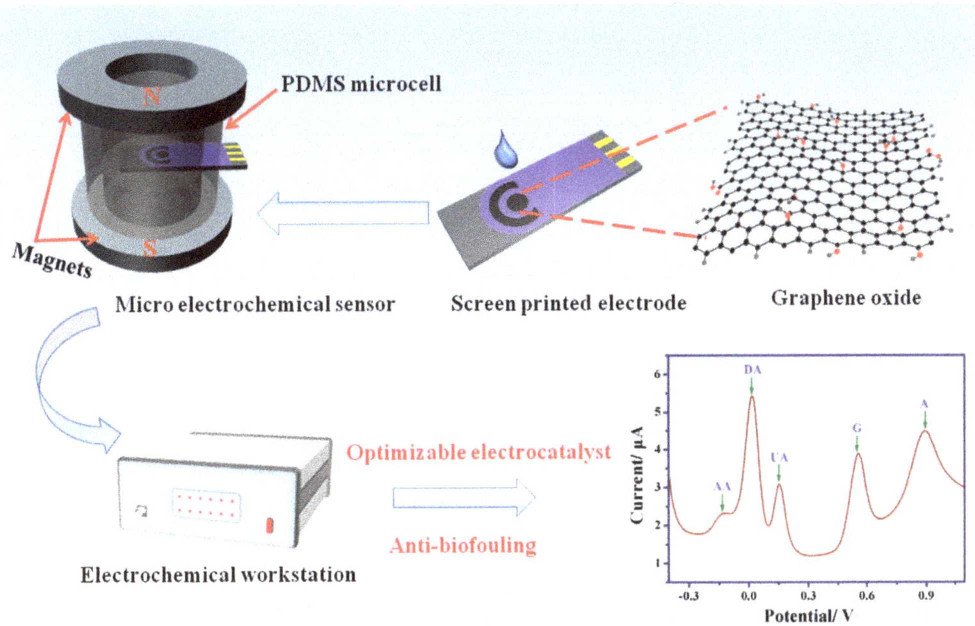

Scheme 1. Electrochemical sensing of multi-analyte based on electrochemically tuned GO electrode.

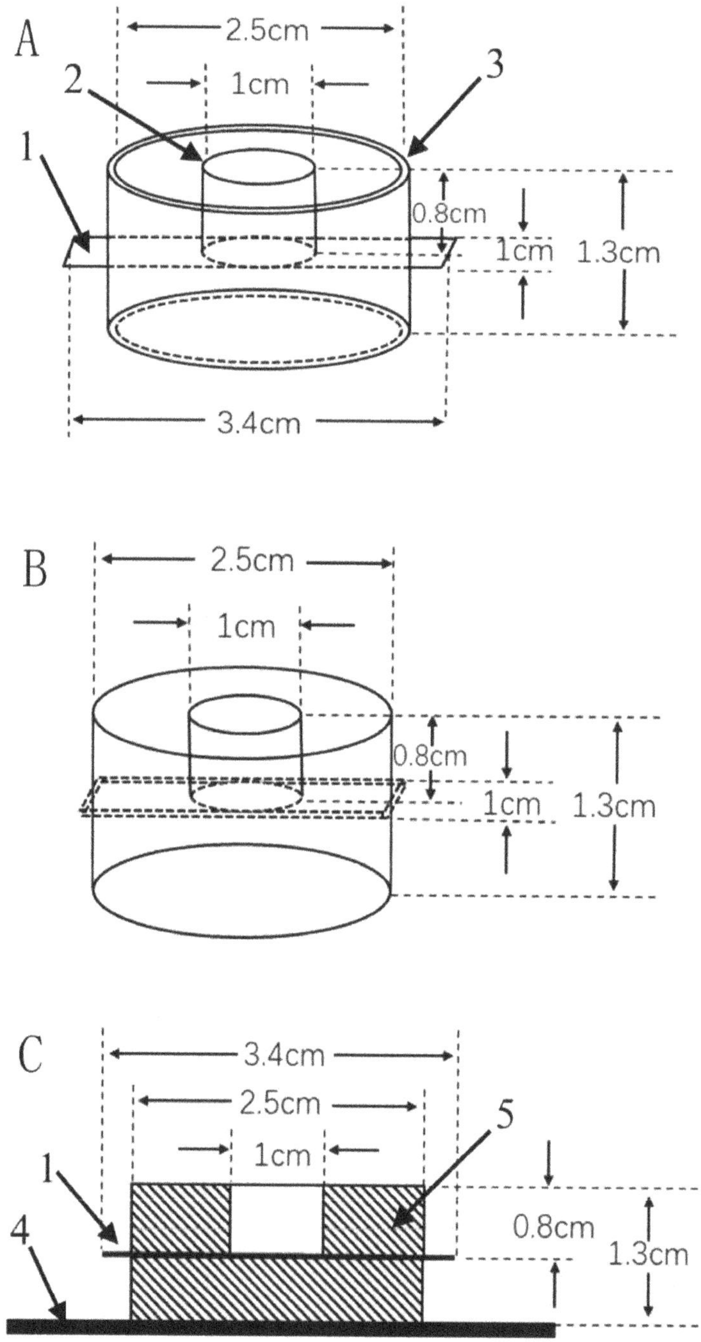

Scheme 2. Preparation of the PDMS-based micro-cell. (**A**) Mold construction: (1) screen-printed electrode template, (2, 3) two cut plastic tubes. (**B**) Schematic view of obtained PDMS micro-cell. (**C**) A cross-sectional view of PDMS micro-cell used for detection: (1) screen-printed electrode template, (4) glass plate, (5) PDMS cell body.

2.3. Electrode Preparation and Modification

Prior to modification, the GCE (3 mm diameter, 0.07 cm^2) was successively polished with 1, 0.3, and 0.05 μm alumina paste to a mirror finish and then rinsed with water followed by an ultrasonic treatment in water and ethanol, respectively. The GO-modified GCE (GO/GCE), or screen-printed electrode, was prepared by dropping 5 μL or 7 μL GO suspension (1 mg/mL) on a cleaned GCE and left to dry at room temperature.

2.4. Interaction between Electrode and Proteins

In order to study the interaction between the electrode and BSA, GO/GCE or $GO_{-0.75V}$/GCE was immersed in 10 mg/mL BSA solution for 30 min, and the dipped electrode was signed with GO-BSA or $GO_{-0.75V}$-BSA. In addition, GO/GCE or $GO_{potential}$ was immersed in 100% serum for 30 min (GO-serum/GCE, $GO_{potential}$-serum/GCE) to investigate the antibiofouling property. The interaction between BSA and UA was tested by immersing the electrode in 10 mg/mL BSA solution for 30 min and then rinsing with water, followed by dipping in 200 μM UA for 30 min ($GO_{-0.75V}$-BSA-UA/GCE).

2.5. Serum Sample Analysis for UA

An original human serum sample was diluted with PBS at a 1:40 v/v ratio for measurements without any other pretreatment.

3. Results and Discussion

3.1. XPS Characterization of the Oxo Functionalities Present on $GO_{potential}$

Over-reduction of GO will lose oxo functionalities, which have been proved to be beneficial for electrocatalysis. Good balance of conductivity and oxo functionalities of GO would play an important part in nonenzymatic electrochemical sensing. XPS was used for the analysis of oxo functionalities present on GO which was electrochemically treated at different potentials from −0.6 to −1 V. Figure 1 shows the recorded C1s spectra of XPS for electrochemically treated GO/GC. Different functional species, including C=C, C-C, C-OH, C-O-C, C=O, and O-C=O, were observed [19]. In order to quantify the change in different components, the content of different functionalities present on $GO_{potential}$ versus potential is shown in Figure 2. It could be seen in Figure 2A, with the negative shift of the potential, the content of both C=O and C-O-C showed a downward trend, and C=O species disappeared when the potential was much lower than −0.8 V. The content of both C-OH and O-C=O species increased until −0.85 V, which might be attributed to the reduction of C=O, and then decreased when the potential was much lower than −0.85 V, indicating the reduction of the C-OH and O-C=O species [19].

3.2. Electrochemical Sensing of Multiple-Analyte

Electrochemically treated GO at different potentials was explored for the electrocatalytic oxidation of UA, DA and AA. These three molecules are electrooxidizable constituents that are commonly present in physiological fluids. It is hard to achieve efficient discrimination from each other by voltammetry due to their overlapped oxidation potential. Figure 3 shows the DPV of the mixture of 200 μM AA, 20 μM DA, and 20 μM UA in 0.1 M pH 7.2 PBS at different electrodes. Only one oxidation peak was observed for bare GCE, and two oxidation waves appeared for GO/GCE. However, when GO was electrochemically treated even at −0.6 V, three oxidation peaks with high selectivity occurred, indicating improved electrocatalytic activity. With the negative shift of the potential, the oxidation current increased until −0.75 V. Nevertheless, electrochemical treatment at more negative potential than −0.75 V would lead to decreased selectivity. This might be caused by the loss of oxo functionalities.

Although the over-electro-reduction of GO can improve the conductivity, useful oxo functionalities will be lost, which may be beneficial for electrocatalysis and selectivity. Conductivity, electro-active sites and binding interaction play an important part in electrochemical sensing. At pH = 7.2, UA exists in the anionic form (pKa = 5.4), which is more hydrophobic than AA and DA due to its very limited solubility in water [20]. AA exists as an anion (pKa = 4.10) while DA is in the cationic form (pKb = 8.87) [21]. Among these oxo functionalities present on GO, ionizable O=C-OH species distributes at the edges, while phenol hydroxyl and epoxide groups mainly present on the basal plane, leading to an edge-to-center distribution of hydrophilic and hydrophobic domains. Pristine GO/GCE exhibited much higher selectivity than bare GCE, suggesting that the oxo functionalities might interact with the analyte and consequently lead to enhanced selectivity, even though GCE has much higher conductivity than pristine GO. In summary, we speculated that the enhanced selectivity was attributed to the hydrophobic interaction between the electrode and UA and the electrostatic force between O=C-OH species and DA. In addition, conductivity and electrocatalytic activity from C=O species might also play an important role in the electrochemical behavior of UA, DA and AA.

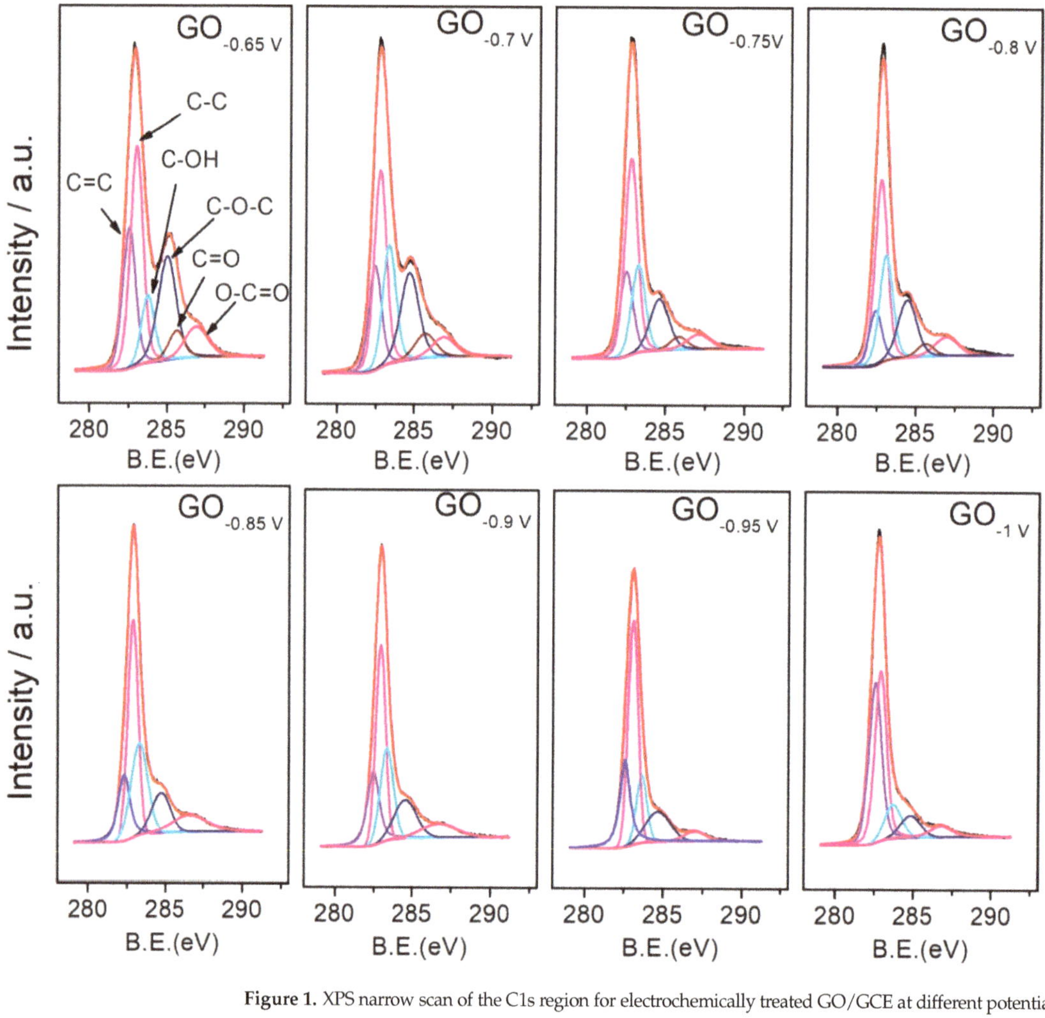

Figure 1. XPS narrow scan of the C1s region for electrochemically treated GO/GCE at different potentials.

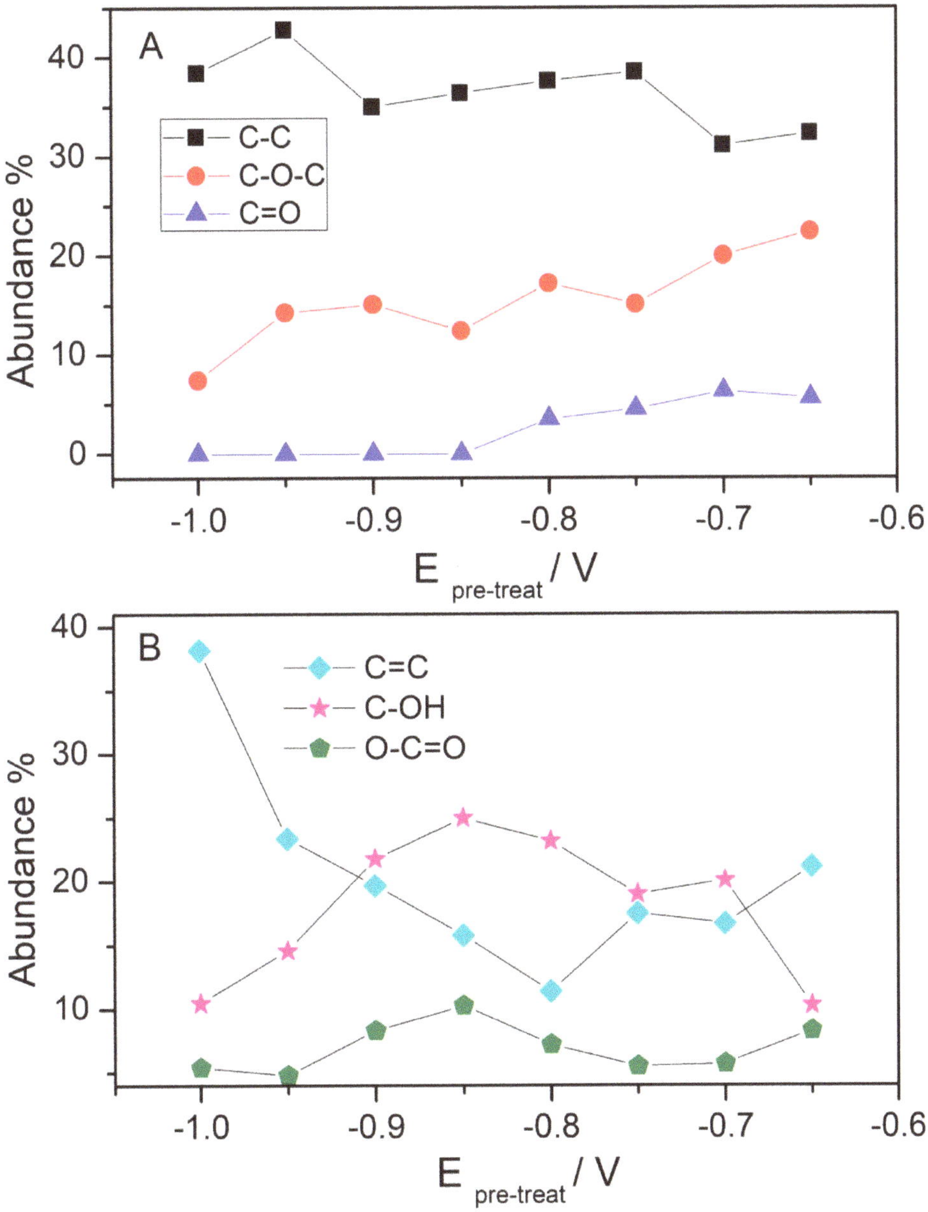

Figure 2. Abundance amount of different functionalities present on GO electrochemically treated at different potentials. (**A**): C-C, C-O-C and C=O; (**B**): C=C, C-OH and O-C=O.

Under optimum conditions, $GO_{-0.75V}$/GCE was employed for the electrochemical detection of UA in the presence of AA and DA in 0.1 M pH 7.2 PBS (Figure S1). Linear calibration curves are obtained over the range of 0.2–20 μM with the calculated detection limit of 0.12 μM. Similarly, Figure S2 demonstrates the electrochemical detection of DA in the presence of AA and UA and gives a linear dependence of the peak current for DA with concentrations ranging from 1 to 100 μM with the calculated detection limit of 0.5 μM.

Electrochemically treated GO at different potentials was also demonstrated for the electrocatalytic oxidation of G and A (Figure S3). GCE showed two anodic waves located at 0.63 and 0.97 V, respectively. For pristine GO-modified electrodes, the two peaks shift to 0.61 and 0.93 V, respectively. When the potential shifted negatively, the oxidation potential of G and A decreased gradually until −0.75 V. The oxidation peak of G and A shifted to 0.55 and 0.85 V at $GO_{-0.75V}$/GCE, respectively. The decreased overpotential of G and A indicated the enhanced electrocatalytic activity, which is the result of the balance of conductivity and some oxo functionalities as seen in C=O species. Upon the potential beyond −0.75 V, the background noise increased. Therefore, $GO_{-0.75V}$/GCE was selected for the electrochemical measurements. Figures S4 and S5 present the electrochemical detection of G and A in 0.1 M pH 7.2 PBS, respectively. The calibration plot shows that the peak current of A (or G) exhibits a linear relationship (R^2 = 0.998 or 0.999) over a concentration range from 2 μM to 100 μM (0.5 to 100 μM) with a detection limit of 0.3 μM (0.2 μM). Table 1 lists the electrochemical sensing performances for five biological molecules based on $GO_{-0.75V}$ and other electrocatalysts. The results showed that this method exhibited a comparable linear range and detection limit with other electrocatalysts.

Next, this method was demonstrated to selectively discriminate UA, DA, AA, G, and A by using a $GO_{-0.75V}$-modified screen-printed electrode couple with a homemade PDMS micro electrochemical cell (Figure 4). The results showed that the electrode exhibited high-resolution discrimination for five biological molecules.

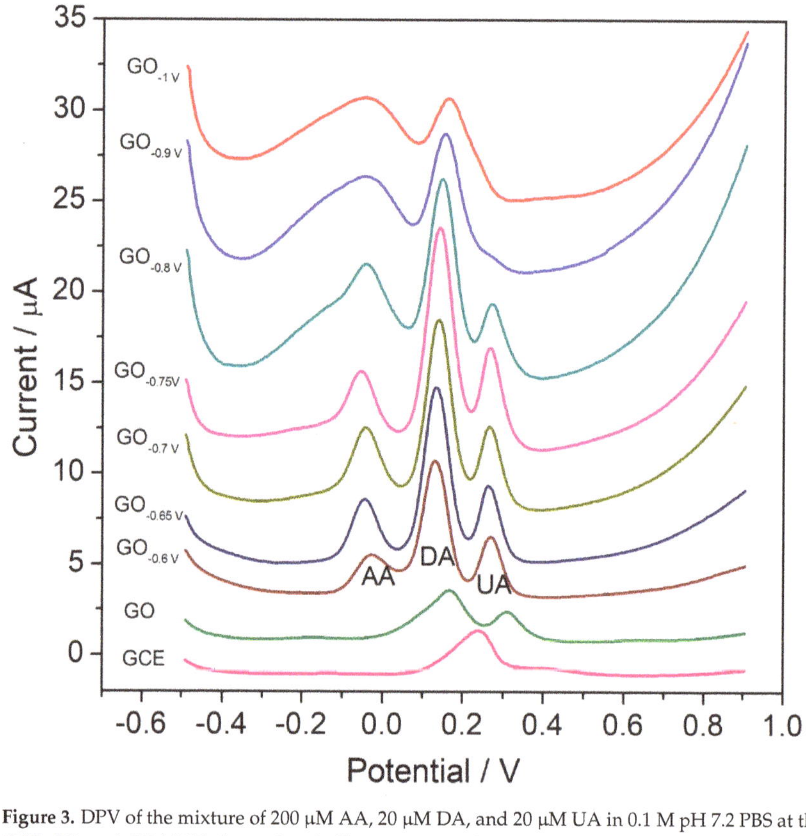

Figure 3. DPV of the mixture of 200 μM AA, 20 μM DA, and 20 μM UA in 0.1 M pH 7.2 PBS at the GCE, GO, and GO/GCE electrochemically treated at different potentials. Conditions: Inc E, 4 mV; Amplitude, 0.05 V; Pulse width, 0.2 s; Sample width, 0.0167 s; Pulse period, 0.5 s.

Table 1. Electrochemical sensing performances for five biological molecules based on $GO_{-0.75V}$ and other electrocatalysts.

Electrocatalysts	Linear Range	Detection Limit	pH	Samples	Ref.
PrGO/PB		AA: 34.7 μM DA: 26.2 μM UA: 8.0 μM	7.4		[22]
TiN-rGO	AA: 50–100 μM DA: 5–175 μM UA: 30–215 μM	AA: DA: 0.159 μM UA: 0.350 μM	7.0	spiked in urine	[23]
PVP-GR	AA: 4.0–1000 μM DA: 0.02–100 μM UA: 0.04–100 μM	AA: 0.8 μM DA: 0.002 μM UA: 0.02 μM	6.0	urine	[24]
h-BN	AA: 30–1000 μM DA: 0.5–150 μM UA: 1–300 μM	AA: 3.77 μM DA: 0.02 μM UA: 0.15 μM	6.0	spiked in urine	[25]
PrGO/MnO_2	AA: 1–800 μM DA: 0.03–45 μM UA: 0.3–80 μM	AA: 1.00 μM DA: 0.02 μM UA: 0.05 μM	6.0	spiked in urine	[26]
CNP	AA: 20–250 μM DA: 10–480 μM UA: 25–2500 μM	AA: 4.25 μM DA: 0.86 μM UA: 0.20 μM	7.0	spiked in serum	[27]
N-rGO	AA: 0.1–4 μM DA: 1–60 μM UA: 1–30 μM	AA: 9.6 μM DA: 0.1 μM UA: 0.2 μM	7.0		[28]
RGO/AuNPs		AA: 9.5 μM DA: 0.14 μM UA: 25 μM	7.4		[29]
MNP/rGO/GCE	AA: 10–100 μM DA: 15 nm–100 μM UA: 10–100 μM	AA: 0.22 μM DA: 0.19 nM UA: 45 nM			[30]
ERGO-pEBT	AA: 10–900 μM DA: 0.5–20 μM UA: 2–70 μM	AA: 0.53 μM DA: 0.009 μM UA: 0.046 μM	6.0	urine	[31]
N-rGO	AA: 550–1500 μM DA: 3–100 μM UA: 20–1000 μM	AA:58 μM DA: 2.4 μM UA: 8.7 μM	7.4		[32]
HCONS	AA: 12.5–1382 μM DA: 1.6–23.6 μM UA: 46.5–806.5 μM	AA: 12.5 μM DA: 0.012 μM UA: 1.5 μM	7.4	serum	[33]
Ni@CHS/rGO	AA: 2–4000 μM DA: 0.25–126 μM UA: 0.25–126 μM	AA: 0.37 μM DA: 0.05 μM UA: 0.05 μM	7.0	spiked in urine	[34]
MoS_2-PGE	G: 15–120 μM A: 15–120 μM	G:0.76 μM A:2.38 μM	7.4		[35]
PPyox/MWNTs-MoS_2/GCE	G: 5–30 μM A: 30–120 μM	G: 1.6 μM A: 1.7 μM	7.0	salmon sperm DNA	[36]
Cu@Ni/MWCNTs	G: 5–180 μM A: 8–150 μM	G: 0.35 μM A: 0.56 μM	4.0	dsDNA from brain tissue of mice	[37]
ds-DNA/ p(L-Cys)/Fe_3O_4 NPs-GO/CPE	G: 0.01–25 μM A: 0.01–30 μM	G: 1.59 μM A: 3.48 μM	6.5	blood DNA	[38]

Table 1. *Cont.*

Electrocatalysts	Linear Range	Detection Limit	pH	Samples	Ref.
p-GLY/GO	G: 0.15–48 μM A: 0.09–103 μM	G: 0.026 μM A: 0.03 μM	3.7	urine and fish sperm DNA	[39]
PDA/MnO_2/IL-G	G: 10–300 μM A: 10–300 μM	G: 0.25 μM A: 0.15 μM	4.0	mouse whole blood	[40]
CuO NPs	G: 1–80 μM A: 1–80 μM	G: 0.687 μM A: 0.472 μM	7.0		[41]
COFS/NH_2-rG/MoS_2	G: 0.5–150 μM A: 1–280 μM	G: 0.51 μM A: 0.44 μM	2.0	herring sperm DNA	[42]
Au/HG	G: 6–500 μM A: 0.6–40 μM	G: 2.5 μM A: 0.42 μM	7.0		[43]
CQD/PAPox	G: 1–65 μM A: 2–70 μM	G: 0.51 μM A: 0.39 μM	3.7	fish sperm DNA	[44]
Cu-CeO_2/ MWCNT/GCE	G: 0.2–6 μM A: 0.1–8 μM	G: 0.128 μM A: 0.062 μM	7.0	a calf thymus DNA	[45]
ZnS NPs/CPE	G: 1–15 μM A: 15–120 μM	G: 0.036 μM A: 0.031 μM	7.4		[46]
βCD/IRGO	G: 0.03–10 μM A: 0.02–7 μM	G: 0.01 μM A: 0.01 μM	7.0		[47]
poly(dopamine)/GR	G: 1–58 μM A: 1–58 μM	G: 0.34 μM A: 0.31 μM	7.0	a calf thymus DNA	[48]
$GO_{-0.75V}$	DA: 1–100 μM UA: 0.2–20 μM G: 0.5–100 μM A: 2–100 μM	DA: 0.5 μM UA: 0.12 μM G: 0.2 μM A: 0.3 μM	7.2		This work

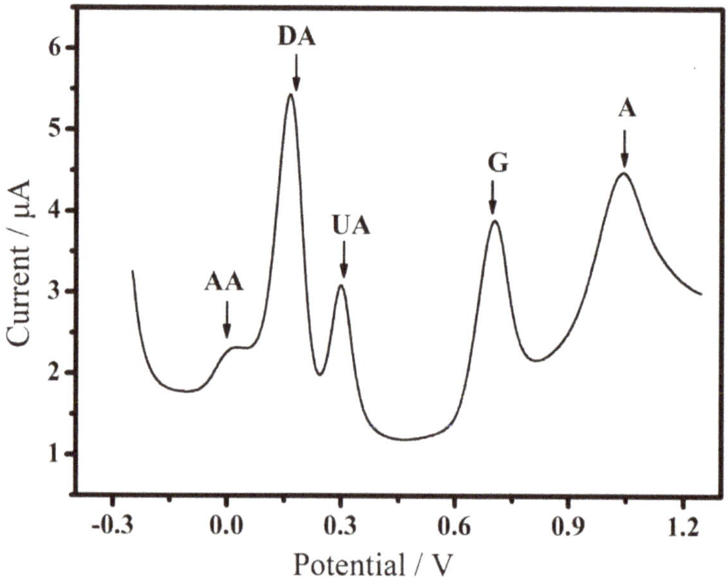

Figure 4. DPV of the mixture of five biological molecules including 200 μM AA, 10 μM DA, 20 μM UA, 20 μM G, and 20 μM A at screen-printed electrode modified with $GO_{-0.75V}$. Conditions are the same as in Figure 3.

3.3. Antibiofouling Property

UA derives from the final metabolism of purine, which is a constituent of nucleic acids, and is finally excreted into the urine by the kidney [49]. Evaluated serum UA level (also called hyperuricemia) could lead to gout and nephrolithiasis, which has recently been found to be related to hypertension, coronary heart disease, heart failure, atrial fibrillation, insulin resistance, and nonalcoholic fatty liver disease [50]. In order to achieve the practical detection of UA in serum, the electrode should possess not only high selectivity, but also excellent antibiofouling ability. First, an $Fe(CN)_6^{3-/4-}$ redox couple was utilized to investigate the binding of electrodes with BSA and antibiofouling property (Figure 5). The electrodes before and after immersing in BSA exhibited distinct differences in $Fe(CN)_6^{3-/4-}$ solution. It is amazing to find that the current of GO-BSA/GCE is about 25 times higher than that of GO/GCE, suggesting that the BSA coating on the electrode surface does not hinder the electron transfer and mass transport. However, the current of $GO_{-0.75V}$-BSA/GCE is only about 1.8 times higher than that of $GO_{-0.75V}$/GCE. The findings indicate that the oxo functionalities are beneficial for the binding of BSA, which may be caused by hydrophilic interaction and hydrogen-bonding interaction. The isoionic point of BSA is about 5, which suggests that BSA exhibits a negative charge at pH 7.2 because of the loss of protons from the imidazole groups of amino acids in BSA [51]. It was considered that the binding of BSA improves the electroactive area of the electrode, and the $Fe(CN)_6^{3-/4-}$ redox couple can transport freely. As a result, the redox current increased dramatically when BSA was attached to the electrode. EIS was carried out to investigate the electrochemical activity of the electrodes, as shown in Figure S6. For GO/GCE and $GO_{-0.75V}$/GCE, the equivalent circuit fitted from the EIS includes a series combination of R1, R2, Ws, CPE1 and CPE2. The electrode surface can be considered as a graphene surface that is covered by the oxygen-containing functional species [52]. In this case, R2 and CPE1 correspond to the graphene surface without oxo functionalities, and CPE2 and Ws correspond to the oxidized surface. R1 and R2 represent the electrolyte resistance and the charge transfer resistance, respectively. Ws is a finite-length Warburg element that represents the diffusion through an oxidized graphene surface. However, additional (R3, CPE3) was observed for GO-BSA/GCE and $GO_{-0.75V}$-BSA/GCE, which is linked with the attached BSA layer. The fitted parameters for different electrodes are listed in Table S1. There is no significant change in the charge transfer resistance (R2) before and after the modification of BSA. However, diffusion resistance (W1-R) decreased when BSA was attached to the surface, indicating that the loading of BSA improved the diffusion of $Fe(CN)_6^{3-/4-}$.

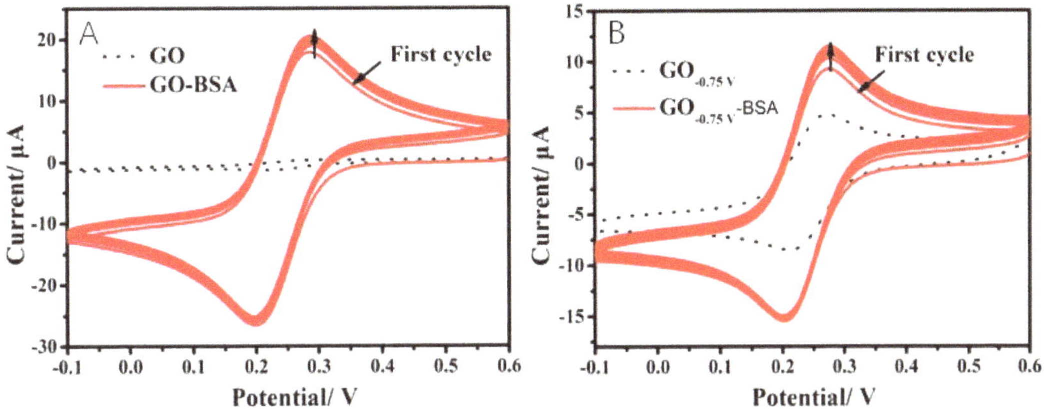

Figure 5. CVs of GO/GCE ((**A**), dotted line), GO-BSA/GCE ((**A**), solid line), $GO_{-0.75V}$/GCE ((**B**), dotted line), and $GO_{-0.75V}$-BSA/GCE ((**B**), solid line) in 5 mM $Fe(CN)_6^{3-/4-}$ redox couple at the scan rate of 25 mV/s.

Similar results were found at 100% serum modified electrode due to the binding of serum proteins (Figure S7). When the potential is more negative than −0.85 V, the current of $GO_{potential}$-serum/GCE is much lower than that of $GO_{potential}$ /GCE. The surface morphology and composition of the electrodes and BSA-modified electrodes were also investigated by XPS (Figure 6) and SEM (Figure S8). The carbon (1 s) spectra for GO-BSA and $GO_{-0.75V}$-BSA can be well fitted to three peaks, suggesting three components of carbon. The peak at 284.7 eV was the bulk C peak ascribed to the alkyl chain. The second peak at approximately 286.4 eV was attributed to the C-S bond, indicating that BSA might be attached to the electrode by thiol groups or disulfide (S-S) presented on BSA. The third peak located at approximately 287.9 eV indicated amide bonds or carboxyl groups. A more clear wrinkled structure was observed for $GO_{-0.75V}$ than $GO_{-0.75V}$-BSA/GCE, which also indicated the attachment of BSA film (Figure S8). The antibiofouling property of the electrodes was also investigated by testing the DPV of the mixture of AA, DA, and UA at the $GO_{-0.75V}$/GCE, $GO_{-0.75V}$-BSA/GCE and $GO_{-0.75V}$-serum/GCE (Figure 7). It was found that the attached BSA did not interfere with the detection of AA, DA, and UA. However, $GO_{-0.75V}$-serum/GCE exhibited a much lower current than that of $GO_{-0.75V}$/GCE, which might be caused by other species that existed in the serum. Figure S9 shows the DPV of $GO_{-0.75V}$/GCE, $GO_{-0.75V}$-BSA/GCE, $GO_{-0.75V}$-serum/GCE and $GO_{-0.75V}$-BSA-UA/GCE in background electrolyte (0.1 M pH 7.2 PBS). An obvious oxidation peak of UA was observed, indicating that UA could be attached to BSA [53].

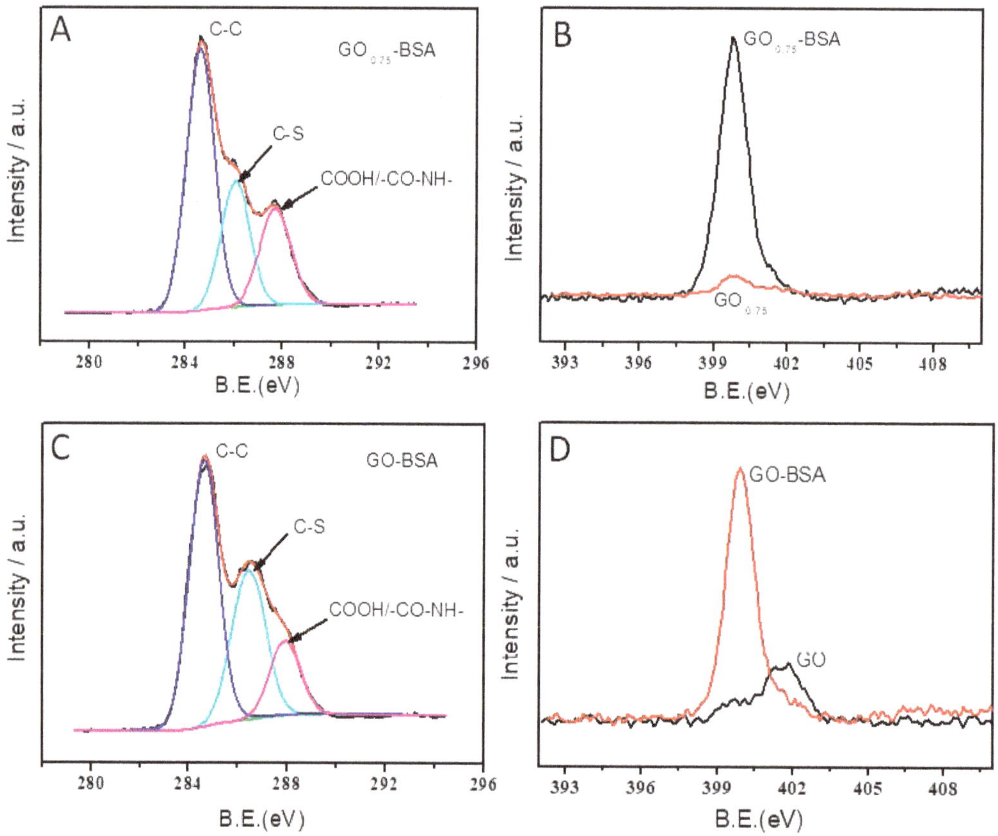

Figure 6. XPS narrow scan of the C1s (**A**,**C**) and N1s (**B**,**D**) region for GO (**D**), $GO_{-0.75V}$ (**B**), $GO_{-0.75V}$-BSA (**A**,**B**) and GO–BSA (**C**,**D**).

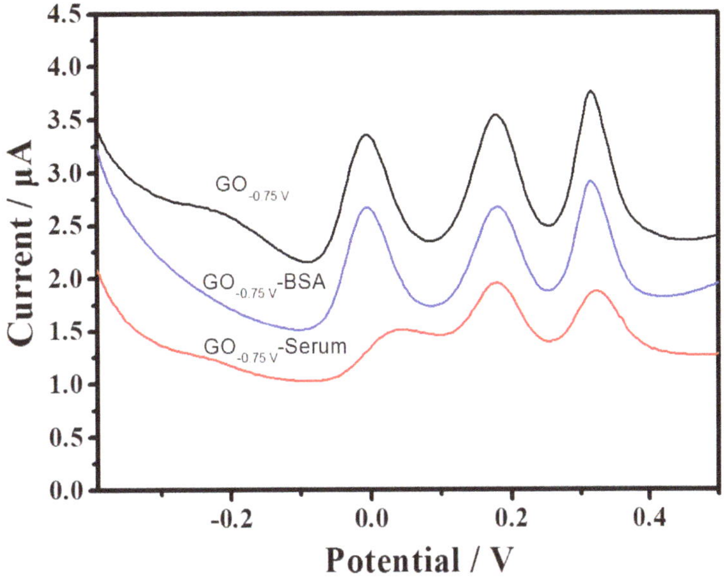

Figure 7. DPV of the mixture of AA, DA, and UA in 0.1 M pH 7.2 PBS at the $GO_{-0.75V}$/GCE, $GO_{-0.75V}$-BSA/GCE and $GO_{-0.75V}$-serum/GCE. Conditions are the same as in Figure 3.

3.4. Analysis of UA in Serum

The GO-based electrochemical sensor was tested to demonstrate the capacity for the detection of UA in practical human serum samples through a standard addition method (Figure S10). The serum sample was diluted with PBS at a 1:40 v/v ratio for direct measurement without any other pretreatment. The detected content of UA in real serum was 160 μM, which is much lower than that (220 μM) obtained by the commercial clinical method.

4. Conclusions

A micro electrochemical sensor was developed for multi-analyte detection based on the electrochemically tuned GO electrode. The key problem of biofouling was investigated and discussed in detail. The results showed that the oxo functionalities present on GO played an important role in electrocatalysis, selectivity, and antibiofouling performance. In this work, we presented a low-cost and simple electrode that possessed both excellent antibiofouling performance and high electrocatalytic activity toward the multi-analyte. Electroactive probe investigation indicated that the attachment of BSA did not hinder the electron transfer but enhanced the diffusion of probes. This electrode was demonstrated to detect the real content of UA in a serum sample. This simple method will provide a promising potential in clinical application.

Supplementary Materials: The following supporting information can be downloaded at: https://www.mdpi.com/article/10.3390/nano12040711/s1, Figure S1: Electrochemical detection of UA in the presence of AA and DA at GO-0.75 V/GCE in 0.1 M pH 7.2 PBS. Inset: calibration plot for the peak current versus UA concentrations. Conditions: Inc E, 4 mV; Amplitude, 0.05 V; Pulse width, 0.2 s; Sample width, 0.0167 s; Pulse period, 0.5 s; Figure S2: Electrochemical detection of DA in the presence of AA and UA at GO-0.75 V/GCE in 0.1 M pH 7.2 PBS. Inset: calibration plot for the peak current versus DA concentrations; Figure S3: DPV of the mixture of G and A in 0.1 M pH 7.2 PBS at the GCE, GO, and GO/GCE electrochemically treated at different potentials; Figure S4: Electrochemical detection of G in the presence of A at GO-0.75 V/GCE in 0.1 M pH 7.2 PBS. Inset: calibration plot for the peak current versus G concentrations; Figure S5: Electrochemical detection of A in the presence of G at GO-0.75 V/GCE in 0.1 M pH 7.2 PBS. Inset: calibration plot for the peak current versus A concentrations; Figure S6: Nyquist diagrams (A and C) and phase angle diagrams vs log of frequency of Bode plots (B and D) for different electrodes. The symbols and solid lines present the experimental and the fitted data, respectively. The insets in phase angle diagrams are the corresponding electrical equivalent circuits. EIS was carried out in 0.1 M KCl containing 5 mM $Fe(CN)_6{}^{3-/4-}$ in the frequency range from 1 MHz to 0.1 Hz at 0.24 V; Figure S7: CVs of 5 mM $Fe(CN)_6{}^{3-/4-}$ redox couple at different GOpotential /GCE (dotted line) and GOpotential-serum /GCE (solid line) at the scan rate of 25 mV/s. A: GO, B: GO-0.6 V, C: GO-0.65 V, D: GO-0.75 V, E:GO-0.85 V, F:GO-0.95 V; Figure S8: SEM images of GO-0.75 V/GCE (A) and GO-0.75 V-BSA/GCE (B); Figure S9: DPV of GO-0.75 V/GCE, GO-0.75 V-BSA/GCE, GO-0.75 V-serum/GCE and GO-0.75 V-BSA-UA/GCE in background electrolyte (0.1 M pH 7.2 PBS); Figure S10: Standard addition method for the measurement of UA in 40-folds diluted serum; Table S1: Obtained parameters from fitting for different electrodes.

Author Contributions: Conceptualization, B.Y.; Data curation, B.Y., L.G. and C.X.; Formal Analysis, B.Y.; Funding acquisition, B.Y.; Investigation, L.G., B.Y. and G.L. (Gang Li); Methodology, B.Y.; Project administration, B.Y. and C.X.; Supervision, B.Y.; Validation, B.Y. and L.G.; Visualization, B.Y. and L.G.; Writing—original draft, B.Y.; Writing—review & editing, B.Y. and G.L. (Gang Liu). All authors have read and agreed to the published version of the manuscript.

Funding: This research was funded by the Natural Science Foundation of Shandong Province, China (No.ZR2020MB058).

Institutional Review Board Statement: Not applicable.

Informed Consent Statement: Not applicable.

Data Availability Statement: The authors confirm that the data supporting the findings of this study are available within the article [and/or] its supplementary materials.

Acknowledgments: This work was funded by the Natural Science Foundation of Shandong Province (No.ZR2020MB058).

Conflicts of Interest: The authors declare no conflict of interest.

Abbreviations

PB	Prussian blue
TiN-rGO	titanium nitride-reduced graphene oxide
N-rGO	nitrogen-doped reduced graphene oxide
PVP	polyvinylpyrrolidone
GR	graphene
h-BN	hexagonal boron nitride
PrGO/MnO_2	poly(3,4-ethylenedioxythiophene)/reduced graphene oxide/manganese dioxide
AuNPs	gold nanoparticles
MNP	magnetic nanoparticles
ERGO-pEBT	electrochemically reduced graphene oxide-poly(eriochrome black T)
HCONS	hexagonal cobalt oxide nanosheets
Ni@CHS/rGO	Ni@carbon hollow nanosphere/reduced graphene oxide
PGE	pencil graphite electrode
PPyox	polypyrrole
p-GLY/GO	poly(glycine)/graphene oxide composite based sensor
CQD/PAPox	carbon quantum dots (CQDs) and overoxidized poly(2-aminopyridine) (PAPox)
βCD/IRGO	β-cyclodextrin functionalized reduced graphite oxide with the aid of ionic liquid
CTME	carbon-thread microelectrode

References

1. Del Río, J.S.; Henry, O.Y.; Jolly, P.; Ingber, D.E. An antifouling coating that enables affinity-based electrochemical biosensing in complex biological fluids. *Nat. Nanotechnol.* **2019**, *14*, 1143–1149. [CrossRef] [PubMed]
2. Jiang, C.; Wang, G.; Hein, R.; Liu, N.; Luo, X.; Davis, J.J. Antifouling strategies for selective in vitro and in vivo sensing. *Chem. Rev.* **2020**, *120*, 3852–3889. [CrossRef] [PubMed]
3. Barfidokht, A.; Gooding, J.J. Approaches toward allowing electroanalytical devices to be used in biological fluids. *Electroanalysis* **2014**, *26*, 1182–1196. [CrossRef]

4. Russo, M.J.; Han, M.; Desroches, P.E.; Manasa, C.S.; Dennaoui, J.; Quigley, A.F.; Kapsa, R.M.I.; Moulton, S.E.; Guijt, R.M.; Greene, G.W.; et al. Antifouling strategies for electrochemical biosensing: Mechanisms and performance toward point of care based diagnostic applications. *ACS Sens.* **2021**, *6*, 1482–1507. [CrossRef]
5. Silva, T.A.; Khan, M.R.K.; Fatibello-Filho, O.; Collinson, M.M. Simultaneous electrochemical sensing of ascorbic acid and uric acid under biofouling conditions using nanoporous gold electrodes. *J. Electroanal. Chem.* **2019**, *846*, 113160. [CrossRef]
6. Brisebois, P.P.; Siaj, M. Harvesting graphene oxide–years 1859 to 2019: A review of its structure, synthesis, properties and exfoliation. *J. Mater. Chem. C* **2020**, *8*, 1517–1547. [CrossRef]
7. Kim, J.; Cote, L.J.; Kim, F.; Yuan, W.; Shull, K.R.; Huang, J. Graphene oxide sheets at interfaces. *J. Am. Chem. Soc.* **2010**, *132*, 8180–8186. [CrossRef]
8. Joshi, D.J.; Koduru, J.R.; Malek, N.I.; Hussain, C.M.; Kailasa, S.K. Surface modifications and analytical applications of graphene oxide: A review. *Trac-Trend. Anal. Chem.* **2021**, *144*, 116448. [CrossRef]
9. Agarwal, V.; Zetterlund, P.B. Strategies for reduction of graphene oxide–A comprehensive review. *Chem. Eng. J.* **2021**, *405*, 127018. [CrossRef]
10. Lavin-Lopez, M.D.P.; Paton-Carrero, A.; Sanchez-Silva, L.; Valverde, J.L.; Romero, A. Influence of the reduction strategy in the synthesis of reduced graphene oxide. *Adv. Powder Technol.* **2017**, *28*, 3195–3203. [CrossRef]
11. Renteria, J.A.Q.; Ruiz-Garcia, C.; Sauvage, T.; Chazaro-Ruiz, L.F.; Rangel-Mendez, J.R.; Ania, C.O. Photochemical and electrochemical reduction of graphene oxide thin films: Tuning the nature of surface defects. *Phys. Chem. Chem. Phys.* **2020**, *22*, 20732–20743. [CrossRef] [PubMed]
12. Kim, H.W.; Ross, M.B.; Kornienko, N.; Zhang, L.; Guo, J.; Yang, P.; McCloskey, B.D. Efficient hydrogen peroxide generation using reduced graphene oxide-based oxygen reduction electrocatalysts. *Nat. Catal.* **2018**, *1*, 282–290. [CrossRef]
13. Tan, X.; Tahini, H.A.; Smith, S.C. Understanding the high activity of mildly reduced graphene oxide electrocatalysts in oxygen reduction to hydrogen peroxide. *Mater. Horiz.* **2019**, *6*, 1409–1415. [CrossRef]
14. Kim, H.W.; Park, H.; Roh, J.S.; Shin, J.E.; Lee, T.H.; Zhang, L.; McCloskey, B.D. Carbon defect characterization of nitrogen-doped reduced graphene oxide electrocatalysts for the two-electron oxygen reduction reaction. *Chem. Mater.* **2019**, *31*, 3967–3973. [CrossRef]
15. Yuan, B.; Zeng, X.; Xu, C.; Liu, L.; Ma, Y.; Zhang, D.; Fan, Y. Electrochemical modification of graphene oxide bearing different types of oxygen functional species for the electro-catalytic oxidation of reduced glutathione. *Sens. Actuators B-Chem.* **2013**, *184*, 15–20. [CrossRef]
16. Zhang, D.; Xu, C.; Li, S.; Zhang, R.; Yan, H.; Miao, H.; Yong, F.; Yuan, B. Electrochemically controlling oxygen functional groups in graphene oxide for the optimization in the electro-catalytic oxidation of dihydroxybenzene isomers and L-methionine. *J. Electroanal. Chem.* **2014**, *717*, 219–224. [CrossRef]
17. Uhm, S.; Tuyen, N.H.; Lee, J. Controlling oxygen functional species of graphene oxide for an electro-oxidation of L-ascorbic acid. *Electrochem. Commun.* **2011**, *13*, 677–680. [CrossRef]
18. Hui, K.H.; Ambrosi, A.; Pumera, M.; Bonanni, A. Improving the analytical performance of graphene oxide towards the assessment of polyphenols. *Chem.-Eur. J.* **2016**, *22*, 3830–3834. [CrossRef]
19. Maccaferri, G.; Zanardi, C.; Xia, Z.Y.; Kovtun, A.; Liscio, A.; Teizi, F.; Palermo, V.; Seeber, R. Systematic study of the correlation between surface chemistry, conductivity and electrocatalytic properties of graphene oxide nanosheets. *Carbon* **2017**, *120*, 165–175. [CrossRef]
20. Zheng, L.; Wu, S.; Lin, X.; Nie, L.; Rui, L. Selective determination of uric acid by using a β-cyclodextrin modified electrode. *Electroanalysis* **2001**, *13*, 1351–1354. [CrossRef]
21. Zen, J.M.; Chen, P.J. A selective voltammetric method for uric acid and dopamine detection using clay-modified electrodes. *Anal. Chem.* **1997**, *69*, 5087–5093. [CrossRef]
22. Dos Santos, P.L.; Katic, V.; Toledo, K.C.F.; Bonacin, J.A. Photochemical one-pot synthesis of reduced graphene oxide/Prussian blue nanocomposite for simultaneous electrochemical detection of ascorbic acid, dopamine, and uric acid. *Sens. Actuators B-Chem.* **2018**, *255*, 2437–2447. [CrossRef]
23. Feng, J.; Li, Q.; Cai, J.P.; Yang, T.; Chen, J.H.; Hou, X.M. Electrochemical detection mechanism of dopamine and uric acid on titanium nitride-reduced graphene oxide composite with and without ascorbic acid. *Sens. Actuators B-Chem.* **2019**, *298*, 126872. [CrossRef]
24. Wu, Y.; Deng, P.; Tian, Y.; Feng, J.; Xiao, J.; Li, J.; Liu, J.; Li, G.; He, Q. Simultaneous and sensitive determination of ascorbic acid, dopamine and uric acid via an electrochemical sensor based on PVP-graphene composite. *J. Nanobiotechnol.* **2020**, *18*, 112. [CrossRef]
25. Li, Q.; Huo, C.R.; Yi, K.; Zhou, L.L.; Su, L.; Hou, X.M. Preparation of flake hexagonal BN and its application in electrochemical detection of ascorbic acid, dopamine and uric acid. *Sens. Actuators B-Chem.* **2018**, *260*, 346–356. [CrossRef]
26. Tukimin, N.; Abdullah, J.; Sulaiman, Y. Electrodeposition of poly(3,4-ethylenedioxythiophene)/reduced graphene oxide/manganese dioxide for simultaneous detection of uric acid, dopamine and ascorbic acid. *J. Electroanal. Chem.* **2018**, *820*, 74–81. [CrossRef]
27. Huang, Y.Q.; Zang, Y.; Ruan, S.J.; Zhang, Y.Y.; Gao, P.P. A high efficiency N, P doped porous carbon nanoparticles derived from lotus leaves for simultaneous electrochemical determination of ascorbic acid, dopamine, and uric acid. *Microchem. J.* **2021**, *165*, 106152. [CrossRef]

28. Zhang, H.Y.; Liu, S. Electrochemical sensors based on nitrogen-doped reduced graphene oxide for the simultaneous detection of ascorbic acid, dopamine and uric acid. *J. Alloys Compd.* **2020**, *842*, 155873. [CrossRef]
29. Lee, C.S.; Yu, S.H.; Kim, T.H. One-step electrochemical fabrication of reduced graphene oxide/gold nanoparticles nanocomposite-modified electrode for simultaneous detection of dopamine, ascorbic acid, and uric acid. *Nanomaterials* **2017**, *8*, 17. [CrossRef]
30. Rosli, A.R.M.; Yusoff, F.; Loh, S.H.; Yusoff, H.M.; Jamil, M.M.A.; Shamsudin, S.H. Simultaneous electrochemical detection of ascorbic acid, dopamine, and uric acid at magnetic nanoparticles/reduced graphene oxide modified electrode. *J. Teknol.* **2021**, *83*, 85–92. [CrossRef]
31. Mohammed, N.; Abdullah, J.; Kamaruzaman, S.; Saiman, M.; Sulaiman, Y. Electrochemical reduced graphene oxide-poly(eriochrome black T)/gold nanoparticles modified glassy carbon electrode for simultaneous determination of ascorbic acid, dopamine and uric acid. *Arab. J. Chem.* **2018**, *11*, 1301–1312. [CrossRef]
32. Minta, D.; Gonzalez, Z.; Wiench, P.; Gryglewicz, S.; Gryglewicz, G. N-Doped reduced graphene oxide/gold nanoparticles composite as an improved sensing platform for simultaneous detection of dopamine, ascorbic acid, and uric acid. *Sensors* **2020**, *20*, 4427. [CrossRef] [PubMed]
33. Zahed, M.A.; Barman, S.C.; Toyabur, R.M.; Sharifuzzaman, M.; Xuan, X.; Nah, J.; Park, J.Y. Ex situ hybridized hexagonal cobalt oxide nanosheets and RGO@MWCNT based nanocomposite for ultra-selective electrochemical detection of ascorbic acid, dopamine, and uric Acid. *J. Electrochem. Soc.* **2019**, *166*, B304–B311. [CrossRef]
34. Zhang, L.; Zhang, J. 3D hierarchical bayberry-like Ni@carbon hollow nanosphere/rGO hybrid as a new interesting electrode material for simultaneous detection of small biomolecules. *Talanta* **2018**, *178*, 608–615. [CrossRef] [PubMed]
35. Vishnu, N.; Badhulika, S. Single step grown MoS_2 on pencil graphite as an electrochemical sensor for guanine and adenine: A novel and low cost electrode for DNA studies. *Biosens. Bioelectron.* **2018**, *124*, 122–128. [CrossRef] [PubMed]
36. Geng, X.T.; Bao, J.; Huang, T.; Wang, X.J.; Hou, C.J.; Hou, J.Z.; Samalo, M.; Yang, M.; Huo, D.Q. Electrochemical Sensor for the Simultaneous Detection of Guanine and Adenine Based on a PPyox/MWNTs-MoS_2 Modified Electrode. *J. Electrochem. Soc.* **2019**, *166*, B498–B504. [CrossRef]
37. Wang, D.; Huang, B.; Liu, J.; Guo, X.; Abudukeyoumu, G.; Zhang, Y.; Ye, B.C.; Li, Y. A novel electrochemical sensor based on Cu@Ni/MWCNTs nanocomposite for simultaneous determination of guanine and adenine. *Biosens. Bioelectron.* **2018**, *102*, 389–395. [CrossRef]
38. Arvand, M.; Sanayeei, M.; Hemmati, S. Label-free electrochemical DNA biosensor for guanine and adenine by ds-DNA/poly(L-cysteine)/Fe3O4 nanoparticles-graphene oxide nanocomposite modified electrode. *Biosens. Bioelectron.* **2018**, *102*, 70–79. [CrossRef]
39. He, S.; He, P.; Zhang, X.; Zhang, X.; Liu, K.; Jia, L.; Dong, F. Poly(glycine)/graphene oxide modified glassy carbon electrode: Preparation, characterization and simultaneous electrochemical determination of dopamine, uric acid, guanine and adenine. *Anal. Chim. Acta* **2018**, *1031*, 75–82. [CrossRef]
40. Zhang, S.; Zhuang, X.; Chen, D.; Luan, F.; He, T.; Tian, C.; Chen, L. Simultaneous voltammetric determination of guanine and adenine using MnO2 nanosheets and ionic liquid-functionalized graphene combined with a permeation-selective polydopamine membrane. *Microchim. Acta* **2019**, *186*, 1–10. [CrossRef]
41. Siddegowda, K.; Mahesh, N.B.; Kumara, S. Fabrication of copper oxide nanoparticles modified carbon paste electrode and its application in simultaneous electroanalysis of guanine, adenine and thymine. *Sens. Actuators A-Phys.* **2018**, *280*, 277–286.
42. Zhao, X.; Guo, H.; Xue, R.; Wang, M.Y.; Guan, Q.X.; Fan, T.; Yang, W.H.; Yang, W. Electrochemical sensing and simultaneous determination of guanine and adenine based on covalent organic frameworks/NH2-rG/MoS2 modified glassy carbon electrode. *Microchem. J.* **2021**, *160*, 105759. [CrossRef]
43. Gao, S.H.; Li, H.J.; Li, M.J.; Li, C.P.; Qian, L.R.; Yang, B.H. A gold-nanoparticle/horizontal-graphene electrode for the simultaneous detection of ascorbic acid, dopamine, uric acid, guanine, and adenine. *J.Solid State Electrochem.* **2018**, *22*, 3245–3254. [CrossRef]
44. He, S.; He, P.; Zhang, X.; Zhang, X.; Dong, F.; Jia, L.; Du, L.; Lei, H. Simultaneous voltammetric determination of guanine and adenine by using a glassy carbon electrode modified with a composite consisting of carbon quantum dots and overoxidized poly(2-aminopyridine). *Microchim. Acta* **2018**, *185*, 107. [CrossRef]
45. Aktürk, M.; Karabiberoğlu, S.U.; Dursun, Z. Fabrication of Cu−CeO_2 coated multiwall carbon nanotube composite electrode for simultaneous determination of guanine and adenine. *Electroanalysis* **2018**, *30*, 238–249. [CrossRef]
46. Mahanthappa, M.; Kottam, N.; Yellappa, S. Electrocatalytic performance of a zinc sulphide nanoparticles-modified carbon paste electrode for the simultaneous determination of acetaminophen, guanine and adenine. *Anal. Methods* **2018**, *10*, 1362–1371. [CrossRef]
47. Wang, M.L.; Cui, M.Z.; Liu, W.F.; Liu, X.G.; Xu, B.S. Facile Synthesis of Cyclodextrin Functionalized Reduced Graphite Oxide with the Aid of Ionic Liquid for Simultaneous Determination of Guanine and Adenine. *Electroanalysis* **2018**, *30*, 842–851. [CrossRef]
48. Zhang, N.; Zhong, W.L.; Zhang, Z.Q.; Fan, Y.W. Electrochemical behaviours of guanine and adenine and their simultaneous determination using a three-dimensional porous poly(dopamine)/Reduced Graphene Oxide-Modified Electrode. *Int. J. Electrochem. Sci.* **2020**, *15*, 5287–5295. [CrossRef]
49. Kim, I.; Im Kim, Y.; Lee, S.W.; Jung, H.G.; Lee, G.; Yoon, D.S. Highly permselective uric acid detection using kidney cell membrane–functionalized enzymatic biosensors. *Biosens. Bioelectron.* **2021**, *190*, 113411. [CrossRef]
50. Lee, S.J.; Oh, B.K.; Sung, K.C. Uric acid and cardiometabolic diseases. *Clin. Hypertens.* **2020**, *26*, 1–7. [CrossRef]

51. Zhou, L.; Wang, K.; Wu, Z.; Dong, H.; Sun, H.; Cheng, X.; Zhang, H.; Zhou, H.; Jia, C.; Jin, Q.; et al. Investigation of controllable nanoscale heat-denatured bovine serum albumin films on graphene. *Langmuir* **2016**, *32*, 12623–12631. [CrossRef] [PubMed]
52. Casero, E.; Parra-Alfambra, A.M.; Petit-Domínguez, M.D.; Pariente, F.; Lorenzo, E.; Alonso, C. Differentiation between graphene oxide and reduced graphene by electrochemical impedance spectroscopy (EIS). *Electrochem. Commun.* **2012**, *20*, 63–66. [CrossRef]
53. Makarska-Bialokoz, M.; Lipke, A. Study of the binding interactions between uric acid and bovine serum albumin using multiple spectroscopic techniques. *J. Mol. Liq.* **2019**, *276*, 595–604. [CrossRef]

nanomaterials

Article

Fabrication and Conductive Mechanism Analysis of Stretchable Electrodes Based on PDMS-Ag Nanosheet Composite with Low Resistance, Stability, and Durability

Chengwei Li, Kai Huang, Tingkang Yuan, Tianze Cong, Zeng Fan and Lujun Pan *

School of Physics, Dalian University of Technology, No. 2 Linggong Road, Ganjingzi District, Dalian 116024, China; chengweili@dlut.edu.cn (C.L.); hk987664186@gmail.com (K.H.); 1424067395@mail.dlut.edu.cn (T.Y.); ctz@mail.dlut.edu.cn (T.C.); fanzeng@dlut.edu.cn (Z.F.)
* Correspondence: lpan@dlut.edu.cn

Abstract: A flexible and stretchable electrode based on polydimethylsiloxane (PDMS)-Ag nanosheet composite with low resistance and stable properties has been investigated. Under the synergistic effect of the excellent flexibility and stretchability of PDMS and the excellent electrical conductivity of Ag nanosheets, the electrode possesses a resistivity as low as 4.28 Ωm, a low resistance variation in the 0–50% strain range, a stable electrical conductivity over 1000 cycles, and a rapid recovery ability after failure caused by destructive large stretching. Moreover, the conductive mechanism of the flexible electrode during stretching is explained by combining experimental tests, theoretical models of contact point-tunneling effect, and finite element simulation. This research provides a simple and effective solution for the structure design and material selection of flexible electrodes, and an analytical method for the conductive mechanism of stretchable electrodes, which has potential for applications in flexible electronic devices, smart sensing, wearable devices, and other fields.

Keywords: stretchable electrodes; PDMS; Ag; contact point; tunneling effect

Citation: Li, C.; Huang, K.; Yuan, T.; Cong, T.; Fan, Z.; Pan, L. Fabrication and Conductive Mechanism Analysis of Stretchable Electrodes Based on PDMS-Ag Nanosheet Composite with Low Resistance, Stability, and Durability. *Nanomaterials* **2022**, *12*, 2628. https://doi.org/10.3390/nano12152628

Academic Editor: Cheol-Min Park

Received: 2 July 2022
Accepted: 28 July 2022
Published: 30 July 2022

1. Introduction

In recent years—with the rapid development of the flexible electronics [1–3], smart wearable devices [4–7], and smart robotics [8,9]—flexible electronic devices have been receiving more and more attention [10–13]. As one of the components of these smart electronic devices, stretchable and flexible electrodes are of great importance in structural design, material selection, and electrical conductivity. In terms of structural design, some scientific studies have fabricated stretchable electrodes using materials with poor stretching properties by specific pattern design or structure construction [14–17], which can be stretched through the deformation of the internal structure of the electrodes, and in this process, the conductive properties of the internal material will not change greatly. Tang et al. developed an Ag wrinkled thin film through the method of pre-stretching and vacuum deposition, and made it into a stretchable electrode. The electrode can maintain a stable electrical conductivity within 100% strain value [18]. Fan et al. proposed a structural strategy that can improve the stretching properties of the material. The research uses the fractal design of the serpentine shapes to topologize the one-dimensional linear structures, which is not stretchable, into loop structures and then into branch-like meshes. As a result, the structure is capable of stretching in multiple directions [19]. Jeong et al. designed a biocompatible metal-patterned porous polydimethylsiloxane (PDMS) electrode with high flexibility and stretchability. In this special structure of the electrode, a large number of pores can disperse the external stress, allowing the electrode to maintain good electrical conductivity during the stretching process [20].

In terms of material selection, some stretchable conductive materials are used to fabricate stretchable electrodes [21–24]. Wang et al. combined poly (3,4-ethylenedioxythiophene):

poly (styrenesulfonate) (PEDOT:PSS) with ionic additives to produce a stretchable electrode. Taking advantage of the good electrical conductivity of PEDOT:PSS and the high fracture strain of the ionic additives, the electrode maintains good electrical conductivity under strain [25]. Zhu et al. fabricated a stretchable fiber-shaped electrode by injecting eutectic gallium–indium (EGaln) into the hollow elastic yarns. Utilizing the property of EGaln as liquid alloy at room temperature, the composite yarn has good conductivity even when deformation occurs [26].

In the above two kinds of strategies, conductive materials are required—such as carbon nanotubes, graphene, and other carbon nanomaterials—as well as gold, silver, and other metal materials. As an electrode material, the material is required to own excellent electrical conductivity, not just the ability to conduct electricity. Moreover, the treatment of conductive materials and structures requires some post-processing methods—such as photolithography, ion implantation, thin film deposition, electroplating, UV treatment, etc.—which increases the cost and difficulty of electrode preparation. As a common flexible electrode material, Ag nanomaterials play an important role in many studies, such as 0-dimensional Ag nanoparticles [27–30], 1-dimensional Ag nanowires [31–36], etc. In some studies related to Ag nanomaterial electrodes, it has been found that the contact resistance at junctions between nanomaterials is one of the main issues affecting their electrical conductivity [37]. Aiming at this issue, some post-treatment strategies have been proposed, such as applying heat [38], pressure [39], capillary force [40], chemical reactions [41], laser nano-welding [32,42], nano-joining at the junctions [43], etc.

Herein, in order to fabricate a flexible and stretchable electrode, a composite conductive structure combining Ag nanosheets and PDMS has been investigated. As a metal material with excellent electrical conductivity, Ag is regarded as one of the candidates for electrode material. Compared with Ag nanoparticles and Ag nanowires, Ag nanosheets show the advantages of physical contact stability as electrode materials due to their surface-to-surface contact with each other, which makes them superior to point-to-point and line-to-line contacts. Using PDMS as an elastic carrier for Ag nanosheets can greatly increase the stretchability, allowing them to retain the good conductivity during stretching. In addition, the contact point theory and the tunneling effect theory were applied to reveal the conductive mechanism of the composite electrode during the stretching process. Combined with the method of finite element simulation, the relationship between the structural parameters of the Ag nanosheets inside the electrode and its conductive properties was explored. This study not only designs a flexible and stretchable electrode structure, but also reveals the structure–effect relationship between PDMS-Ag composite and the conductivity, providing a simple and effective solution for the field of flexible electrodes.

2. Experimental Section

2.1. Materials

The silver glue used in this study was purchased from the Taobao online store of Alibaba Network Technology Co., Ltd. (Hangzhou, China). The silver glue (Model: DJ-F4) has a silver content of 68 wt %, a sheet resistance of less than 0.01 Ω/sq, a density of 1.46–1.48 g/cm^3 at 25 °C, a silver monomer diameter of less than 10 μm, a viscosity of more than 20,000 CPS, and a weight dilution ratio of 1:0.7–0.8 (conductive agent to diluent). For the preparation of the elastic substrate, a mixture of the base and curing agent for polydimethylsiloxane (PDMS, Dow Corning Sylgard 184) with a mass ratio of 10:1 was prepared.

2.2. Fabrication of the Stretchable Electrode

For the fabrication of the stretchable electrode, the PDMS film and a mixture of PDMS and silver glue were used as the flexible substrate and the conductive material, respectively. The detailed preparation process is as follows. The PDMS mixed with the curing agent in advance was dripped into a square mold, and its thickness was controlled to 1 mm, then the air bubbles were removed in a vacuum chamber and the PDMS was dried in an oven at 60 °C for 2 h to obtain a solid PDMS film. Then, the film was cut into a size of 2 cm × 1 cm × 1 mm to obtain a flexible substrate. In order to mix PDMS with Ag nanosheets uniformly, the colloid in silver glue was used as a dispersion medium, thus improving the dispersion of Ag nanosheets in PDMS. In detail, the silver glue and the prepared PDMS were physically mixed in the mass ratios of 1:0, 1:0.3, 1:0.4, 1:0.5, 1:0.6, 1:0.7, and 1:0.8, respectively. The composite PDMS-Ag conductor can be obtained after stirring uniformly. After that, the composite conductor was smeared on the surface of the PDMS substrate, and after removing the air bubbles and drying at 60 °C for 2 h, a composite electrode with a size of about 1 cm × 0.5 cm × 0.2 mm was formed. The schematic of the stretchable electrode is shown in Figure 1.

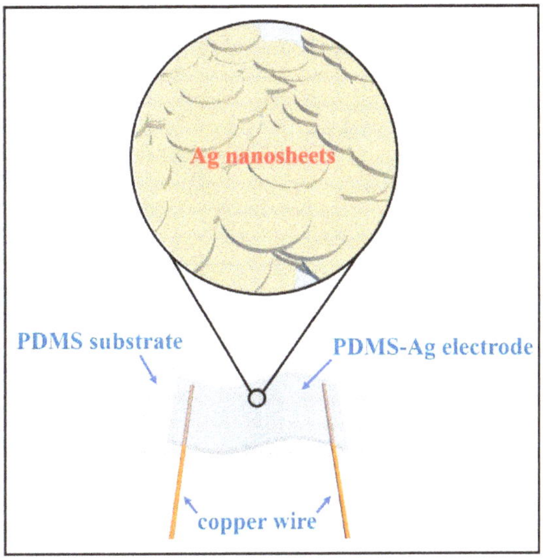

Figure 1. Schematic of the stretchable electrode based on PDMS-Ag nanosheets.

2.3. Characterization and Measurement of the Stretchable Electrode

Optical microscopy (SemiShare SE-4, Shenzhen SemiShare Technology Co., Ltd., Shenzhen, China) and scanning electron microscopy (JCM-5000, JEOL Beijing Technology and Trade Co., Ltd., Beijing, China) were used to observe the micro-structure of the stretchable electrode. A universal material testing machine (YL-S70, Guangzhou Aipeisen Instruments Co., Ltd., Guang Zhou, China) was used to apply a programmed stretching process. A high precision source meter (Agilent B2902A, Agilent Technologies Co., Ltd., Santa Clara, CA, USA) was used to measure the resistance change of the electrode.

2.4. Simulation Methods

In order to investigate the electrical conductivity of the composite electrode containing different ratios of Ag and PDMS to help understand the conductive mechanism, the simulation of finite element analysis was performed using the software of COMSOL Multiphysics (version 5.4). Specifically, the potential distribution and current distribution inside the electrodes with different ratios were simulated by constructing a random distribution model of Ag nanosheets.

3. Results and Discussion

3.1. Morphology Characterization and Basic Electrical Properties of the Stretchable Electrodes

In this study, six stretchable electrodes with different mass ratios of silver glue and PDMS were selected as the research objects, under the premise that the mass of silver glue was fixed at 30 mg, and the mass ratios between silver glue and PDMS were set as 1:0, 1:0.3, 1:0.4, 1:0.5, 1:0.6, and 1:0.8, respectively. The morphology characterization of the six stretchable electrodes were observed by SEM (Figure 2) and optical microscopy (Figure S1). It can be seen that the Ag nanosheets are uniformly dispersed in the PDMS. With the gradual increase in the PDMS content, the density of the Ag nanosheets gradually decreases, and the phenomenon that the Ag nanosheets are separated from each other by the gully-like PDMS becomes more and more obvious.

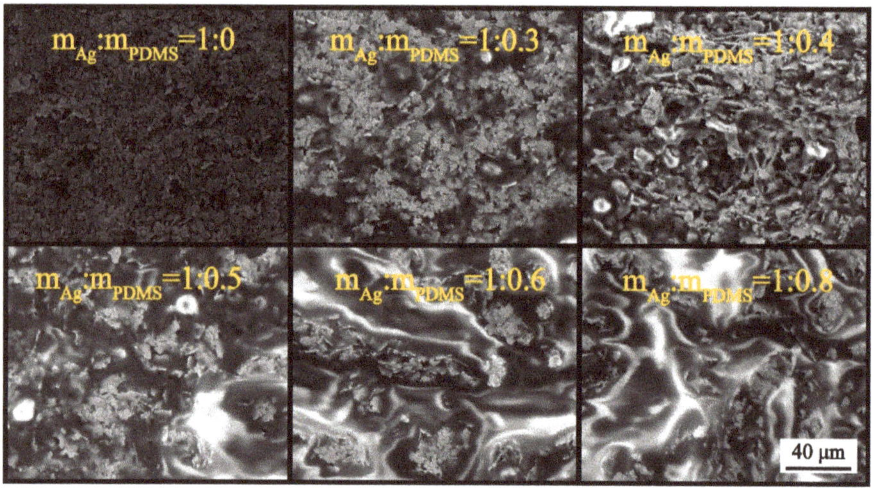

Figure 2. Morphology characterization of the stretchable electrodes based on PDMS-Ag nanosheets with different mass ratios.

According to the statistics of the size of PDMS-Ag composite electrode (Table S1) and the measurement of the initial resistance (Figure 3a), the resistivity of the electrodes with different mass ratios can be calculated, as shown in Figure 3b. It is observed that when the mass ratio of silver glue to PDMS varies from 1:0 to 1:0.6, the initial resistances are all within 6 Ω, and the resistivity increases linearly. When the mass ratio of silver glue to PDMS is 1:0.8, the initial resistance can reach 21.2 Ω, and the resistivity produces a sudden change to 21.3×10^{-4} Ω·m. It indicates that with the increase in the PDMS content, PDMS causes a negative effect on the silver conductive pathway, which hinders the conductivity of Ag nanosheets.

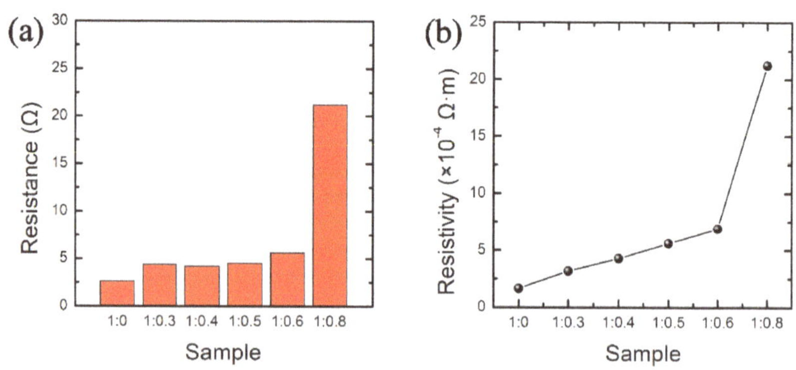

Figure 3. Comparison of (**a**) initial resistance and (**b**) resistivity of PDMS-Ag electrodes with different mass ratios.

3.2. Conductive Performance Test of PDMS-Ag Electrodes during Stretching

The conductive performance of the PDMS-Ag electrodes with different mass ratios during stretching is shown in Figure 4. It is observed that the resistance of all these electrodes increases with the strain variable. By comparing the conductive performances of different electrodes during the stretching process, it can be found that the maximum strain value of the electrodes with a mass ratio of silver glue to PDMS of 1:0 and 1:0.8 is less than 15%, and the resistance shows a rapidly increasing trend. This is because when the mass ratio of silver glue to PDMS is 1:0, the electrode is completely composed of Ag nanosheets, and a large number of Ag nanosheets form an Ag nanosheet film, whose mechanical properties are similar to that of the Ag thin film, and the stretchability is very poor. Therefore, the conductivity of the electrode can be greatly influenced by a small strain. When the mass ratio of silver glue to PDMS is 1:0.8, the PDMS content inside the composite electrode is very high, and the large amount of PDMS separates the limited number of Ag nanosheets from each other, so that the number of initial conductive pathways is very small. When the strain occurs, the original limited number of conductive pathways are further damaged, resulting in the loss of conductivity within 15% of the strain value. Similarly, low contents of PDMS cause the composite electrode to behave Ag film-like property, resulting in the process from integrity to destruction within a relatively small strain range. High contents of PDMS makes for poor conductive ability of the composite electrode, resulting in an easily damaged conductive ability when subjected to a small strain. Importantly, when the mass ratio of silver glue to PDMS is 1:0.4, the conductive performance during stretching is the most stable, and the resistance change is controlled to about 20 Ω within a strain range of more than 50%. Therefore, the composite electrode with this ratio is the optimal electrode under this structural system.

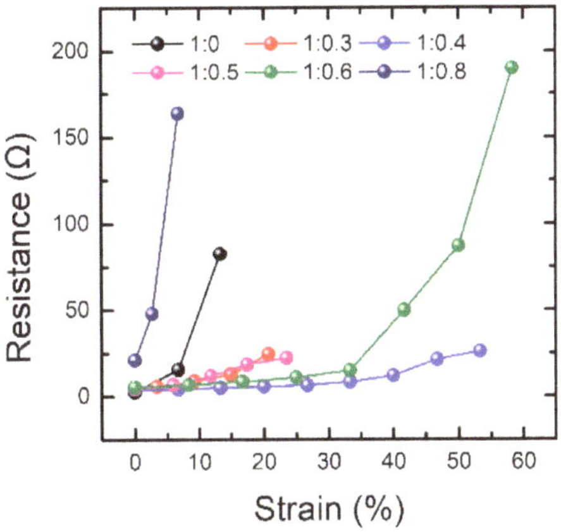

Figure 4. Comparison of the conductive performance of the electrodes with different ratios of silver glue and PDMS during stretching.

For the optimal electrode with a mass ratio of silver glue to PDMS of 1:0.4, the related tests on the stability of the conductivity during the stretching process were performed, as shown in Figure 5. Figure 5a demonstrates the resistance of this electrode at a sustained strain of 20%. It can be seen that the resistance of the electrode increases by only about 4 Ω under a 20% strain, and the resistance is stable and remains about 10 Ω during the process of holding it under the strain of 20% for 25 s. When it stops stretching, the resistance returns to the initial state. It indicates that the electrode ensures stable conductivity under constant static stretching. Figure 5b shows the resistance of the electrode under cyclic stretching. The stress–strain curve in Figure S2 illustrates that the stretching and releasing process of the composite electrode during cyclic stretching is very stable, which provides a guarantee for the resistance measurement during the cyclic test. During 1000 cycles of stretching, the resistance of this electrode consistently varied between 5 and 10 Ω. In addition, the inset shows that the resistance change caused by each stretching is stable. It reflects that the electrode has an excellent fatigue durability to withstand repeated stretching. Figure 5c shows the resistance of the same electrode under intermittent stretching. During the first to fifth stretching tests, the strain was 20% for each stretching process, and the time interval between each test was 4 h. It is illustrated that the resistance change is about 4 Ω during the whole stretching process in each test. In the same stretching test at a fixed interval time, the behavior of the resistance change caused by each stretch is similar. It indicates that the long periods of non-use state do not affect the performance of this electrode, which has excellent stability and reliability in a single stretching test.

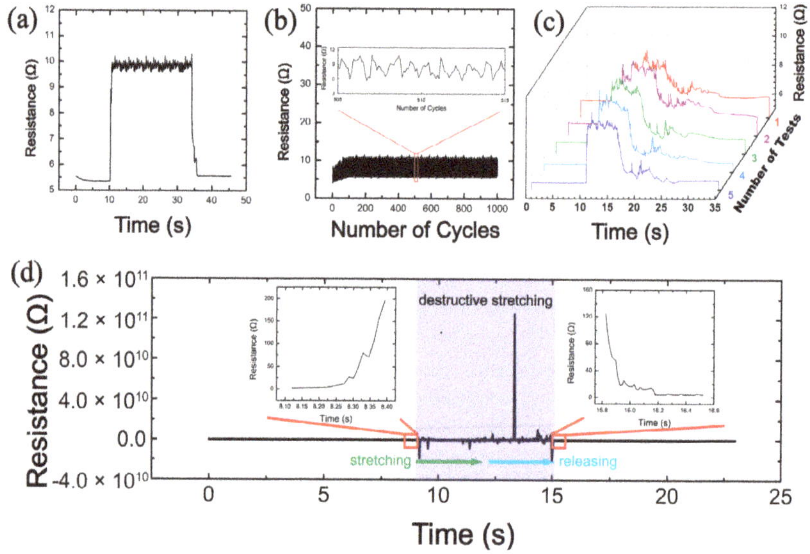

Figure 5. Conductive stability tests of the optimal electrode with a mass ratio of silver glue to PDMS of 1:0.4 during stretching process. Under a strain of 20%, (**a**) sustained stretching test, (**b**) 1000 times cyclic stretching test, and (**c**) intermittent stretching test. (**d**) Destructive stretching-releasing test under a strain of 200%.

It is observed from Figure 5a–c that when a tolerable strain occurs, a small change in resistance appears with the strain. This small resistance change is supposed to originate from the slippage between adjacent Ag nanosheets and a small amount of separation. The slippage and the degree of separation are related to the density and size of the Ag nanosheets. The detailed information will be described later in the finite element analysis.

Figure 5d shows the influence of a destructive large strain on the performance of this electrode. Before the destructive stretching begins, the initial resistance of the electrode is 6 Ω. During the time period of 8–16 s, the electrode undergoes a destructive stretching process with a large strain of 200%, during which the conductive path inside the electrode is completely destroyed, resulting in an open-circuit state of the electrode. After 16 s, the electrode was fully released and the resistance is restored to 6 Ω, ensuring that the electrode could continue to operate normally. The destructive stretching and releasing process under an optical microscopy is shown in Figure S3. The inset of Figure 5d shows the resistance change in a period of time before the start and after the end of the destructive stretching, from which it can be clearly observed that the resistance increases rapidly at the initial state of destructive stretching and decreases rapidly after the electrode is fully released. These results show that this electrode has good robustness and will not permanently fail as a result of a destructive stretch. Furthermore, as a practical flexible sheet, the stability for bending and temperature variation is important. Therefore, the stability tests for the situations of bending operation and temperature variation were performed, as shown in Figure S4. It is observed that the electrode has low resistance both in bending operation and temperature variation, and its resistance changes are small (4 Ω for bending, 2 Ω for temperature variation), confirming the possibility that the electrode can be used under a variety of external stimuli.

3.3. Conductive Mechanism Analysis of PDMS-Ag Electrodes

In order to explain the conductive mechanism of the PDMS-Ag electrodes, the number of Ag nanosheets was counted according to the SEM images in Figure S5 and the statistical results are listed in Table S2. Based on above results, the density of the Ag nanosheets and the conductivity of the electrodes with different ratios of Ag and PDMS are shown in Figure 6. It is observed that, with the increase in PDMS content, the conductivity of the electrode and the density of Ag nanosheets inside it gradually decrease. This is because the higher the PDMS content is, the more the Ag nanosheets are separated from each other, resulting in a decrease in the number of Ag conductive paths and a decrease in the conductivity. Moreover, the decreasing trends of conductivity and Ag nanosheet density are similar, indicating a positive correlation between them.

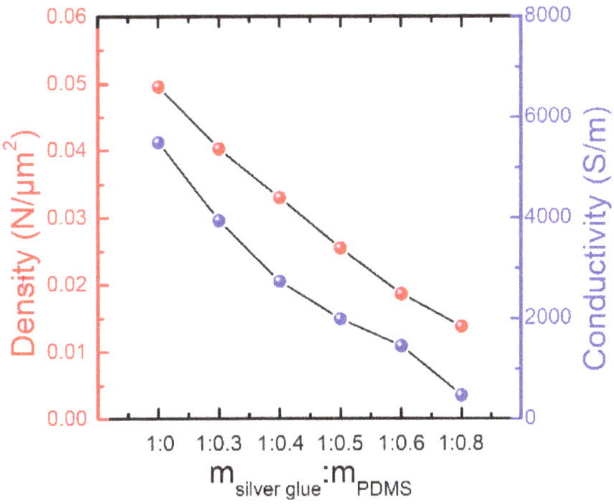

Figure 6. Conductivity of the electrode and density of Ag nanosheets in the electrode containing different ratios of silver glue and PDMS.

According to the above analysis, it is clear that the conductivity of the electrode is directly related to the number of contact points between Ag nanosheets. Therefore, the relationship between the conductivity of electrode and Ag nanosheets can be expressed by the contact point theory as shown in the Supplementary Information [44,45]. After a series of theoretical derivations, the relationship between the resistance and the density of Ag nanosheets can be obtained as

$$\ln \frac{R}{R_0} = \alpha' \frac{D}{D_0} + b' \quad (1)$$

where D and D_0 are the density and initial density of Ag nanosheets, respectively. Thus, the contact point theory suggests a linear relationship between the relative density (D/D_0) of the Ag nanosheets and the logarithm of the relative resistance (R/R_0) of the electrode.

The resistance of the electrode is the smallest when the mass ratio of silver glue to PDMS is 1:0, and the resistance increases gradually with the increasing mass ratio of PDMS, which is consistent with the phenomenon that the resistance increases gradually with the increasing strain during the stretching process. Therefore, the density of Ag nanosheets and the resistance of electrode with a mass ratio of silver glue and PDMS of 1:0 are selected as the initial parameters, and the relative density and the relative resistance of electrodes with different mass ratios are calculated. Based on the above analysis, a relationship between relative resistance and relative density is statistically obtained for the electrodes with different mass ratios, as shown in Figure 7a. It can be seen that relative density and relative resistance are consistent with the fitting result (red line) of contact point theory. It indicates that the number of contact points between Ag nanosheets gradually decreases with the increasing of PDMS content, which reflects the conductive state of the electrodes with different mass ratios.

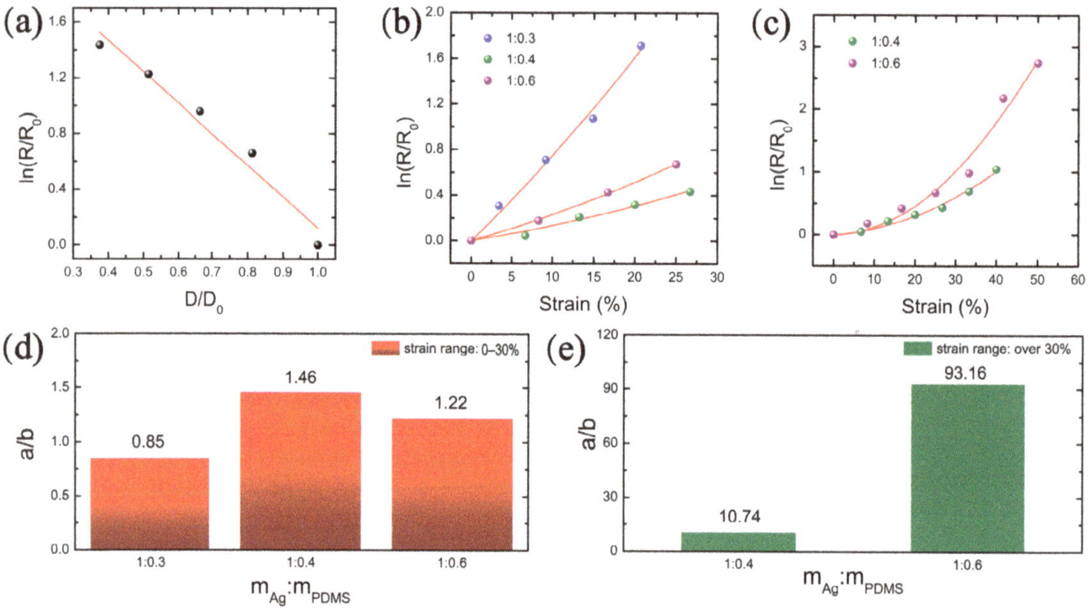

Figure 7. Conductive mechanism analysis of the PDMS-Ag electrodes. (**a**) Contact point theoretical fitting. Tunneling effect theoretical fitting under (**b**) small and (**c**) large strains. Statistics of the proportional coefficients a/b of the tunneling theoretical equation under (**d**) small strain and (**e**) large strain.

When the stretching process is in progress, the change of electrode resistance not only comes from the change of the contact resistance, but also from the change of the tunneling resistance between the incompletely contacted Ag nanosheets. Therefore, taking the change of tunneling resistance into account, the conductive mechanism of the electrode during stretching can be explained. According to the tunneling theoretical model [46–49], combined with the actual situation in this study, after a series of theoretical derivations (Supplementary Information), the relationship between tunneling resistance and strain can be obtained as

$$\ln(\frac{R_t}{R_0}) = a\varepsilon^2 + b\varepsilon \quad (2)$$

Different from the contact point theory, which reflects the state quantity of resistance, the tunneling effect theory reflects the resistance change during the stretching process. In Equation (2), $a\varepsilon^2$ represents the tunneling resistance term during stretching process, and $b\varepsilon$ represents the contact resistance term of the electrode. Therefore, the total resistance of the electrode can reflect both the contact state inside the electrode and the change process of resistance. The ratio of a/b can reflect the influence degree of contact resistance and tunneling resistance on the change of electrode resistance, so as to judge the dominant role of them in the contribution of resistance change. When the ratio of a/b is small, the contact resistance is dominant. Conversely, the tunneling resistance is dominant.

Based on this conclusion, the tunneling effect theory is used to fit the resistance changes of electrodes with a mass ratio of silver glue and PDMS of 1:0.3, 1:0.4, and 1:0.6 during the stretching process within the strain range of 0–30%. The results are shown in Figure 7b. After fitting, the fitting degree (R^2) of the above three electrodes all reach 0.99, indicating that the tunneling theory is in good agreement with the experimental data. According to the fitting results, the proportional coefficients a/b of the tunneling theoretical equations for these three electrodes are 0.85, 1.46, and 1.22, respectively, as shown in Figure 7d. It indicates that the contact resistance dominates in the electrode with the mass ratio of silver glue to PDMS of 1:0.3. This is because the PDMS content in the electrode is relatively small, and the overall stretchability of the electrode is poor, a small strain can undergo the process from complete to destruction of the internal conductive network of the electrode. Therefore, the resistance change of this electrode is mainly manifested by the contact resistance. The proportional coefficients a/b of the electrodes with the mass ratio of silver glue to PDMS of 1:0.4 and 1:0.6 are larger than that of 1:0.3, indicating that the tunneling effect on the resistance gradually increases with the decrease in the density of Ag nanosheets.

Since the maximum strain value of electrode with a mass ratio of silver glue and PDMS of 1:0.3 is only 20%, the tunneling theory is fitted for the electrodes with a mass ratio of 1:0.4 and 1:0.6 in the wider range of 0–50%, and the fitting results and the statistics of the proportional coefficients a/b of the tunneling theoretical equation are shown in Figure 7c,e respectively. It is observed that in the strain range of 0–50%, the resistance changes of both electrodes during stretching are consistent with the tunneling effect theory, and the proportional coefficients a/b of the tunneling theoretical equation of the two electrodes are 10.74 and 93.16, respectively. In the stretching process from small to large, the proportional coefficients a/b of the electrode with a mass ratio of 1:0.6 increases from 1.22 to 93.16, which is much larger than that of the electrode with a mass ratio of 1:0.4 from 1.46 to 10.74. It indicates that, with the further increase in the strain, the influence of tunneling resistance becomes more obvious, which is caused by the separation of a large number of Ag nanosheets from each other under large strains. For the proportional coefficients a/b of the tunneling theoretical equation, the electrode with a mass ratio of 1:0.6 is significantly more than that with a mass ratio of 1:0.4. This is because the electrode with a mass ratio of 1:0.6 has a high PDMS content, resulting in a low number of initial conductive pathways. The separation between Ag nanosheets caused by stretching is obvious, leading to a high tunneling resistance. However, the electrode with a mass ratio of 1:0.4 keeps a large number of Ag nanosheets in contact with each other during the stretching process, thus maintaining a relatively stable conductivity under a large strain. Therefore, the electrode with a mass ratio of 1:0.4 is the optimal ratio for the PDMS-Ag electrode.

3.4. Finite Element Analysis of the PDMS-Ag Electrodes

In order to obtain a deeper and comprehensive understanding of the conductive mechanism and the influence factors, the PDMS-Ag electrode is modeled by the method of finite element analysis. Due to the fact that the thickness of the electrodes is much smaller than the size of the surface region, it can be considered that the Ag nanosheets are less distributed in the direction perpendicular to the plane and mainly concentrated in the surface region. Moreover, in the direction perpendicular to the surface, the Ag nanosheets in contact with each other can be regarded as a whole. Therefore, for the purpose of simplifying the model, the three-dimensional conductive model can be replaced by the two-dimensional in-plane conductive model, which plays a dominant role. Firstly, the size of the Ag nanosheets in the SEM image of the electrode is counted, as shown in Figure S6. According to the statistical results, the average radius of the Ag nanosheets is 3.912 μm. Therefore, the radius of the Ag nanosheet is set to 4 μm during modeling. Then, the physical model of the PDMS-Ag composite film is established by combining the statistics of the density and size of the Ag nanosheets in the electrode, and an electric potential of 0.1 V is applied to the model, the potential distribution is shown in Figure S7. Based on this model, the current density distribution inside the composite electrode was simulated under the mass ratios of Ag nanosheets to PDMS of 1:0.3, 1:0.4, 1:0.5, and 1:0.6, respectively, as shown in Figure 8a. It is clear that the current density decreases gradually with the increase in the PDMS content, this trend is consistent with the experimental phenomenon. The simulation data under different relative densities of Ag nanosheets are fitted using the contact point theory, and the fitting results are found to be consistent with the theory, as shown in Figure 8b. Comparing the simulation result with the experimental data (Figure 8c), it is found that the relative resistance change of the simulation result is large than that of the experimental result, and the fitting slope of the simulation result is 4.58, which is larger than that of the experimental data of 1.44. The reason for the difference between the simulation and experimental results may come from the fact that some measured Ag nanosheets are covered by other adjacent Ag nanosheets during the counting process, resulting in a smaller measured value of the Ag nanosheets than the actual value.

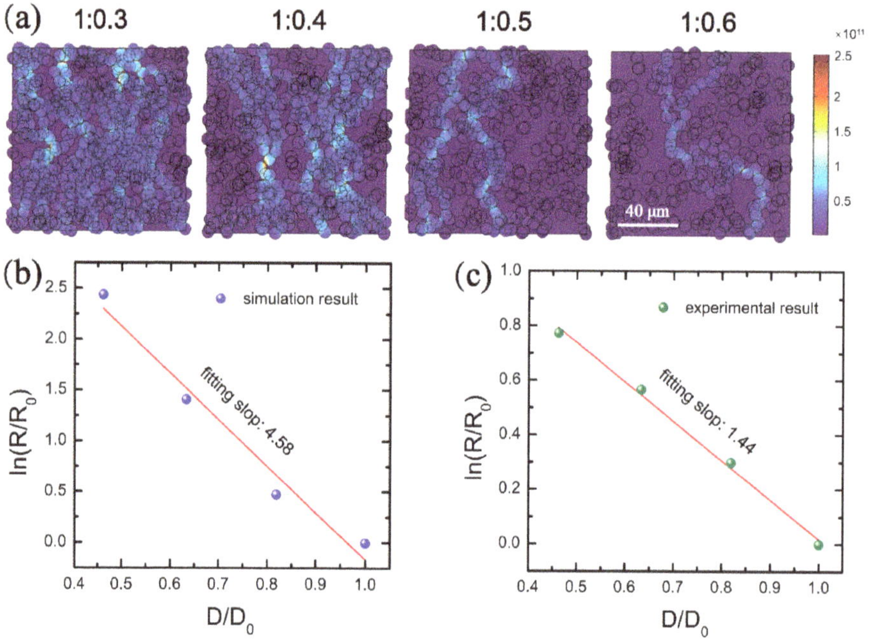

Figure 8. Comparison of simulation and experimental results. (**a**) The current density distribution inside the composite electrode under the mass ratios of Ag nanosheets to PDMS of 1:0.3, 1:0.4, 1:0.5, and 1:0.6, respectively. Contact point theoretical fitting of (**b**) simulation and (**c**) experimental results.

Considering the deviation in the measurement process of Ag nanosheet size, the conductivity of the electrode with different Ag nanosheet sizes was simulated to modify the model based on the contact point theory. Therefore, the radius of the Ag nanosheets was increased from the initial 4 μm to 5 μm and 6 μm, respectively, and the current density distribution was simulated, as shown in Figure 9a. It can be seen from the simulation results that with the increase in the Ag nanosheet size, the current density and conductive pathways in the electrodes with the same mass ratio increase gradually, and correspondingly, the resistance gradually decreases. Contact point theory was used to fit the Ag nanosheets with a radius of 5 μm and 6 μm, and the fitting results are shown in Figure 9b,c. According to the fitting results of these two situations, the fitting slopes are 2.34 and 0.82, the logarithm of the relative resistances are 1.26 and 0.42, respectively. This is because with the increase in the Ag nanosheet size, the contact area between the mutually stacked Ag nanosheets becomes larger. During the stretching process, Ag nanosheets are more difficult to separate from each other, resulting in a decrease in the change rate of the number of conductive pathways and a smaller resistance change in the electrode. Therefore, a large Ag nanosheet size is beneficial to improve the conductive stability of the electrode during stretching. From the above discussion and the experimental results in Figure 8c, it can be seen that the fitting slope of the contact point theory of the experimental data is 1.44, and the change of logarithm of relative resistance is 0.77. Comparing the experimental results in Figure 8c with the simulation results in Figure 9b,c, it can be seen that both the fitting slope and the change of logarithm of relative resistance of the experimental results are between the fitting results for the Ag nanosheet radius of 5 μm and 6 μm. Following this trend, an Ag nanosheet radius between 5 and 6 μm would exist. Under this radius, the slope and the change of logarithm of relative resistance fitted by the contact point theory are similar to the experimental results in Figure 8c. By simulating and fitting different Ag nanosheet radii between 5 and 6 μm, it is found that when the Ag nanosheet radius is 5.5 μm, the fitting slope and the change of logarithm of relative resistance are 1.39 and 0.75, respectively, which are very close to the experimental results. Therefore, the average radius of the Ag nanosheets used in the experiment can be inferred to be around 5.5 μm through this method. The simulation results modify the measured values of the Ag nanosheet size based on SEM images.

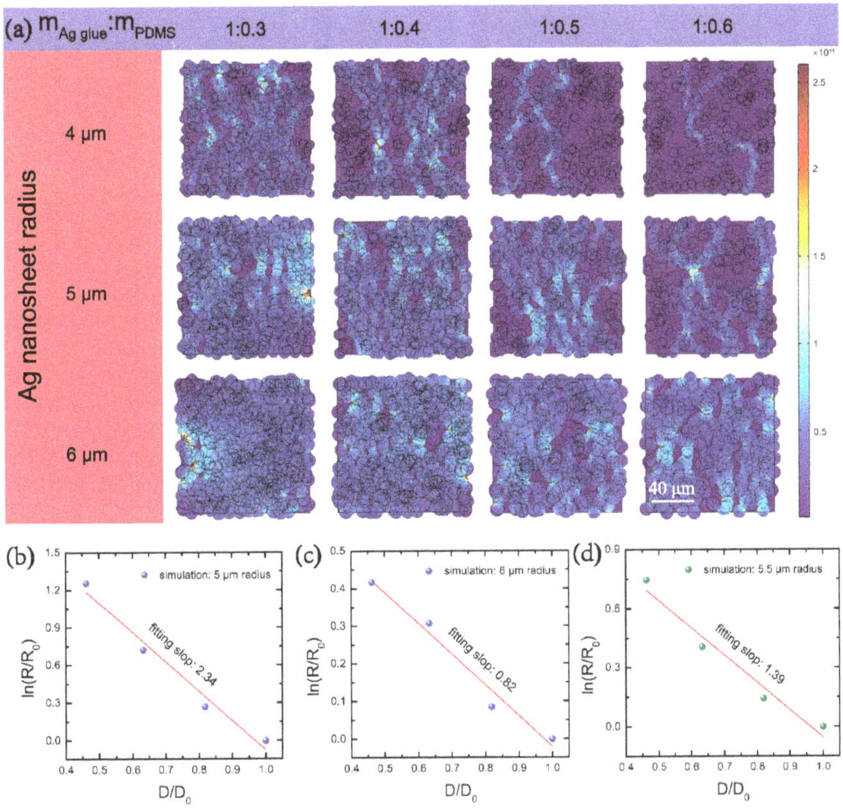

Figure 9. Simulation of the conductivity of electrodes with different Ag nanosheet radii. (**a**) Comparison of the current density distribution when the radius of Ag nanosheets is 4, 5, and 6 μm, respectively. Contact point theoretical fitting for the Ag nanosheet radius of (**b**) 5 μm, (**c**) 6 μm, and (**d**) 5.5 μm, respectively.

Therefore, the flexible and stretchable electrodes can be fabricated by mixing PDMS with Ag nanosheets, and the conductive stability of the composite electrodes during stretching can be regulated by adjusting the density and size of the Ag nanosheets.

4. Conclusions

In summary, a PDMS-Ag-based composite electrode was prepared. Taking advantage of the properties of good flexibility and stretchability of PDMS and good conductivity of Ag, the electrode has the ability to maintain good conductivity during stretching. By comparing the conductivity of different electrodes, it is found that the electrode with a mass ratio of PDMS and silver glue of 1:0.4 has the best stability during stretching. The electrode has an initial resistance of only about 5 Ω, a maximum strain value of over 50% in normal operation, and a limited resistance change within 20 Ω during the whole stretching process. In addition, it can maintain stable conductivity during constant stretching process; withstand more than 1000 cycles of stretching tests; and under the state of the electrode failure caused by destructive large stretching, the conductivity can be rapidly recovered after the stress is released. The conductive mechanism of the electrode is explained by combining the contact point theory and the tunneling theory. The change rule of the conductivity under different Ag nanosheet densities and the conductivity of the electrodes under different mass ratios during stretching process are investigated. The dominant role of contact resistance and tunneling resistance is judged by the value of the proportional coefficients a/b of the tunneling theoretical equations. Moreover, the simulation method is used to explain the effect of Ag nanosheet size on the conductivity, and the average radius of the Ag nanosheet inside the electrode is inferred to be 5.5 μm by simulation. This study provides a simple and effective method to fabricate a flexible and stretchable electrode, and has potential in flexible electronics, wearable devices, and other fields.

Supplementary Materials: The following supporting information can be downloaded at: https://www.mdpi.com/article/10.3390/nano12152628/s1, Figure S1: optical microscopy observation of the stretchable electrodes with different mass ratios silver glue and PDMS; Table S1: The statistics of the size of PDMS-Ag composite electrodes; Figure S2: The stress-strain curves of 100 continuous cyclic stretching-releasing test; Figure S3: The surface morphology of the composite electrode during the process of destructive stretching and releasing under an optical microscopy; Figure S4: The stability of the electrode under different external stimuli, (a) bending test, (b) temperature variation test; Figure S5: The statistics on the number of Ag nanosheets by counting method; Table S2: The statistics on the density of Ag nanosheets; Figure S6: The statistical size of the Ag nanosheet. (a) SEM image of the statistical area. (b) The statistical results of the Ag nanosheet radii; Figure S7: Modeling process of the electrode through the finite element simulation. The schematic of (a) model building and (b) potential distribution.

Author Contributions: Conceptualization, C.L.; methodology, C.L.; software, K.H.; validation, C.L. and L.P.; formal analysis, C.L. and K.H.; investigation, C.L., K.H. and T.Y.; resources, K.H. and T.Y.; data curation, K.H. and T.Y.; writing—original draft preparation, C.L. and K.H.; writing—review and editing, C.L.; visualization, C.L., K.H. and T.C.; supervision, L.P. and Z.F.; project administration, L.P. and Z.F.; funding acquisition, L.P., Z.F. and C.L. All authors have read and agreed to the published version of the manuscript.

Funding: This work was supported by the China Postdoctoral Science Foundation (no. 2021M700658), the National Natural Science Foundation of China (no. 51972039, 51661145025, 51803018), and LiaoNing revitalization Talent Program (no. XLYC1902122).

Conflicts of Interest: The authors declare no conflict of interest.

References

1. Wang, C.Y.; Xia, K.L.; Wang, H.M.; Liang, X.P.; Yin, Z.; Zhang, Y.Y. Advanced Carbon for Flexible and Wearable Electronics. *Adv. Mater.* **2019**, *31*, 1801072. [CrossRef]
2. Zhao, J.; Li, N.; Yu, H.; Wei, Z.; Liao, M.Z.; Chen, P.; Wang, S.P.; Shi, D.X.; Sun, Q.J.; Zhang, G.Y. Highly Sensitive MoS2 Humidity Sensors Array for Noncontact Sensation. *Adv. Mater.* **2017**, *29*, 1702076. [CrossRef]

3. Li, C.W.; Zhang, Y.F.; Yang, S.T.; Zhao, H.T.; Guo, Y.; Cong, T.Z.; Huang, H.; Fan, Z.; Liang, H.W.; Pan, L.J. A flexible tissue-carbon nanocoil-carbon nanotube-based humidity sensor with high performance and durability. *Nanoscale* **2022**, *14*, 7025–7038. [CrossRef]
4. Li, C.W.; Yang, S.T.; Guo, Y.; Huang, H.; Chen, H.; Zuo, X.Q.; Fan, Z.; Liang, H.W.; Pan, L.J. Flexible, multi-functional sensor based on all-carbon sensing medium with low coupling for ultrahigh-performance strain, temperature and humidity sensing. *Chem. Eng. J.* **2021**, *426*, 130364. [CrossRef]
5. Liao, M.; Ye, L.; Zhang, Y.; Chen, T.Q.; Peng, H.S. The Recent Advance in Fiber-Shaped Energy Storage Devices. *Adv. Electron. Mater.* **2019**, *5*, 1800456. [CrossRef]
6. Lv, T.; Yao, Y.; Li, N.; Chen, T. Wearable fiber-shaped energy conversion and storage devices based on aligned carbon nanotubes. *Nano. Today* **2016**, *11*, 644–660. [CrossRef]
7. Tadesse, M.G.; Mengistie, D.A.; Chen, Y.; Wang, L.C.; Loghin, C.; Nierstrasz, V. Electrically conductive highly elastic polyamide/lycra fabric treated with PEDOT:PSS and polyurethane. *J. Mater. Sci.* **2019**, *54*, 9591–9602. [CrossRef]
8. Wang, Z.P.; Wang, Y.S.; Chen, Y.J.; Yousaf, M.; Wu, H.S.; Cao, A.Y.; Han, R.P.S. Reticulate Dual-Nanowire Aerogel for Multifunctional Applications: A High-Performance Strain Sensor and a High Areal Capacity Rechargeable Anode. *Adv. Funct. Mater.* **2019**, *29*, 1807467. [CrossRef]
9. Zhang, C.; Liu, S.Y.; Huang, X.; Guo, W.; Li, Y.Y.; Wu, H. A stretchable dual-mode sensor array for multifunctional robotic electronic skin. *Nano. Energy* **2019**, *62*, 164–170. [CrossRef]
10. Yu, L.T.; Yeo, J.C.; Soon, R.H.; Yeo, T.; Lee, H.H.; Lim, C.T. Highly Stretchable, Weavable, and Washable Piezoresistive Microfiber Sensors. *Acs. Appl. Mater. Inter.* **2018**, *10*, 12773–12780. [CrossRef] [PubMed]
11. Lv, Z.S.; Luo, Y.F.; Tang, Y.X.; Wei, J.Q.; Zhu, Z.Q.; Zhou, X.R.; Li, W.L.; Zeng, Y.; Zhang, W.; Zhang, Y.Y.; et al. Editable Supercapacitors with Customizable Stretchability Based on Mechanically Strengthened Ultralong MnO2 Nanowire Composite. *Adv. Mater.* **2018**, *30*, 1704531. [CrossRef] [PubMed]
12. Trung, T.Q.; Kim, C.; Lee, H.B.; Cho, S.M.; Lee, N.E. Toward a Stretchable Organic Light-Emitting Diode on 3D Microstructured Elastomeric Substrate and Transparent Hybrid Anode. *Adv. Mater. Technol.* **2020**, *5*, 1900995. [CrossRef]
13. An, B.W.; Gwak, E.J.; Kim, K.; Kim, Y.C.; Jang, J.; Kim, J.Y.; Park, J.U. Stretchable, Transparent Electrodes as Wearable Heaters Using Nanotrough Networks of Metallic Glasses with Superior Mechanical Properties and Thermal Stability. *Nano. Lett.* **2016**, *16*, 471–478. [CrossRef] [PubMed]
14. Hwang, H.; Kim, D.G.; Jang, N.S.; Kong, J.H.; Kim, J.M. Simple method for high-performance stretchable composite conductors with entrapped air bubbles. *Nanoscale Res. Lett.* **2016**, *11*, 14. [CrossRef]
15. Hong, H.C.; Chen, C.M. Design, Fabrication and Failure Analysis of Stretchable Electrical Routings. *Sens. Basel* **2014**, *14*, 11855–11877. [CrossRef]
16. Qaiser, N.; Damdam, A.N.; Khan, S.M.; Bunaiyan, S.; Hussain, M.M. Design Criteria for Horseshoe and Spiral-Based Interconnects for Highly Stretchable Electronic Devices. *Adv. Funct. Mater.* **2021**, *31*, 2007445. [CrossRef]
17. Webb, R.C.; Bonifas, A.P.; Behnaz, A.; Zhang, Y.H.; Yu, K.J.; Cheng, H.Y.; Shi, M.X.; Bian, Z.G.; Liu, Z.J.; Kim, Y.S.; et al. Ultrathin conformal devices for precise and continuous thermal characterization of human skin. *Nat. Mater.* **2013**, *12*, 938–944. [CrossRef]
18. Tang, J.; Guo, H.; Zhao, M.M.; Yang, J.T.; Tsoukalas, D.; Zhang, B.Z.; Liu, J.; Xue, C.Y.; Zhang, W.D. Highly Stretchable Electrodes on Wrinkled Polydimethylsiloxane Substrates. *Sci. Rep.* **2015**, *5*, 16527. [CrossRef]
19. Fan, J.A.; Yeo, W.H.; Su, Y.W.; Hattori, Y.; Lee, W.; Jung, S.Y.; Zhang, Y.H.; Liu, Z.J.; Cheng, H.Y.; Falgout, L.; et al. Fractal design concepts for stretchable electronics. *Nat. Commun.* **2014**, *5*, 3266. [CrossRef]
20. Jeong, G.S.; Baek, D.H.; Jung, H.C.; Song, J.H.; Moon, J.H.; Hong, S.W.; Kim, I.Y.; Lee, S.H. Solderable and electroplatable flexible electronic circuit on a porous stretchable elastomer. *Nat. Commun.* **2012**, *3*, 977. [CrossRef]
21. Chen, W.; Li, Y.D.; Li, R.Q.; Thean, A.V.Y.; Guo, Y.X. Bendable and Stretchable Microfluidic Liquid Metal-Based Filter. *IEEE Microw. Wirel. Co.* **2018**, *28*, 203–205. [CrossRef]
22. Cao, Y.; Morrissey, T.G.; Acome, E.; Allec, S.I.; Wong, B.M.; Keplinger, C.; Wang, C. A Transparent, Self-Healing, Highly Stretchable Ionic Conductor. *Adv. Mater.* **2017**, *29*, 1605099. [CrossRef] [PubMed]
23. Park, M.; Park, J.; Jeong, U. Design of conductive composite elastomers for stretchable electronics. *Nano. Today* **2014**, *9*, 244–260. [CrossRef]
24. Ladd, C.; So, J.H.; Muth, J.; Dickey, M.D. 3D Printing of Free Standing Liquid Metal Microstructures. *Adv. Mater.* **2013**, *25*, 5081–5085. [CrossRef] [PubMed]
25. Wang, Y.; Zhu, C.X.; Pfattner, R.; Yan, H.P.; Jin, L.H.; Chen, S.C.; Molina-Lopez, F.; Lissel, F.; Liu, J.; Rabiah, N.I.; et al. A highly stretchable, transparent, and conductive polymer. *Sci. Adv.* **2017**, *3*, e1602076. [CrossRef]
26. Zhu, S.; So, J.H.; Mays, R.; Desai, S.; Barnes, W.R.; Pourdeyhimi, B.; Dickey, M.D. Ultrastretchable Fibers with Metallic Conductivity Using a Liquid Metal Alloy Core. *Adv. Funct. Mater.* **2013**, *23*, 2308–2314. [CrossRef]
27. Hong, G.B.; Luo, Y.H.; Chuang, K.J.; Cheng, H.Y.; Chang, K.C.; Ma, C.M. Facile Synthesis of Silver Nanoparticles and Preparation of Conductive Ink. *Nanomaterials-Basel* **2022**, *12*, 171. [CrossRef]
28. Kim, K.K.; Choi, J.; Kim, J.H.; Nam, S.; Ko, S.H. Evolvable Skin Electronics by In Situ and In Operando Adaptation. *Adv. Funct. Mater.* **2022**, *32*, 2106329. [CrossRef]
29. Nguyen, H.L.; Jo, Y.K.; Cha, M.; Cha, Y.J.; Yoon, D.K.; Sanandiya, N.D.; Prajatelistia, E.; Oh, D.X.; Hwang, D.S. Mussel-Inspired Anisotropic Nanocellulose and Silver Nanoparticle Composite with Improved Mechanical Properties, Electrical Conductivity and Antibacterial Activity. *Polymers-Basel* **2016**, *8*, 102. [CrossRef]

30. Yeo, J.; Kim, G.; Hong, S.; Kim, M.S.; Kim, D.; Lee, J.; Lee, H.B.; Kwon, J.; Suh, Y.D.; Kang, H.W.; et al. Flexible supercapacitor fabrication by room temperature rapid laser processing of roll-to-roll printed metal nanoparticle ink for wearable electronics application. *J. Power Sources.* **2014**, *246*, 562–568. [CrossRef]
31. Wu, X.L.; Wang, S.Y.; Luo, Z.W.; Lu, J.X.; Lin, K.W.; Xie, H.; Wang, Y.H.; Li, J.Z. Inkjet Printing of Flexible Transparent Conductive Films with Silver Nanowires Ink. *Nanomaterials-Basel* **2021**, *11*, 1571. [CrossRef] [PubMed]
32. Lee, P.; Lee, J.; Lee, H.; Yeo, J.; Hong, S.; Nam, K.H.; Lee, D.; Lee, S.S.; Ko, S.H. Highly Stretchable and Highly Conductive Metal Electrode by Very Long Metal Nanowire Percolation Network. *Adv. Mater.* **2012**, *24*, 3326–3332. [CrossRef] [PubMed]
33. Vo, T.T.; Lee, H.J.; Kim, S.Y.; Suk, J.W. Synergistic Effect of Graphene/Silver Nanowire Hybrid Fillers on Highly Stretchable Strain Sensors Based on Spandex Composites. *Nanomaterials-Basel* **2020**, *10*, 2063. [CrossRef] [PubMed]
34. Won, P.; Park, J.J.; Lee, T.; Ha, I.; Han, S.; Choi, M.; Lee, J.; Hong, S.; Cho, K.J.; Ko, S.H. Stretchable and Transparent Kirigami Conductor of Nanowire Percolation Network for Electronic Skin Applications. *Nano. Lett.* **2019**, *19*, 6087–6096. [CrossRef] [PubMed]
35. Jung, J.; Cho, H.; Yuksel, R.; Kim, D.; Lee, H.; Kwon, J.; Lee, P.; Yeo, J.; Hong, S.; Unalan, H.E.; et al. Stretchable/flexible silver nanowire electrodes for energy device applications. *Nanoscale* **2019**, *11*, 20356–20378. [CrossRef]
36. Cho, J.H.; Kang, D.J.; Jang, N.S.; Kim, K.H.; Won, P.; Ko, S.H.; Kim, J.M. Metal Nanowire-Coated Metal Woven Mesh for High-Performance Stretchable Transparent Electrodes. *Acs. Appl. Mater. Inter.* **2017**, *9*, 40905–40913. [CrossRef]
37. Ge, Y.J.; Duan, X.D.; Zhang, M.; Mei, L.; Hu, J.W.; Hu, W.; Duan, X.F. Direct Room Temperature Welding and Chemical Protection of Silver Nanowire Thin Films for High Performance Transparent Conductors. *J. Am. Chem. Soc.* **2018**, *140*, 193–199. [CrossRef]
38. Song, T.B.; Chen, Y.; Chung, C.H.; Yang, Y.; Bob, B.; Duan, H.S.; Li, G.; Tu, K.N.; Huang, Y.; Yang, Y. Nanoscale Joule Heating and Electromigration Enhanced Ripening of Silver Nanowire Contacts. *Acs. Nano.* **2014**, *8*, 2804–2811. [CrossRef]
39. Tokuno, T.; Nogi, M.; Karakawa, M.; Jiu, J.T.; Nge, T.T.; Aso, Y.; Suganuma, K. Fabrication of silver nanowire transparent electrodes at room temperature. *Nano. Res.* **2011**, *4*, 1215–1222. [CrossRef]
40. Celano, T.A.; Hill, D.J.; Zhang, X.; Pinion, C.W.; Christesen, J.D.; Flynn, C.J.; McBride, J.R.; Cahoon, J.F. Capillarity-Driven Welding of Semiconductor Nanowires for Crystalline and Electrically Ohmic Junctions. *Nano. Lett.* **2016**, *16*, 5241–5246. [CrossRef] [PubMed]
41. Lu, H.F.; Zhang, D.; Cheng, J.Q.; Liu, J.; Mao, J.; Choy, W.C.H. Locally Welded Silver Nano-Network Transparent Electrodes with High Operational Stability by a Simple Alcohol-Based Chemical Approach. *Adv. Funct. Mater.* **2015**, *25*, 4211–4218. [CrossRef]
42. Lee, J.; Lee, P.; Lee, H.; Lee, D.; Lee, S.S.; Ko, S.H. Very long Ag nanowire synthesis and its application in a highly transparent, conductive and flexible metal electrode touch panel. *Nanoscale* **2012**, *4*, 6408–6414. [CrossRef]
43. Lee, J.; Lee, P.; Lee, H.B.; Hong, S.; Lee, I.; Yeo, J.; Lee, S.S.; Kim, T.S.; Lee, D.; Ko, S.H. Room-Temperature Nanosoldering of a Very Long Metal Nanowire Network by Conducting-Polymer-Assisted Joining for a Flexible Touch-Panel Application. *Adv. Funct. Mater.* **2013**, *23*, 4171–4176. [CrossRef]
44. Li, C.W.; Pan, L.J.; Deng, C.H.; Wang, P.; Huang, Y.Y.; Nasir, H. A flexible, ultra-sensitive strain sensor based on carbon nanocoil network fabricated by an electrophoretic method. *Nanoscale* **2017**, *9*, 9872–9878. [CrossRef] [PubMed]
45. Li, C.W.; Pan, L.J.; Deng, C.H.; Cong, T.Z.; Yin, P.H.; Wu, Z.L. A highly sensitive and wide-range pressure sensor based on a carbon nanocoil network fabricated by an electrophoretic method. *J. Mater. Chem. C* **2017**, *5*, 11892–11900. [CrossRef]
46. Xu, S.; Rezvanian, O.; Peters, K.; Zikry, M.A. The viability and limitations of percolation theory in modeling the electrical behavior of carbon nanotube-polymer composites. *Nanotechnology* **2013**, *24*, 155706. [CrossRef]
47. Dang, Z.M.; Jiang, M.J.; Xie, D.; Yao, S.H.; Zhang, L.Q.; Bai, J.B. Supersensitive linear piezoresistive property in carbon nanotubes/silicone rubber nanocomposites. *J. Appl. Phys.* **2008**, *104*, 024114. [CrossRef]
48. Chen, L.; Chen, G.H.; Lu, L. Piezoresistive behavior study on finger-sensing silicone rubber/graphite nanosheet nanocomposites. *Adv. Funct. Mater.* **2007**, *17*, 898–904. [CrossRef]
49. Li, C.W.; Pan, L.J.; Deng, C.H.; Wang, P. CNC-Al2O3-Ti: A new unit for micro scale strain sensing. *Rsc. Adv.* **2016**, *6*, 107683–107688. [CrossRef]

nanomaterials

Article

Disposal Immunosensor for Sensitive Electrochemical Detection of Prostate-Specific Antigen Based on Amino-Rich Nanochannels Array-Modified Patterned Indium Tin Oxide Electrode

Liang Yan [1,2,*], Shuai Xu [3] and Fengna Xi [3,*]

1 Shanxi Bethune Hospital, Shanxi Academy of Medical Sciences, Tongji Shanxi Hospital, Third Hospital of Shanxi Medical University, Taiyuan 030032, China
2 Tongji Hospital, Tongji Medical College, Huazhong University of Science and Technology, Wuhan 430030, China
3 Department of Chemistry, Zhejiang Sci-Tech University, Hangzhou 310018, China
* Correspondence: yanliangyl2022@126.com (L.Y.); fengnaxi@zstu.edu.cn (F.X.)

Abstract: Sensitive detection of prostate-specific antigens (PSA) in serum is essential for the prevention and early treatment of prostate cancer. Simple and disposable electrochemical immunosensors are highly desirable for screening and mobile detection of PSAs in high-risk populations. Here, an electrochemical immunosensor was constructed based on amino-rich nanochannels array-modified patterned, inexpensive, and disposable indium tin oxide (ITO) electrodes, which can be employed for the sensitive detection of PSA. Using an amino-group-containing precursor, a vertically ordered mesoporous silica nanochannel film (VMSF) containing amino groups (NH_2-VMSF) was rapidly grown on ITO. When NH_2-VMSF contained template surfactant micelle (SM), the outer surface of NH_2-VMSF was directionally modified by aldehyde groups, which enabled further covalent immobilization of the recognitive antibody to prepare the immuno-recognitive interface. Owing to the charge-based selective permeability, NH_2-VMSF can electrostatically adsorb negatively charged redox probes in solution ($Fe(CN)_6^{3-/4-}$). The electrochemical detection of PSA is realized based on the mechanism that the antigen–antibody complex can reduce the diffusion of redox probes in solution to the underlying electrode, leading to the decrease in electrochemical signal. The constructed immunosensor can achieve sensitive detection of PSA in the range from 10 pg/mL to 1 μg/mL with a limit of detection (LOD) of 8.1 pg/mL. Sensitive detection of PSA in human serum was also achieved. The proposed disposable immunosensor based on cheap electrode and nanochannel array is expected to provide a new idea for developing a universal immunosensing platform for sensitive detection of tumor markers.

Keywords: disposable immunosensor; electrochemical detection; nanochannel array; patterned ITO electrode; prostate-specific antigen

Citation: Yan, L.; Xu, S.; Xi, F. Disposal Immunosensor for Sensitive Electrochemical Detection of Prostate-Specific Antigen Based on Amino-Rich Nanochannels Array-Modified Patterned Indium Tin Oxide Electrode. *Nanomaterials* **2022**, *12*, 3810. https://doi.org/10.3390/nano12213810

Academic Editor: Camelia Bala

Received: 25 September 2022
Accepted: 24 October 2022
Published: 28 October 2022

1. Introduction

As one of the most common malignant tumors in men worldwide, prostate cancer, an epithelial malignancy of the prostate gland, is regarded as the "invisible killer" of middle-aged and elderly men [1,2]. This is because patients with early-stage prostate cancer can achieve good therapeutic effects or be cured by radical surgery or radiotherapy. However, the early symptoms of prostate cancer are not obvious, and the fatality rate is high when found in the late stage. Therefore, early prevention, early diagnosis, and early treatment are the keys to improving the survival rate of prostate cancer. The clinical diagnosis of prostate cancer mainly relies on digital rectal examination, detection of prostate-specific antigen (PSA) in serum, prostate ultrasound (transrectal), and pelvic magnetic resonance imaging (MRI). Amongst these, the detection of PSA in serum has the advantages of non-invasiveness and easy operation, demonstrating great potential in screening of high-risk groups, early diagnosis, and monitoring of curative effect [3,4]. As a tumor biomarker secreted by prostate epithelial cells, PSA exists in prostate tissue and semen, and is present in very low levels in normal human serum. PSA in normal human serum is generally less than 4 ng/mL. A higher PSA level of 4-10 ng/mL is called the gray zone for prostate cancer diagnosis. Prostate cancer patients commonly have a PSA level higher than 10 ng/mL [5]. As PSA has extremely high tissue specificity, it has become the preferred biomarker for the diagnosis of prostate cancer. The development of a simple and low-cost method for sensitive detection of PSA in serum is of great significance for non-invasive screening and early diagnosis of prostate cancer.

Until now, immunoassays, including enzyme-linked immunoassay (ELISA), chemiluminescence immunoassay, and electrochemiluminescence (ECL) immunoassay based on magnetic bead, were mainly applied for the detection of PSA [6–9]. However, these strategies mostly employ sandwich immunoassays by forming primary antibody (Ab_1)/antigen (Ag, analyte)/labeled secondary antibody (labeled-Ab_2) complexes (Ab_1/Ag/labeled-Ab_2), leading to a cumbersome operating process. In addition, the detection sensitivity of the first two methods is low. Although the latter has high detection sensitivity, it suffers from high detection costs because of the use of streptavidin-coated magnetic microbeads and ECL ruthenium complex-labeled antibodies. The above detection methods are also difficult to adapt to on-site detection. Electrochemical sensors have the advantages of rapid detection, high sensitivity, simple instrument, easy integration, and portability [10–16]. Therefore, electrochemical immunoassays offer advantages for sensitive, convenient, and even in-situ or point-of-care detection of PSA in serum.

A suitable supporting electrode is the basis for constructing electrochemical immunosensors. Since it is difficult to maintain excellent detection performance after regeneration in most immunosensors, the construction of disposable electrochemical immunosensors with inexpensive and one-use only electrodes is highly desirable [17,18]. This is attributed to the disadvantages of electrochemical immunosensors based on reusable electrochemical electrodes, such as high cost, complex fabrication, and difficulty in batch fabrication. For example, the most common renewable electrodes are noble metal electrodes (e.g., Au, Pt electrodes) and carbon electrodes (e.g., glassy carbon electrode (GCE), carbon paste electrodes, carbon fiber electrodes, etc.). However, these electrodes are expensive and often need to be polished with particulate slurries (such as 1 μm, 0.3 μm, and 0.05 μm Al_2O_3 slurry) before use, which is complicated to operate [19,20]. Disposable screen-printed electrodes (SPCE) are inexpensive and could be mass-manufactured. Several SPCE-based immunosensors were developed for PSA detection [21–23]. However, the fabrication of the recognitive interface in these sensors is usually complicated, and it is easy to contaminate the electrode surface. Recently, patterned indium tin oxide (ITO) electrodes attracted much attention as inexpensive and disposable electrodes [24–26]. ITO is prepared by doping high-valent Sn^{4+} into In_2O_3. A large number of free electrons resulting from the doping structure endow ITO with excellent electrical conductivity as an n-type semiconductor. Until now, many techniques (e.g., magnetron sputtering, chemical vapor deposition, sol–gel, electron beam evaporation, etc.) were used to prepare ITO films on different substrates (such as rigid glass, or flexible polyethylene terephthalate (PET), polyimide (PI), and so on) with firm bonding with the substrate, and scratch resistance. Therefore, ITO electrodes have the advantages of rigid or flexible structure, easy patterning, mass production, low cost, and excellent electrochemical performance, demonstrating great potential for constructing disposable electrochemical immunosensors.

The detection modes of electrochemical immunosensors include two categories [27,28]. One is to the directly or indirectly electrochemical signals generated by labels in the formed Ab_1/Ag/labeled-Ab_2 sandwich complex. The other type is to achieve label-free detection by redox probes in solution or immobilized on the electrode surface [29–31]. Briefly, the binding of antibodies towards the detected antigen on the immuno-recognitive interface changes the interface resistance of the electrode, which, in turn, leads to a change in the electrical signal of the probe. Amongst these, the detection based on solution-state probes has the advantages of convenient operation and simple electrode construction. The improvement in the detection sensitivity of this solution–probe-based immunosensor by introducing nanomaterials with signal amplification is crucial. Very recently, the remarkable signal amplification effect by vertically ordered mesoporous silica nanochannel film (VMSF) attracted much attention [32–34]. VMSF is a nanometer ultrathin film (50~200 nm in thickness) composed of silica nanochannels parallel to each other with high density (up to 3~12 × 10^{12} cm^{-2}) and uniform pore size (usually 2–3 nm in diameter) [35–37]. On the one hand, the open and high-density nanochannel array ensures efficient diffusion of small molecules. On the other hand, the ultra-small nanochannels have an ultra-high specific surface area, showing excellent charge-based permselectivity [38–40]. The variability of VMSF structure endows it with flexible enrichment towards small molecules with different charges [40,41]. For example, the ionization of silanol groups (Si-OH, pK_a~2) on commonly prepared VMSF using tetraethoxysiloxane (TEOS) as the precursor provides negative charge that can repel anions, but shows efficient enrichment towards cations [42–44]. When VMSF with rich amino groups (NH_2-VMSF) is prepared using 3-aminopropyltriethoxysilane-APTES as a precursor, it has a large number of positively charged sites, leading to a significant attraction on negatively charged probes [45]. Therefore, VMSF-modified electrodes can significantly enrich small molecule probes in solution, improving the detection sensitivity of the electrochemical immunosensors. In addition, the size exclusion effect of ultra-small nanochannels can avoid the contamination of the electrode surface by the complex matrix (e.g., proteins) in biological samples [46]. Therefore, VMSF-modified disposable electrodes have great potential in the convenient and sensitive detection of PSA in serum.

Herein, a label-free electrochemical immunosensing platform was fabricated for sensitive detection of PSA in human serum based on modification of miniaturized, integrated, and disposable ITO electrodes with amino-rich nanochannel arrays (NH_2-VMSF). When surfactant micelle (SM) and amino-containing siloxanes were used as template and precursor, respectively, NH_2-VMSF was rapidly grown (<10 s) by electrochemical-assisted self-assembly (EASA) method. To achieve covalent immobilization of the recognitive antibody (Ab), the amino groups on the outer surface of NH_2-VMSF reacted with bifunctional glutaraldehyde to generate aldehyde-based surfaces. The blocking of the nanochannels by SM ensures that the aldehydeylation occurs only at the entrance of the nanochannels and not within the nanochannels. The open nanochannel array after SM was removed exhibited remarkable enrichment towards anionic electrochemical redox probes ($Fe(CN)_6^{3-/4-}$) in the solution. Since the immunocomplex formed by the binding of Ab and PSA on the immuno-recognitive interface hindered the diffusion of $Fe(CN)_6^{3-/4-}$ to the underlying electrode, the immunosensor can realize sensitive detection of PSA. In comparison with the immunosensors with complicated fabrication process, our immunosensor has the advantages of simple fabrication and high sensitivity.

2. Materials and Methods

2.1. Chemicals and Materials

All reagents used in the experiment were of analytical grade without further treatment. Prostate-specific antigen (PSA), mouse anti-human PSA monoclonal antibody (Ab), carcinoembryonic antigen (CEA), and carcinoma antigen 125 (CA125) were purchased from Beijing KEY-BIO Biotech Co., Ltd. (Beijing, China). Bone gamma-carboxyglutamate protein (BGP) was purchased from Nanjing Okay Biotechnology Co., Ltd. (Jiangsu, China). S100 calcium-binding protein β (S 100) was obtained from Proteintech (Wuhan, China). Potassium ferricyanide ($K_3[Fe(CN)_6]$), potassium ferricyanide ($K_4[Fe(CN)_6]$), tetraethyl orthosilicate (TEOS), cetyltrimethylammonium bromide (CTAB), potassium hydrogen phthalate (KHP), glutaraldehyde (GA), and fetal bovine serum (BSA) were all purchased from Aladdin Biochemical Technology Co., Ltd. (Shanghai, China). Sodium nitrate ($NaNO_3$)

was obtained from Prospect Chemical Reagent Co., Ltd. (Wuxi, China). Additionally, 3-aminopropyltriethoxysilane (APTES) was obtained from Macklin biochemical Technology Co., Ltd. (Shanghai, China). Anhydrous ethanol and sodium hydroxide (NaOH) were purchased from Gaojing Fine Chemical Co., Ltd. (Hangzhou, China). Phosphate buffer (PBS, 0.01 M, pH 7) was prepared by Na_2HPO_4 and NaH_2PO_4. Deionized water (18.2 MΩ cm) was prepared by Mill-Q system (Millipore Company, Shanghai, China). ITO-coated glasses (<17 Ω/square, thickness: 100 ± 20 nm) obtained from Zhuhai Kaivo Optoelectronic Technology (Zhuhai, China) were first cleaned by NaOH aqueous solution (1 M), and subsequently sonicated in acetone, ethanol, and ultrapure water prior to use.

2.2. Measurements and Instrumentations

The morphology of NH_2-VMSF was investigated by transmission electron microscope (TEM, JEM-2100, JEOL, Tokyo, Japan) with an acceleration voltage of 200 kV. Before measurement, the NH_2-VMSF on the ITO electrode was scraped off slowly with a blade, dispersed evenly with ethanol, and dripped on the copper net. The morphology and thickness of NH_2-VMSF were characterized by scanning electron microscope (SEM, SU8010, Hitachi, Tokyo, Japan) with an acceleration voltage of 5 kV. Before investigation, the electrode surface was scratched with a glass knife and divided into small pieces, the cross-section was placed upward and stuck on the sample table with conductive adhesive. Then, the sample was observed after spraying with gold. X-ray photoelectron spectroscopy (XPS) analysis was carried out on a PHI5300 electron spectrometer using 250 W, 14 kV, Mg Kα radiation (PE Ltd., Boston, MA, United States). Electrochemical impedance spectroscopy (EIS), cyclic voltammetry (CV), and differential pulse voltammetry (DPV) measurements were performed on an Autolab (PGSTAT302N) electrochemical workstation (Metrohm, Switzerland). A traditional three-electrode system was employed for electrochemicals. Briefly, bare or modified ITO was used as the working electrode, Ag/AgCl was used as the reference electrode, and a platinum wire electrode was used as the counter electrode. The scan rate used in cyclic voltammetry (CV) measurement was 50 mV/s, the parameters for DPV measurements included step potential (0.005 V), pulse amplitude (0.05 V), interval time (0.2 s), and pulse time (0.05 s).

2.3. Preparation of NH_2-VMSF/ITO Electrode

As reported previously, NH_2-VMSF was grown on the bare ITO electrode by using the electrochemically assisted self-assembly (EASA) method [47]. Briefly, APTES (0.318 mL) and 1.585 g CTAB (4.34 mM) were added to a mixture of 20 mL $NaNO_3$ (0.1 M, pH 2.6) and 20 mL ethanol. When TEOS (2.732 mL) was added to the solution, it was necessary to adjust the pH to 3 with concentrated HCl. The solution was aged for 2.5 h under stirring before use. Then, the bare ITO electrode was immersed into the above solution and grown for 10 s with constant current density (−0.70 mA/cm^2). After being quickly washed with ultrapure water, the obtained electrode was dried with nitrogen, aged at 120 °C overnight, and termed as SM@NH_2-VMSF/ITO. SM template extraction was achieved by immersing the film electrode in an ethanol solution containing 0.1 M HCl under moderate stirring for 5 min, and the resulting electrode was termed as NH_2-VMSF/ITO.

2.4. Fabrication of Label-Free Immunosensor

Owing to the presence of amino groups on NH_2-VMSF, glutaraldehyde (GA) was chosen as the bifunctional linker for covalent immobilization of the recognitive bioligands. Firstly, the prepared SM@NH_2-VMSF/ITO electrode was soaked in 5 % GA solution for 30 min at 37 °C in dark. After unlinked GA was washed off, the micelle was removed by stirring for 5 min in 0.1M HCl/ethanol (V: V = 1: 1) solution to obtain GA/NH_2-VMSF/ITO. Then, to fabricate the immune recognition interface, the PSA antibody (50 μL, 10 μg/mL) was drop-coated on the surface of GA-NH_2-VMSF/ITO. After incubation at 37 °C for 90 min, the unbound antibody was rinsed with PBS (0.01 M, pH 7). Finally, the obtained electrode was then incubated with BSA solution (1 %, wt%) for 60 min to block the non-specific sites, followed by rinsing with PBS. The prepared immunosensor was denoted as Ab/GA/NH_2-VMSF/ITO and stored in a refrigerator at 4°C.

2.5. Electrochemical Determination of PSA

The Ab/NH_2-VMSF/ITO immunosensor was incubated with different concentrations of PSA (antigen) at 37 °C for 45 min before electrochemical testing. KCl (0.1 M) containing $Fe(CN)_6^{3-/4-}$ (2.5 mM) was applied as the electrolyte solution. The electrochemical signal of the Fe $(CN)_6^{3-/4-}$ in the electrolyte was measured before and after PSA binding. Healthy human serum was diluted 50-fold with PBS (0.01 M, pH 7) for real sample analysis. To simulate the different PSA concentrations of prostate cancer patients, artificial PSA was added to the serum and then detected with the developed immunosensor.

3. Results and Discussion

3.1. Fabrication of Immunosensor on Amino-Rich Nanochannel Array-Modified Electrode

In recent years, researchers developed electrochemical-assisted growth method (EASA) or Stöber-solution growth method to induce the synergistic occurrence of surfactant molecular self-assembly and organosilane hydrolysis/polycondensation [48,49], which establish the growth of vertically ordered mesoporous silica nanochannel films (VMSF) on solid surfaces. VMSF has excellent properties, including ultra-thin and adjustable thickness, highly uniform pore size and nanochannel distribution, extremely high porosity, excellent mechanical/chemical/thermal stability and biocompatibility, and easy surface functionalization [50,51]. Moreover, the cost is low, and it can be prepared in batches in a large area, so it is an ideal electrode modification material. VMSF can be stably modified on indium tin oxide (ITO) electrodes [52,53]. This is attributed to the covalent bonding between VMSF and the ITO surface through the formation of -Si-O-In- or -Si-O-Sn- bonds. In this work, VMSFs containing amino groups (NH_2-VMSF) were grown on the surface of patterned ITO electrode by the EASA method, which can grow VMSFs in a very short time (within 10 s). Although antibodies, as protein macromolecules, cannot enter the nanochannels of VMSFs, they can be easily immobilized on the outer surface of VMSFs, which is the entrance of the nanochannels. By introducing reactive groups (such as -NH_2 groups) into VMSFs, the conversion of chemical groups can be flexibly achieved through covalent reactions.

As illustrated in Figure 1, the patterned ITO electrode consists of a square area as the working electrode and a thin linear section as the wire. Insulating tape is applied to the intersection of the thin wire and the working electrode area to ensure that different electrodes have a consistent electrode area. When NH_2-VMSF is grown on ITO, SM remains within the nanochannels (SM@NH_2-VMSF/ITO), thereby blocking the nanochannel. Then, aldehyde group derivatization is performed on the outer surface of NH_2-VMSF to obtain a surface with aldehyde groups through reactions between amino groups and glutaraldehyde (GA). After removal of SM, the covalent immobilization of the recognitive antibody (Ab) is subsequently achieved through the reaction between aldehyde groups and amino groups in Ab. The immunosensor Ab/GA/NH_2-VMSF/ITO is obtained after blocking non-specific sites on the electrode surface with bovine serum albumin (BSA). Since Ab can bind of the targeted PSA, the formed immunocomplex prevents the entry of $Fe(CN)_6^{3-/4-}$ probe in solution, resulting in a reduction in the electrochemical signal. Based on this mechanism, electrochemical detection of PSA can be achieved.

3.2. Characterization of Morphology and Structure of NH_2-VMSF

The morphology of NH_2-VMSF was characterized by scanning electron microscopy (SEM) and transmission electron microscopy (TEM). In Figure 2a, the cross-sectional SEM image of the NH_2-VMSF/ITO shows the layered structure, including the glass layer, the ITO layer, and the NH_2-VMSF layer. The thickness of NH_2-VMSF is 78 nm (Figure 2b). Figure 2c is the top-view TEM image of NH_2-VMSF after scraping off the electrode. As seen, NH_2-VMSF has a continuous porous structure. There are no cracks in the larger area. The high-resolution TEM (HRTEM) image reveals a hexagonal packing structure (inset of

Figure 2c). The pore diameter is between 2–3 nm. The pore density is ~$7.5 \times 10^{12}/cm^2$, corresponding to a porosity of ~44 %.

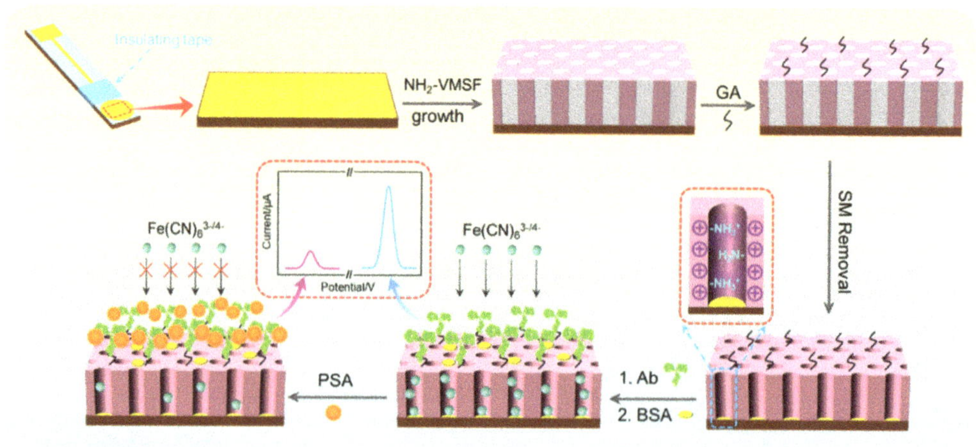

Figure 1. Schematic illustration for the fabrication of label-free immunosensor and the determination of PSA using immunocomplex-gated electrochemical signal.

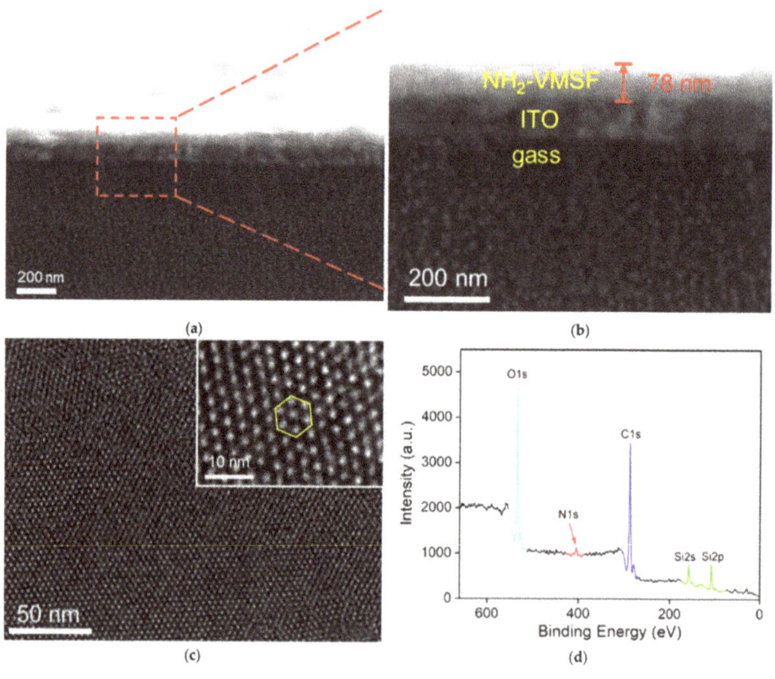

Figure 2. (**a**,**b**) The cross-sectional SEM image of the NH_2-VMSF/ITO at different magnificence. (**c**) Top-view TEM image of NH_2-VMSF at different magnifications. Inset is the HRTEM image. (**d**) X-ray photoelectron spectrum (XPS) of SM@NH_2-VMSF/ITO.

NH_2-VMSF was prepared using TEOS and APTES as mixed siloxanes. The functional groups on the outer surface of NH_2-VMSF were applied for the fabrication of immuno-recognitive interface. To investigate the elemental composition of the outer surface of NH_2-VMSF, X-ray photoelectron spectroscopy (XPS) of SM@NH_2-VMSF/ITO was investigated. As shown in Figure 2d, in addition to C element (from SM), O and Si elements (from SiO_2 structure) and the signal peaks of N also appear, proving that APTES introduces NH_2 groups to NH_2-VMSF. The active NH_2 groups endow NH_2-VMSF/ITO with great potential for further modification and functionalization.

3.3. Charge-Based Selective Permeability or Redox Probe in NH_2-VMSF

In the growth process of NH_2-VMSF on ITO electrode, SM closed and open nanochannel array-modified electrodes are obtained successively. The integrity of NH_2-VMSF and the penetration of probes were investigated by examining the electrochemical behavior of anionic ($Fe(CN)_6^{3-/4-}$, Figure 3a) and cationic ($Ru(NH_3)_6^{3+}$, Figure 3b) standard redox probes on different electrodes. As controls, the supporting ITO electrode and VMSF-modified electrodes without amino groups (VMSF/ITO) were also examined. As shown, the redox peaks of the two probes are obvious on the ITO electrode. Despite the excellent electrochemical properties of ITO, its surface is difficult to derivatize directly. Furthermore, in practical applications, co-existing components in complex matrices tend to adhere to the ITO surface through non-specific adsorption, leading to fouling of the electrode and degraded electrochemical performance. In the case of the SM@NH_2-VMSF/ITO electrode, no electrochemical signals are observed for either redox probes. This is due to the fact that the hydrophobic SM blocks the nanochannels, so that hydrophilic probes in solution cannot diffuse to the underlying electrode. This phenomenon proves that the grown VMSF is intact without cracks. When SM was removed to obtain an open nanochannel array, significant oxidation and reduction peaks of the two probes appear on the NH_2-VMSF/ITO electrode. As shown, VMSF/ITO exhibits higher peak current in $Ru(NH_3)_6^{3+}$ solution in comparison with ITO. In the employed pH of the electrolyte solution (pH 4), the ionization of abundant silanol groups (pK_a~2) in VMSF make a negatively charged surface, leading to electrostatic adsorption effect on cationic probe $Ru(NH_3)_6^{3+}$ and a higher CV signal. Compared with ITO, the electrochemical signal of NH_2-VMSF/ITO recovers to some extent because of the reduced active surface after the growth of non-conductive NH_2-VMSF. It is worth noting that NH_2-VMSF/ITO shows different peak currents for both probes in comparison with that of VMSF/ITO. As is well-known, the ionization of abundant silanol groups (pK_a~2) in VMSF make a negatively charged surface. The introduction of amino groups endows NH_2-VMSF with positively charged sites. Due to the high surface area of the nanochannel arrays, the modified electrodes exhibit remarkable charge-based permselectivity for ion probes. Specifically, the electrostatic repulsion between the positively charged sites of NH_2-VMSF and $Ru(bpy)_3^{2+}$ leads to a lower peak current on NH_2-VMSF/ITO than that of VMSF/ITO. In the negatively charged probe solution, the situation is just the opposite. The signal on NH_2-VMSF/ITO is higher than that of VMSF/ITO electrode due to the electrostatic attraction towards the negatively charge probe. Thus, the modification of NH_2-VMSF facilitates the diffusion of negative Fe $(CN)_6^{3-/4-}$ to the underlying electrode.

3.4. Electrochemical Characterization of the Fabrication Process of Immunosensor

Electrochemical methods including cyclic voltammetry (CV) and electrochemical impedance spectroscopy (EIS) were used to investigate the changes in the electrode surface during the fabrication of the immunosensor. Figure 4a shows the cyclic voltammetric curves of different electrodes. GA/NH_2-VMSF/ITO was obtained by reacting SM@NH2-VMSF/ITO with GA and then removing SM. It can be seen that the peak current of redox probes on GA/NH_2-VMSF is slightly reduced, which may be attributed to the cross-linking of partial amino groups by GA at the entrance of nanochannels. The peak current of the redox probe on the resulting immunosensor Ab/GA/NH_2-VMSF/ITO further reduces when Ab is covalently immobilized on the aldehyde surface and the following blocking of non-specific binding sites by BSA. This is attributed to the fact that proteins act as non-conductive species to increase the interfacial resistance of the electrode surface. When the immuno-electrode is incubated with PSA, the probe signal significantly decreases due to the formation of antigen–antibody immunocomplexes that hinder the entry of the probe into the nanochannel, reducing the electrochemical signal of the probe. Similar conclusions are verified by EIS, as shown in Figure 4b, with the electron transfer resistance (Ret) related to the semicircle diameter of each curve gradually increasing with GA modification, fabrication of immuno-recognitive interface, and PSA binding. The above results validate the efficient development of the immunosensor.

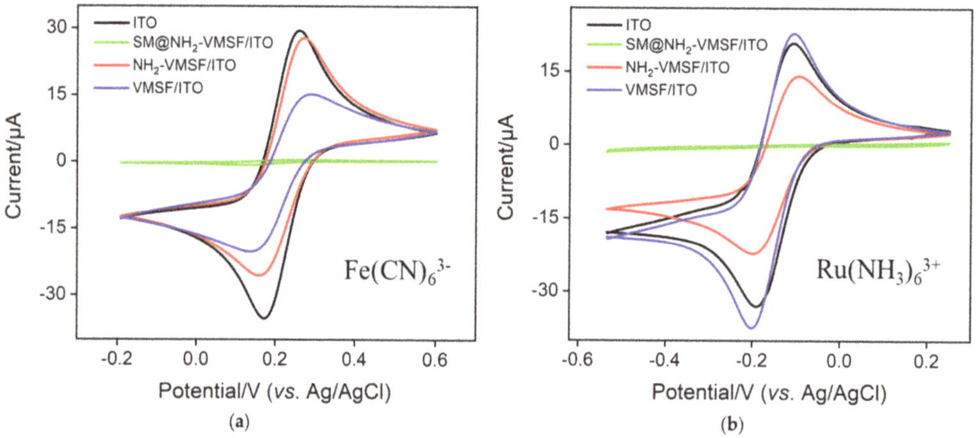

Figure 3. CV curves obtained from different electrodes in 50 mM KHP (pH 4) containing 0.5 mM $Fe(CN)_6^{3-}$ (**a**) and $Ru(NH_3)_6^{3+}$ (**b**).

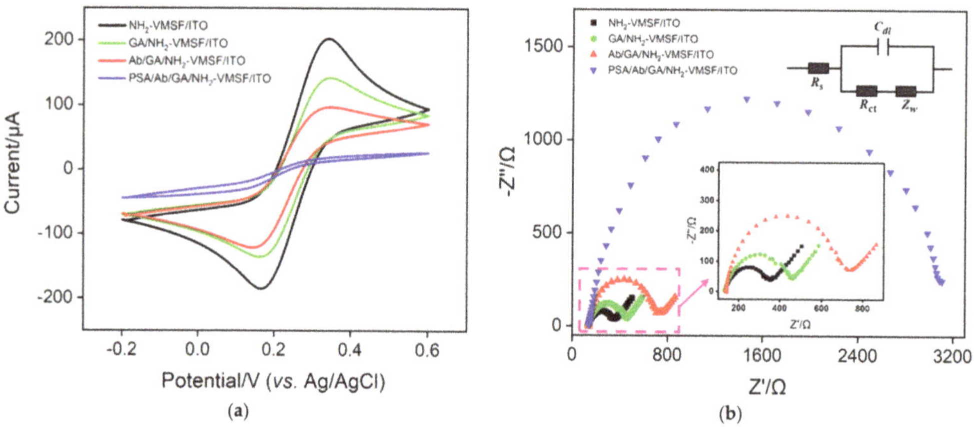

Figure 4. CV (**a**) and EIS (**b**) curves obtained on different electrodes. The electrolyte solution is 0.1 M KCl containing 2.5 mM Fe $(CN)_6^{3-/4-}$. Insets in (**b**) are equivalent circuits of detection (top inset) and the enlarged view of the EIS curves at the high frequency region (bottom inset).

The aldehydeylation of the outer surface of NH_2-VMSF is the basis for the construction of the immunosensing interface. The effect of aldehydeylation on the performance of the constructed immunosensors using open or SM-closed nanochannel arrays was investigated. Amongst these, the former was denoted as GA-NH_2-VMSF after derivatization with GA. The electrochemical signal of $Fe(CN)_6^{3-/4-}$ on different electrode was determined by DPV. Figure 5a shows the relative current value (I/I_0) obtained on different electrodes, where I_0 is the peak current obtained on the electrode before GA modification, and I

is the peak current obtained on the electrode after stepwise modification. As seen, the signal of the redox probe on the GA-NH_2-VMSF/ITO electrode significantly reduces when the open nanochannel is employed for GA modification. This is attributed to the cross-linking of GA to the amino groups inside the nanochannel, which greatly reduces the diffusion of the probe to the underlying electrode. The immunosensor constructed by this strategy has a very low electrochemical signal. This results in the detection of PSA only in a narrow concentration range. In contrast, when the SM-blocked nanochannel array is aldehydeylated, the reaction is directed to the outer surface of NH_2- because the nanochannels are filled with hydrophobic SM (Figure 5b). After SM removal, the open nanochannels effectively ensure the diffusion of probes. In addition, the immunosensor after binding with Ab still has a high electrochemical signal and the binding of PSA significantly reduces the electrochemical signal, which leads to sensitive electrochemical detection and a wide detection range. Therefore, in this paper, SM-blocked nanochannel arrays are used for the introduction of aldehyde groups to construct immunosensors.

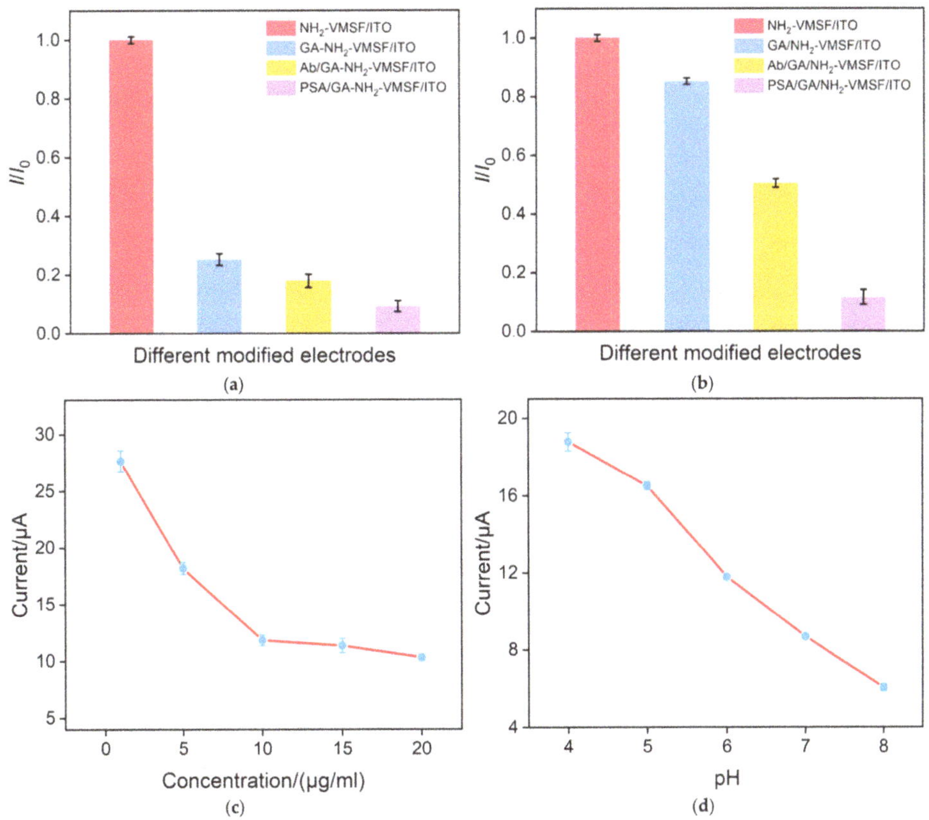

Figure 5. (**a**,**b**) Relative current values (I/I_0) obtained at open (**a**) or SM-closed (**b**) nanochannel array-modified electrodes, where I_0 is the peak current obtained on the electrode before GA modification, and I is the peak current obtained on the electrode after stepwise modification. (**c**,**d**) Peak current obtained in 0.1 M KCl solutions containing 2.5 mM $Fe(CN)^{3-/4-}$ after GA/NH_2-VMSF/ITO was incubated with different concentrations of antibody. (**b**) Peak current obtained on Ab/GA/NH_2-VMSF/ITO in 0.1 M KCl solutions containing 2.5 mM $Fe(CN)^{3-/4-}$ at different pH.

The influence of the amount of antibody or pH value was investigated. Figure 5c reveals the peak current of GA/NH_2-VMSF/ITO after incubation with different concentrations of Ab. When Ab concentration increases, the peak current of redox probe decreases and then reaches an almost stable value. When the antibody is fixed on the outer surface of VMSF, it closes the entrance of some nanochannels because of the large size, leading to reduced diffusion of redox probes into the nanochannels. The appearance of stable signals indicates that the amount of immobilized antibody is close to saturation. Thus, the concentration of Ab was chosen as 10 μg/mL. Figure 5d shows the effect of pH on the peak current of the redox probe on the immunosensor. The peak current increases with the increase in pH value. This is due to the charge change of amino groups under different pH conditions. With the increase in pH, the deprotonation of amino group leads to the gradual decrease in positive sites, weakening the adsorption towards negatively charged probes. Considering the stability of Ab and the neutral environment of serum samples, pH 6 was selected for further experiments.

3.5. Label-Free Electrochemical Determination of PSA

The performance of the constructed sensor in the electrochemical detection of PSA was investigated. The detection principle was the reduction in the electrochemical probe signal in solution after the specific binding of PSA and Ab. The immunosensors were incubated with different concentrations of PSA, and then the DPV curves of the electrodes were determined. As shown in Figure 6a, the DPV peak current of the immunosensor gradually decreases with increasing PSA concentration. The linear regression curve in inset in Figure 6a shows a good linear relationship between the peak current (I) and the logarithmic value of PSA concentration ($\log C_{PSA}$). The detection range is from 10 pg/mL to 1 ug/mL ($I = -1.42\log C_{PSA} + 5.68$, $R^2 = 0.995$). The limit of detection (LOD) calculated based on a three-fold signal-to-noise ratio (S/N=3) is 1.3 pg/mL. The LOD is lower than that obtained based on electrochemical detection using PtCu hollow nanoframes [54], metal-ions-functionalized gold nanoparticles-carbon nanospheres CNSs@AuNPs [55], nafion/graphene oxide/aldehyde methyl pyridine (Nafion/rGO/CHO-MP) [56], or metal organic frame-235/methylene blue (MOF-235/MB) [57]-modified electrode, but higher than that obtained based on palladium-nanoparticles-loaded electroactive amino-zeolitic imidazolate framework-67 (Pd/NH_2-ZIF-67) [58], electrochemiluminescence (ECL) detection based on cadmium sulfide/chitosan/g-C_3N_4 (CdS/Chito/g-C_3N_4) [59], or electrodeposited gold@poly-luminol nanocomposite (Au@PL-NC) [60].

Excellent selectivity is crucial for the practical application of immunosensors. The selectivity of the constructed immunosensor was examined by detecting PSA or other interfering tumor biomarkers, including carcinoembryonic antigen (CEA), bone gamma-carboxyglutamate protein (BGP), S100 calcium-binding protein β (S 100), and carcinoma antigen 125 (CA125). Immunosensors were incubated with a single protein or a mixture of all proteins. As shown in Figure 6b, a significant reduction in electrochemical signal is observed only when the immunosensor is co-incubated with PSA and PSA-containing mixed proteins. The other tested proteins do not cause significant changes in the electrochemical signaling response. This indicates that the constructed immunosensor has high selectivity for PSA. Using the same experimental conditions, we prepared five sensing electrodes in parallel and measured the response to 50 ng/mL PSA to evaluate the reproducibility of the immunosensor construction process. As shown in Figure 6c, the five electrodes show close responses with a relative standard deviation of 3.4 %. Therefore, the electrode preparation process has high reproducibility. The storage stability of the immunoelectrode was also investigated. After storage of the immunosensor at 4 °C for 6 days, the response to 50 ng/mL PSA maintains 91.5 % of the initial signal, indicating the high storage stability of the immunosensor.

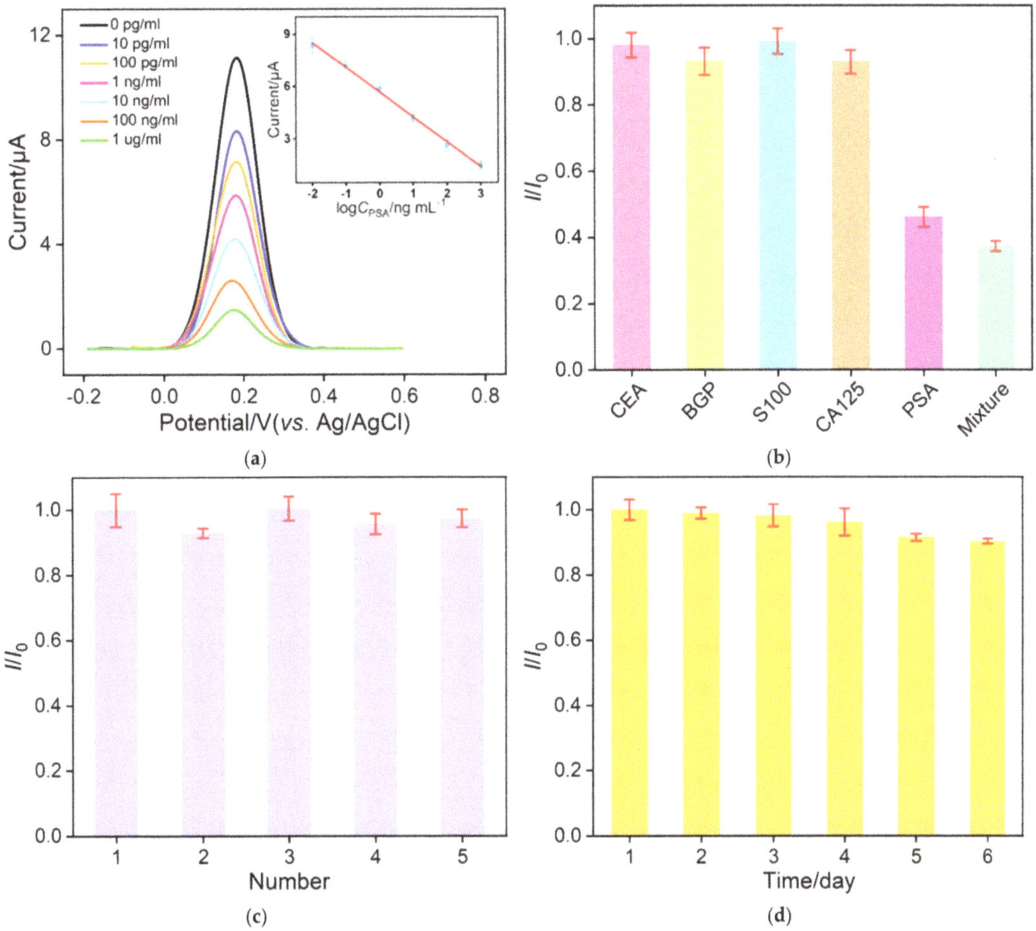

Figure 6. (**a**) DPV response of the immunosensor after incubation with different concentrations of PSA. The inset is the linear relationship between the peak current and the logarithmic value of PSA concentration. (**b**) Relative ratio of DPV peak current before (I_0) and after (I) incubation with CEA (10 ng/ml), BGP (1 ng/mL), S100 (10 ng/ml), CA125 (1 μU/mL), PSA (10 ng/mL), or their mixture. (**c**) Relative ratio of DPV peak current obtained on five parallel electrodes. I0 and I are peak current obtained on the 1st or other electrodes. (**d**) Relative ratio of DPV peak current obtained on immunosensor stored at 4 °C for different times.

3.6. Real Sample Analysis

The practical application of the constructed immunosensor was evaluated by measuring the concentration of PSA in human serum. To investigate the effect of the serum matrix on the detection of PSA, PSA added in serum diluted by different factors was detected. As shown in Figure 7, the detection recovery is poor at a low dilution factor, which may be due to the influence of the viscous matrix on the diffusion of redox probes or the interference of coexisting substances. When the dilution factor is not less than 50 times, the serum matrix has no obvious effect on the detection recovery. The PSA concentration detected by the constructed immunosensor (1.25 ng/mL) is consistent with that (1.32 ng/mL) obtained by the commercial electrochemiluminescence analyzer. In addition, different concentrations of PSA were added to serum samples to mimic cancer patients with higher PSA concentrations. As shown in Table 1, the detected concentration of PSA recovery ranges from 95.7% to 103.2%, indicating good accuracy. Therefore, the immunosensor constructed in this paper has potential application in the clinical detection of PSA.

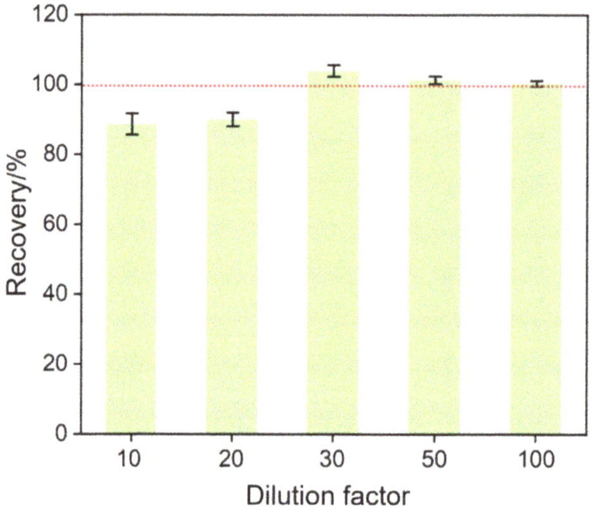

Figure 7. Recovery for the detection of PSA (10 ng/mL) in serum matrix when serum was diluted by different factors.

Table 1. Determination of PSA in human serum.

Sample [a]	Added (ng/mL)	Found (ng/mL)	RSD (%, n = 3)	Recovery (%)
	0.0100	0.0103	2.7	103.2
Human serum [a]	0.100	0.0957	2.8	95.7
	10.0	10.1	3.7	101.4

[a] Samples were diluted 50 times when PSA was added. The indicated concentration of PSA was obtained after dilution.

4. Conclusions

In summary, we constructed a disposable immunosensor for electrochemical detection of PSA in serum based on an array of amino-containing nanochannels modified on patterned ITO electrodes. Using amino-containing siloxane as a precursor, amino-rich VMSF (NH_2-VMSF) was rapidly grown. Directed aldehyde group functionalization of the outer surface of NH_2-VMSF followed by covalent immobilization of the antibody was performed when VMSF contained micelles to close the nanochannels. NH_2-VMSF can facilitate the diffusion of negatively charged redox probes in solution to the underlying electrode. Sensitive electrochemical detection of PSA is achieved based on the hindrance of probe diffusion by immunocomplexes. The immunosensor constructed herein has the advantages of simple construction, sensitive detection, high reproducibility, and good stability. In addition, the ITO electrode used has low preparation cost, a highly adjustable electrode shape and area, and can be mass-produced. Therefore, the provided strategy is expected to be a universal electrochemical immunoassay platform for the convenient and mobile detection of tumor markers.

Author Contributions: Investigation, data curation, S.X.; writing—original draft preparation, L.Y.; writing—review and editing, conceptualization and supervision, F.X. All authors have read and agreed to the published version of the manuscript.

Funding: This research was funded by the National Natural Science Foundation of China (21904117), and the Zhejiang Provincial Natural Science Foundation of China (LY20B050007).

Institutional Review Board Statement: Not applicable.

Informed Consent Statement: Not applicable.

Data Availability Statement: The data presented in this study are available on request from the corresponding author.

Conflicts of Interest: The authors declare no conflict of interest.

References

1. Shokorlou, Y.M.; Heidarzadeh, H. Multispectral plasmonic biosensors based on a Penta-supercell metamaterial for detection of prostate-specific antigen: Ultrasensitive in LC resonance mode. *Biosens. Bioelectron.* **2022**, *217*, 114722. [CrossRef] [PubMed]
2. Mishra, S.; Kim, E.S.; Sharma, P.; Wang, Z.; Yang, S.; Kaushik, A.; Wang, C.; Li, Y.; Kim, N.Y. Tailored biofunctionalized biosensor for the label-free sensing of prostate-specific antigen. *ACS Appl. Bio Mater.* **2020**, *3*, 7821–7830. [CrossRef] [PubMed]
3. Tang, L.; Li, J. Plasmon-based colorimetric nanosensors for ultrasensitive molecular diagnostics. *ACS Sens.* **2017**, *2*, 857–875. [CrossRef] [PubMed]
4. Wang, Y.; Wang, M.; Yu, H.; Wang, G.; Ma, P.; Pang, S.; Jiao, Y.; Liu, A. Screening of peptide selectively recognizing prostate-specific antigen and its application in detecting total prostate-specific antigen. *Sens. Actuators B Chem.* **2022**, *367*, 132009. [CrossRef]
5. Han, L.; Wang, D.; Yan, L.; Petrenko, V.; Liu, A. Specific phages-based electrochemical impedimetric immunosensors for label-free and ultrasensitive detection of dual prostate-specific antigens. *Sens. Actuators B Chem.* **2019**, *297*, 126727. [CrossRef]
6. Dejous, C.; Krishnan, U. Sensors for diagnosis of prostate cancer: Looking beyond the prostate specific antigen. *Biosens. Bioelectron.* **2020**, *173*, 112790. [CrossRef]
7. Ghorbani, F.; Abbaszadeh, H.; Dolatabadi, J.; Aghebati-Maleki, L.; Yousefi, M. Application of various optical and electrochemical aptasensors for detection of human prostate specific antigen: A review. *Biosens. Bioelectron.* **2019**, *142*, 111484. [CrossRef]
8. Negahdary, M.; Sattarahmady, N.; Heli, H. Advances in prostate specific antigen biosensors-impact of nanotechnology. *Clin. Chim. Acta* **2020**, *504*, 43–55. [CrossRef]
9. Ozyurt, C.; Uludag, I.; Ince, B.; Sezginturk, M. Biosensing strategies for diagnosis of prostate specific antigen. *J. Pharm. Biomed. Anal.* **2022**, *209*, 114535. [CrossRef]
10. Zhou, H.; Dong, G.; Sailjoi, A.; Liu, J. Facile pretreatment of three-dimensional graphene through electrochemical polarization for improved electrocatalytic performance and simultaneous electrochemical detection of catechol and hydroquinone. *Nanomaterials* **2022**, *12*, 65. [CrossRef]
11. Gong, J.; Tang, H.; Wang, M.; Lin, X.; Wang, K.; Liu, J. Novel three-dimensional graphene nanomesh prepared by facile electro-etching for improved electroanalytical performance for small biomolecules. *Mater. Design* **2022**, *215*, 110506. [CrossRef]
12. Wang, K.; Yang, L.; Huang, H.; Lv, N.; Liu, J.; Liu, Y. Nanochannel array on electrochemically polarized screen printed carbon electrode for rapid and sensitive electrochemical determination of clozapine in human whole blood. *Molecules* **2022**, *27*, 2739. [CrossRef]
13. Bao, C.; Zhang, R.; Qiao, Y.; Cao, X.; He, F.; Hu, W.; Wei, M.; Lu, W. Au nanoparticles anchored on cobalt boride nanowire arrays for the electrochemical determination of prostate-specific antigen. *ACS Appl. Nano Mater.* **2021**, *4*, 5707–5716. [CrossRef]
14. Chen, M.; Han, R.; Li, Y.; Luo, X. Nonfouling and ratiometric electrochemical detection of prostate specific antigen in whole serum. *Anal. Chim. Acta* **2022**, *1224*, 340191. [CrossRef]
15. He, Y.; Xie, S.; Yang, X.; Yuan, R.; Chai, Y. Electrochemical peptide biosensor based on in situ silver deposition for detection of prostate specific antigen. *ACS Appl. Mater. Interfaces* **2015**, *7*, 13360–13366. [CrossRef]
16. Li, S.; Zhang, J.; Tan, C.; Chen, C.; Hu, C.; Bai, Y.; Ming, D. Electrochemical immunosensor based on hybrid MoS_2/Pt@Au-nanoprism/PDA for simultaneous detection of free and total prostate specific antigen in serum. *Sens. Actuators B Chem.* **2022**, *357*, 131413. [CrossRef]
17. Kabay, G.; Yin, Y.; Singh, C.; Ahmad, N.; Gunasekaran, S.; Mutlu, M. Disposable electrochemical immunosensor for prostate cancer detection. *Sens. Actuators B Chem.* **2022**, *360*, 131667. [CrossRef]
18. Rong, S.; Zou, L.; Li, Y.; Guan, Y.; Guan, H.; Zhang, Z.; Zhang, Y.; Gao, H.; Yu, H.; Zhao, F.; et al. An ultrasensitive disposable sandwich-configuration electrochemical immunosensor based on OMC@AuNPs composites and AuPt-MB for alpha-fetoprotein detection. *Bioelectrochemistry* **2021**, *141*, 107846. [CrossRef]
19. Huang, J.; Zhang, T.; Dong, G.; Zhu, S.; Yan, F.; Liu, J. Direct and sensitive electrochemical detection of bisphenol a in complex environmental samples using a simple and convenient nanochannel-modified electrode. *Front. Chem.* **2022**, *10*, 900282. [CrossRef]
20. Yan, F.; Chen, J.; Jin, Q.; Zhou, H.; Sailjoi, A.; Liu, J.; Tang, W. Fast one-step fabrication of a vertically-ordered mesoporous silica-nanochannel film on graphene for direct and sensitive detection of doxorubicin in human whole blood. *J. Mater. Chem. C* **2020**, *8*, 7113–7119. [CrossRef]
21. Thunkhamrak, C.; Chuntib, P.; Ounnunkad, K.; Banet, P.; Aubert, P.; Saianand, G.; Gopalan, A.I.; Jakmunee, J. Highly sensitive voltammetric immunosensor for the detection of prostate specific antigen based on silver nanoprobe assisted graphene oxide modified screen printed carbon electrode. *Talanta* **2020**, *208*, 120389. [CrossRef] [PubMed]

22. Wan, Y.; Deng, W.; Su, Y.; Zhu, X.; Peng, C.; Hu, H.; Peng, H.; Song, S.; Fan, C. Carbon nanotube-based ultrasensitive multiplexing electrochemical immunosensor for cancer biomarkers. *Biosens. Bioelectron.* **2011**, *30*, 93–99. [CrossRef]
23. Choosang, J.; Khumngern, S.; Thavarungkul, P.; Kanatharana, P.; Numnuam, A. An ultrasensitive label-free electrochemical immunosensor based on 3D porous chitosan-graphene-ionic liquid-ferrocene nanocomposite cryogel decorated with gold nanoparticles for prostate-specific antigen. *Talanta* **2021**, *224*, 121787. [CrossRef] [PubMed]
24. Yan, L.; Zhang, C.; Xi, F. Disposable amperometric label-free immunosensor on chitosan–graphene-modified patterned ITO electrodes for prostate specific antigen. *Molecules* **2022**, *27*, 5895. [CrossRef] [PubMed]
25. Demirbakan, B.; Sezgintürk, M. A novel electrochemical immunosensor based on disposable ITO-PET electrodes for sensitive detection of PAK 2 antigen. *J. Electroanal. Chem.* **2019**, *848*, 113304. [CrossRef]
26. Yadav, A.; Verma, D.; Lakshmi, G.; Eremin, S.; Solanki, P. Fabrication of label-free and ultrasensitive electrochemical immunosensor based on molybdenum disulfide nanoparticles modified disposable ITO: An analytical platform for antibiotic detection in food samples. *Food Chem.* **2021**, *363*, 130245. [CrossRef]
27. Chen, S.; Xu, L.; Sheng, K.; Zhou, Q.; Dong, B.; Bai, X.; Lu, G.; Song, H. A label-free electrochemical immunosensor based on facet-controlled Au nanorods/reduced graphene oxide composites for prostate specific antigen detection. *Sens. Actuators B Chem.* **2021**, *336*, 129748. [CrossRef]
28. Jones, A.; Dhanapala, L.; Kankanamage, R.; Kumar, C.; Rusling, J. Multiplexed immunosensors and immunoarrays. *Anal. Chem.* **2020**, *92*, 345–362. [CrossRef]
29. Zhang, J.; Yang, L.; Pei, J.; Tian, Y.; Liu, J. A reagentless electrochemical immunosensor for sensitive detection of carcinoembryonic antigen based on the interface with redox probe-modified electron transfer wires and effectively immobilized antibody. *Front. Chem.* **2022**, *10*, 939736. [CrossRef]
30. Lin, J.; Li, K.; Wang, M.; Chen, X.; Liu, J.; Tang, H. Reagentless and sensitive determination of carcinoembryonic antigen based on a stable Prussian blue modified electrode. *RSC Adv.* **2020**, *10*, 38316–38322. [CrossRef]
31. Gong, J.; Zhang, T.; Luo, T.; Luo, X.; Yan, F.; Tang, W.; Liu, J. Bipolar silica nanochannel array confined electrochemiluminescence for ultrasensitive detection of SARS-CoV-2 antibody. *Biosens. Bioelectron.* **2022**, *215*, 114563. [CrossRef]
32. Zhou, H.; Ma, X.; Sailjoi, A.; Zou, Y.; Lin, X.; Yan, F.; Su, B.; Liu, J. Vertical silica nanochannels supported by nanocarbon composite for simultaneous detection of serotonin and melatonin in biological fluids. *Sens. Actuators B Chem.* **2022**, *353*, 131101. [CrossRef]
33. Yan, F.; Luo, T.; Jin, Q.; Zhou, H.; Sailjoi, A.; Dong, G.; Liu, J.; Tang, W. Tailoring molecular permeability of vertically-ordered mesoporous silica-nanochannel films on graphene for selectively enhanced determination of dihydroxybenzene isomers in environmental water samples. *J. Hazard. Mater.* **2021**, *410*, 124636. [CrossRef]
34. Ma, X.; Liao, W.; Zhou, H.; Tong, Y.; Yan, F.; Tang, H.; Liu, J. Highly sensitive detection of rutin in pharmaceuticals and human serum using ITO electrodes modified with vertically-ordered mesoporous silica-graphene nanocomposite films. *J. Mater. Chem. B* **2020**, *8*, 10630–10636. [CrossRef]
35. Walcarius, A. Electroinduced surfactant self-assembly driven to vertical growth of oriented mesoporous films. *Acc. Chem. Res.* **2021**, *54*, 3563–3575. [CrossRef]
36. Zhou, H.; Ding, Y.; Su, R.; Lu, D.; Tang, H.; Xi, F. Silica nanochannel array film supported by ß-cyclodextrin-functionalized graphene modified gold film electrode for sensitive and direct electroanalysis of acetaminophen. *Front. Chem.* **2022**, *9*, 812086. [CrossRef]
37. Zhu, X.; Xuan, L.; Gong, J.; Liu, J.; Wang, X.; Xi, F.; Chen, J. Three-dimensional macroscopic graphene supported vertically-ordered mesoporous silica-nanochannel film for direct and ultrasensitive detection of uric acid in serum. *Talanta* **2022**, *238*, 123027. [CrossRef]
38. Liang, R.; Jiang, J.; Zheng, Y.; Sailjoi, A.; Chen, J.; Liu, J.; Li, H. Vertically oriented mesoporous silica film modified fluorine-doped tin oxide electrode for enhanced electrochemiluminescence detection of lidocaine in serum. *RSC Adv.* **2021**, *11*, 34669–34675. [CrossRef]
39. Zheng, W.; Su, R.; Lin, X.; Liu, J. Nanochannel array modified three-dimensional graphene electrode for sensitive electrochemical detection of 2,4,6-trichlorophenol and prochloraz. *Front. Chem.* **2022**, *10*, 954802. [CrossRef]
40. Zhou, P.; Yao, L.; Chen, K.; Su, B. Silica nanochannel membranes for electrochemical analysis and molecular sieving: A comprehensive review. *Crit. Rev. Anal. Chem.* **2019**, *50*, 424–444. [CrossRef]
41. Xi, F.; Xuan, L.; Lu, L.; Huang, J.; Yan, F.; Liu, J.; Dong, X.; Chen, P. Improved adhesion and performance of vertically-aligned mesoporous silica-nanochannel film on reduced graphene oxide for direct electrochemical analysis of human serum. *Sens. Actuators B Chem.* **2019**, *288*, 133–140. [CrossRef]
42. Wang, M.; Lin, J.; Gong, J.; Ma, M.; Tang, H.; Liu, J.; Yan, F. Rapid and sensitive determination of doxorubicin in human whole blood by vertically-ordered mesoporous silica film modified electrochemically pretreated glassy carbon electrodes. *RSC Adv.* **2021**, *11*, 9021–9028. [CrossRef] [PubMed]
43. Ma, K.; Zheng, Y.; An, L.; Liu, J. Ultrasensitive immunosensor for prostate-specific antigen based on enhanced electrochemiluminescence by vertically ordered mesoporous silica-nanochannel film. *Front. Chem.* **2022**, *10*, 851178. [CrossRef] [PubMed]
44. Zhou, Z.; Guo, W.; Xu, L.; Yang, Q.; Su, B. Two orders-of-magnitude enhancement in the electrochemiluminescence of by vertically ordered silica mesochannels. *Anal. Chim. Acta* **2015**, *886*, 48–55. [CrossRef] [PubMed]

45. Ma, K.; Yang, L.; Liu, J.; Liu, J. Electrochemical sensor nanoarchitectonics for sensitive detection of uric acid in human whole blood based on screen-printed carbon electrode equipped with vertically-ordered mesoporous silica-nanochannel film. *Nanomaterials* **2022**, *12*, 1157. [CrossRef]
46. Zhou, L.; Hou, H.; Wei, H.; Yao, L.; Sun, L.; Yu, P.; Su, B.; Mao, L. In vivo monitoring of oxygen in rat brain by carbon fiber microelectrode modified with antifouling nanoporous membrane. *Anal. Chem.* **2019**, *91*, 3645–3651. [CrossRef]
47. Gong, J.; Zhang, T.; Chen, P.; Yan, F.; Liu, J. Bipolar silica nanochannel array for dual-mode electrochemiluminescence and electrochemical immunosensing platform. *Sens. Actuators B Chem.* **2022**, *368*, 132086. [CrossRef]
48. Walcarius, A.; Sibottier, E.; Etienne, M.; Ghanbaja, J. Electrochemically assisted self-assembly of mesoporous silica thin films. *Nat. Mater.* **2007**, *6*, 602–608. [CrossRef]
49. Teng, Z.; Zheng, G.; Dou, Y.; Li, W.; Mou, C.; Zhang, X.; Asiri, A.; Zhao, D. Highly ordered mesoporous silica films with perpendicular mesochannels by a simple Stöber-solution growth approach. *Angew. Chem. Int. Ed.* **2012**, *51*, 2173–2177. [CrossRef]
50. Zou, Y.; Zhou, X.; Xie, L.; Tang, H.; Yan, F. Vertically-ordered mesoporous silica films grown on boron nitride-graphene composite modified electrodes for rapid and sensitive detection of carbendazim in real samples. *Front. Chem.* **2022**, *10*, 939510. [CrossRef]
51. Yan, F.; Ma, X.; Jin, Q.; Tong, Y.; Tang, H.; Lin, X.; Liu, J. Phenylboronic acid-functionalized vertically ordered mesoporous silica films for selective electrochemical determination of fluoride ion in tap water. *Microchim. Acta* **2020**, *187*, 470. [CrossRef]
52. Zhang, M.; Zou, Y.; Zhou, X.; Yan, F.; Ding, Z. Vertically-ordered mesoporous silica films for electrochemical detection of Hg(II) ion in pharmaceuticals and soil samples. *Front. Chem.* **2022**, *10*, 952936. [CrossRef]
53. Luo, X.; Zhang, T.; Tang, H.; Liu, J. Novel electrochemical and electrochemiluminescence dual-modality sensing platform for sensitive determination of antimicrobial peptides based on probe encapsulated liposome and nanochannel array electrode. *Front. Nutr.* **2022**, *9*, 962736. [CrossRef]
54. Chen, Y.; Yuan, P.; Wang, A.; Luo, X.; Xue, Y.; Zhang, L.; Feng, J. A novel electrochemical immunosensor for highly sensitive detection of prostate-specific antigen using 3D open-structured PtCu nanoframes for signal amplification. *Biosens. Bioelectron.* **2019**, *126*, 187–192. [CrossRef]
55. Li, L.; Wei, Y.; Zhang, S.; Chen, X.; Shao, T.; Feng, D. Electrochemical immunosensor based on metal ions functionalized CNSs@Au NPs nanocomposites as signal amplifier for simultaneous detection of triple tumor markers. *J. Electroanal. Chem.* **2021**, *880*, 114882. [CrossRef]
56. Jeong, S.; Barman, S.; Yoon, H.; Park, J. A prostate cancer detection immunosensor based on Nafion/Reduced graphene oxide/aldehyde functionalized methyl pyridine composite electrode. *J. Electrochem. Soc.* **2019**, *166*, B920–B926. [CrossRef]
57. Zhang, M.; Hu, X.; Mei, L.; Zhang, L.; Wang, X.; Liao, X.; Qiao, X.; Hong, C. PSA detection electrochemical immunosensor based on MOF-235 nanomaterial adsorption aggregation signal amplification strategy. *Microchem. J.* **2021**, *171*, 106870. [CrossRef]
58. Dai, L.; Li, Y.; Wang, Y.; Luo, X.; Wei, D.; Feng, R.; Yan, T.; Ren, X.; Du, B.; Wei, Q. A prostate-specific antigen electrochemical immunosensor based on Pd NPs functionalized electroactive Co-MOF signal amplification strategy. *Biosens. Bioelectron.* **2019**, *132*, 97–104. [CrossRef]
59. Huang, B.; Liu, X.; Chen, J.; Mao, C.; Niu, H.; Jin, B. Electrochemiluminescence immunoassay for the prostate-specific antigen by using a CdS/chitosan/g-C3N4 nanocomposite. *Microchim. Acta.* **2020**, *187*, 155. [CrossRef]
60. Paramasivam, S.; Veerapandian, M.; Kumar, S. Rational design of effective solid-state electrochemiluminescence platform of Gold@Polyluminol nanocomposite as an ultrasensitive immuno-probe for selective detection of prostate specific antigen. *Anal. Chim. Acta.* **2022**, *1206*, 339736. [CrossRef]

nanomaterials

MDPI

Review

In Situ Assembly of Nanomaterials and Molecules for the Signal Enhancement of Electrochemical Biosensors

Yong Chang, Ning Xia, Yaliang Huang, Zhifang Sun * and Lin Liu *

College of Chemistry and Chemical Engineering, Anyang Normal University, Anyang 455000, China; 7180610011@stu.jiangnan.edu.cn (Y.C.); xianing82414@csu.edu.cn (N.X.); 182301010@csu.edu.cn (Y.H.)
* Correspondence: allensune@gmail.com (Z.S.); liulin@aynu.edu.cn (L.L.)

Abstract: The physiochemical properties of nanomaterials have a close relationship with their status in solution. As a result of its better simplicity than that of pre-assembled aggregates, the in situ assembly of nanomaterials has been integrated into the design of electrochemical biosensors for the signal output and amplification. In this review, we highlight the significant progress in the in situ assembly of nanomaterials as the nanolabels for enhancing the performances of electrochemical biosensors. The works are discussed based on the difference in the interactions for the assembly of nanomaterials, including DNA hybridization, metal ion–ligand coordination, metal–thiol and boronate ester interactions, aptamer–target binding, electrostatic attraction, and streptavidin (SA)–biotin conjugate. We further expand the range of the assembly units from nanomaterials to small organic molecules and biomolecules, which endow the signal-amplified strategies with more potential applications.

Keywords: electrochemical biosensors; self-assembly; nanomaterials; hybridization; peptide; streptavidin

Citation: Chang, Y.; Xia, N.; Huang, Y.; Sun, Z.; Liu, L. In Situ Assembly of Nanomaterials and Molecules for the Signal Enhancement of Electrochemical Biosensors. *Nanomaterials* **2021**, *11*, 3307. https://doi.org/10.3390/nano11123307

Academic Editors: Dong Liu, Baiqing Yuan and Camelia Bala

Received: 4 November 2021
Accepted: 4 December 2021
Published: 6 December 2021

Publisher's Note: MDPI stays neutral with regard to jurisdictional claims in published maps and institutional affiliations.

1. Introduction

During the past several decades, miscellaneous nanomaterials with various elements and different morphologies have been designed and synthesized. Owing to the amazing and powerful properties, they have been widely utilized in chemical, physical, and biological-related fields. Generally, the properties of nanomaterials are closely related to their physicochemical parameters, including composition, shape, and size. For instance, silver nanoparticles (AgNPs) can generate a well-defined and amplified electrochemical peak based on the highly characteristic solid-state Ag/AgCl process [1], which have been widely used as the electrochemical tracers for the detection of various targets [2]. The status of nanomaterials in solution (monodispersion and aggregation) may have an important influence on their performances. For example, gold nanoparticles (AuNPs) and AgNPs exhibit different local surface plasmon resonance (LSPR) adsorption and endow the solution with different color. The inherent enzyme-mimetic catalytic activity can also be reversibly regulated by modulating the status of nanozymes, such as MoS_2, quantum dots (QDs), $Cu_{2-x}Se$ NPs, and AuNPs [3–6]. Traditional fluorescent dyes show aggregation-induced quenching properties, but on the contrary, aggregation-induced emission (AIE) phenomenon was also observed in various organic molecules and nanomaterials [7–10]. Therefore, lots of optical methods (fluorescence and colorimetric) have been precisely explored to analyze targets of interest based on the aggregation-induced effect, including metal ions, small molecules, DNA, and enzymes [11,12]. Up to now, several mechanisms for stimulating the aggregation of nanomaterials have been demonstrated, including DNA hybridization, antigen–antibody association, aptamer–target binding, electrostatic attraction, streptavidin (SA)–biotin interaction, metal ion–ligand coordination, and covalent bond formation [13–18]. However, the homogeneous methods based on target-recognition-induced aggregation in solution are less sensitive.

As a result of the merits of low cost, high sensitivity, and ease of calibration and miniaturization, electrochemical biosensors have received broad research interest for applications from environmental monitoring to food safety and disease diagnosis. To further improve their sensitivity, plenty of signal-amplified strategies have been designed and applied in electrochemical assays. Usually, the signal-amplified strategies contain enzyme catalysis (e.g., horseradish peroxidase and alkaline phosphatase), DNA assembly techniques such as catalytic hairpin assembly (CHA) and hybridization chain reaction (HCR), and various functional nanomaterials [19–21]. Among these protocols, nanomaterials have shown great promise for improving the sensitivity and selectivity of electrochemical assays because of their excellent characteristics, such as good conductivity, high surface to volume ratio, and ease of functionalization. Up to now, nanomaterials, including carbon nanotubes, graphene, semiconductor quantum dots (QDs), metal nanoparticles (NPs), and 2D layered nanosheets, have been exploited for the fabrication of electrochemical biosensors with enhanced performances [22]. During various electrochemical assays, nanomaterials mainly play four important roles for enhancing the detection performances: electrode substrate modifiers, nanoelectrocatalysts, nanocarriers for enzymes and recognition elements, and electroactive tracers. For example, carbon nanomaterials are frequently utilized in electroanalytical and electrocatalytic sensing fields because of their high electrical conductivity, excellent surface to volume ratio, and chemical stability [23]. Gold nanoparticles (AuNPs) are easily incorporated with enzymes and biorecognition elements, perfectly combining the catalytic properties of enzymes, recognition abilities of biorecognition elements with excellent electrochemical properties of AuNPs [24].

It is a promising signal-amplified strategy to integrate the organized assembly or unordered aggregation of nanomaterials into electrochemical methods [25]. Moreover, the excellent conductivity and electrocatalytic activity of nanomaterials can significantly enhance the conductivity and redox current. For instance, Chen et al. employed melamine to induce the formation of the PdPt nanodendrites–melamine networks based on the firm interactions between the nanodendrites and the three amino groups of each melamine molecule [26]. Then, the formed networks with excellent catalytic ability were utilized as the labels to increase the current for gene mutation detection. AgNPs aggregates induced by the hybridization of DNA on AgNPs were successfully applied for multiplexed DNA target detection [27]. However, these aggregate tags were synthesized through hybridization prior to the sandwich assays, which may affect the uniformity of the size of the aggregates. In 2015, Dai et al. introduced the concept of re-creation of the existing platforms, which transferred the NPs-based colorimetric assay into the electrochemical analysis with Hg^{2+} as the model analyte [28]. After that, a number of attempts have been put into the construction of various detection techniques based on the conversion of aggregation-based colorimetirc assays to interfacial analytical assays [29–33]. In addition to acting as the aggregation triggers, biomacromolecules with nanoscale sizes can also be used as the self-assembling building blocks to form diverse nanostructures for drug delivery and biosensing. Meanwhile, the low conductivity of biomacromolecules facilitates the development of electrochemical impedance biosensors.

In this review, we mainly discuss the design and application of electrochemical biosensors based on the in situ assembly of nanomaterials on the electrode surface for signal readout and amplification. To facilitate the readability and comprehension, the works are briefly discussed based on the difference in the interactions for the assembly of nanomaterials. Furthermore, we expand the range of assembly units from nanomaterials to small molecules and biomacromolecules.

2. In Situ Assembly of Nanomaterials for Signal Amplification

2.1. DNA Hybridization

Due to the specificity of base-pairing hybridization, a variety of DNA assembly nanotechnologies have been elaborately designed and versatilely applied in biosensing for signal amplification [34,35]. It is a promising signal amplification strategy to combine the assembly technologies with nanomaterials. Moreover, the excellent conductivity and electrocatalytic activity of nanomaterials can significantly transform the intrinsic electron inert polymeric DNA into a conductive DNA nanostructure. Among them, hybridization-inducing aggregation without the use of enzymes is simple and can be conducted at mild conditions [36]. Song et al. developed a disposable electrochemical aptasensor array for multiplied proteins detection by in situ DNA hybridization-induced AgNPs aggregation for signal amplification [37]. As shown in Figure 1A, AgNPs were modified with two complementary DNA sequences and two kinds of aptamers against platelet-derived growth factor (PDGF-BB) and thrombin, respectively. After the capture of the target and the sandwich-type reaction, DNA-labeled AgNPs were captured to form the AgNPs aggregates on the electrode surface through the in situ hybridization of DNA. A remarkably amplified electrochemical signal was observed by differential pulse stripping voltammetry (DPSV). The sensitivity of this in situ hybridization-induced formation of AgNPs aggregates was calculated to be 10 orders of magnitude higher than that of the single AgNP nanolabel. Moreover, the DNA-induced assembly of AuNPs was employed for protein kinase activity analysis (Figure 1B) [38]. In this paper, the Zr^{4+}-labeled phosphorylated peptide could capture DNA-modified AuNPs (DNA-AuNPs) via the coordination interaction between the phosphate groups in DNA and Zr^{4+} ions Then, DNA-AuNPs polymeric networks were formed in situ by DNA hybridization on the electrode surface. The conductive and negative charged networks could accommodate a large amount of $[Ru(NH_3)_6]^{3+}$ ions by the electrostatic interactions. The current intensity was dramatically enhanced, and a low detection limit and a wide linear range were achieved.

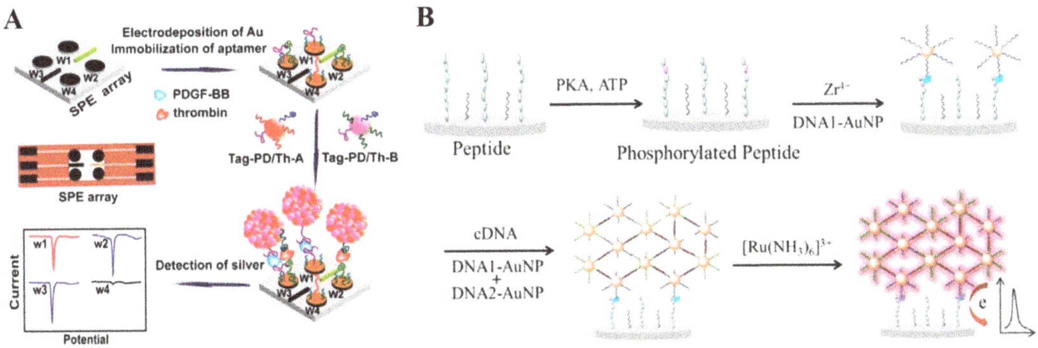

Figure 1. (**A**) Schematic representation of the principle of the dnGOPs-based electrochemical detection of DNA [37]. Copyright 2014 American Chemical Society. (**B**) Schematic representation of DNA-AuNPs assembled polymeric network amplified electrochemical biosensor for kinase activity detection [38]. Copyright 2014 American Chemical Society.

At the same time, other DNA-based assembly nanotechnologies including enzyme-aided and enzyme-free methods have also gained attractive attention. For example, Yu et al. developed a cascade signal amplification platform through integrating duplex-specific nuclease (DSN)-assisted target recycling with CHA reaction for the detection of microRNA-141 (miR-141) (Figure 2) [39]. During the DSN-assisted target recycling amplification, one miR-141 extracted from human breast cancer cells could induce the production of massive DNA connectors, which would trigger the next CHA reaction. Then, the AuNP hot spots were self-assembled into networks on the H_2-immobilized electrode surface. Numerous positively charged RuHex ions were captured by the anionic phosphate backbone of DNA duplex, finally resulting in a significant amplification in the electrochemical signal. Moreover, two-input AND and INHIBIT (INH) molecular logic gates were fabricated to analyze miRNAs. As one enzyme-free isothermal alternative, HCR avoids the restriction of precise dependence of pH and temperature in enzyme-mediated methods and has been used to detect miRNA and others [40,41]. For example,

Yuan et al. developed an electrochemical biosensor for the simultaneous detection of multiple miRNAs, in which the DNA-modified magnetic nanoprobes loaded with two different electroactive molecules were bound with the products of HCR [42]. However, in this method, one copy of miRNA only triggered one copy of polymeric HCR product. To further improve the sensitivity of the HCR-based method, Miao et al. reported an electrochemical method for miRNA detection based on the analyte-triggered nanoparticle localization on the electrode in combination with HCR amplification [43]. In this work, miRNA induced the opening of a hairpin on the electrode-immobilizing tetrahedral DNA for the capture of HCR-H0-modified AuNPs. Free HCR-H0 strands on the surface of AuNPs could further induce in situ hairpin polymerization. Subsequently, numerous AgNPs were assembled on the electrode, generating a sharp stripping current peak during the solid-state Ag/AgCl reaction. Although DNA-modified NPs as the sensing units have been widely used in electrochemical assays, extensive and complicated conjugation steps increase the complexity and cost of the assays.

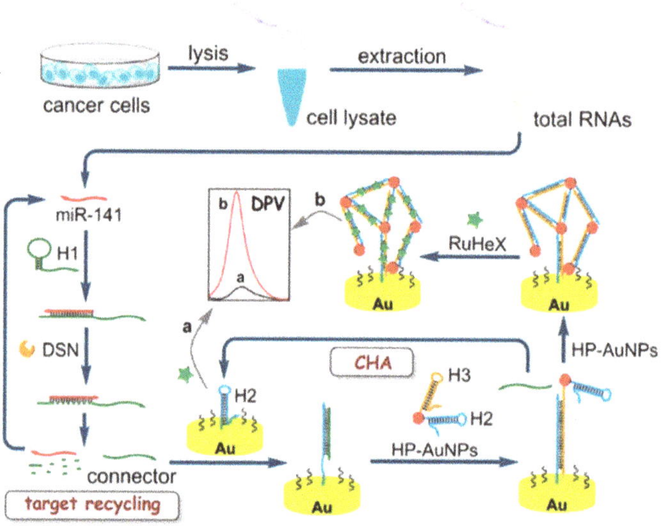

Figure 2. Schematic representation of cascade amplification of DSN-assisted target recycling and CHA reaction and in situ self-assembly of AuNP networks on electrodes for label-free electrochemical detection of miR-141 in signal-on mode [39]. Copyright 2018 American Chemical Society.

To overcome the shortcoming of DNA label-based experiments, label-free and nanomaterials-based signal amplification strategies have attracted more interest. As is known to us, the polymeric products generated by DNA self-assembly can be used as the templates for the assembly of nanomaterials through the electrostatic interaction or in situ metallization [44,45]. For this view, Li et al. demonstrated that the positively charged AuNPs could electrostatically assemble onto the double-helix of HCR products to amplify the electrochemical signal [46]. It was reported that a cytosine (C)-rich DNA sequence can be used as the template to prepare silver nanoclusters (AgNCs) that showed excellent electrocatalytic ability and redox property [47,48]. Yang et al. developed a label-free electrochemical method for the detection of miRNA (miRNA-199a) based on the in situ DNA-templated synthesis of AgNCs [49]. As displayed in Figure 3, when miRNA-199a was hybridized with the template probe, the target-assisted polymerization nicking reaction (TAPNR) amplification was initiated, and massive intermediate sequences were generated to bind with the secondary DNA probes on the electrode. Then, the HCR amplification was triggered by the surface-tethered intermediate sequences. In this process, numerous C-rich loop DNAs were formed in the dsDNA polymers. In the presence of $AgNO_3$ and $NaBH_4$, a large amount of AgNCs were produced by using C-rich loop DNAs as the templates. Similar to AgNPs, the AgNCs could generate a dramatically enhanced current response. Up to now, integrating HCR with the in situ formation of AgNCs has also been used to develop electrochemical biosensors for the detection of methyltransferase activity, Pb^{2+} and Type b3a2 [50–52]. In addition, AgNCs generated in the polymeric HCR products exhibit excellent ECL property and have been utilized to detect HATs activities [53]. In addition, double-stranded DNA (dsDNA) can be employed as the template for the preparation of copper nanoclusters (CuNCs). Zhao et al. reported an electrochemical method for protein detection based on the HCR-assisted formation of CuNCs [54]. The formed CuNCs could release numerous Cu^{2+} ions by acid dissolution, thus catalyzing the oxidation of o-phenylenediamine by O_2 and leading to the strong electrochemical signal. With exonuclease T7 triggered targets recycling and HCR amplification, Wang et al. prepared tree-like overlapping and branching Y-shaped dsDNA for the precise in situ growth of CuNCs [55].

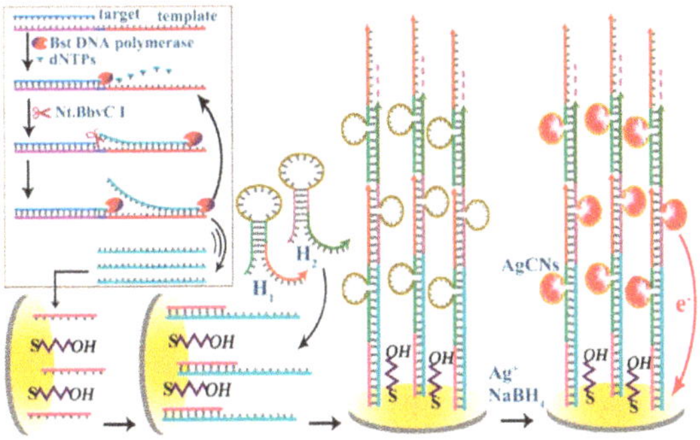

Figure 3. Schematic representation of ultrasensitive and label-free electrochemical detection of miRNA-199a based on in situ generated AgNCs by coupling TAPNR with HCR amplifications [49]. Copyright 2015 American Chemical Society.

2.2. *Metal Ion–Ligand Coordination*

Organic ligands modified on the surface of NPs can bind metal ions with different stability constants. The formation of the ligand–metal–ligand complex can induce the aggregation of NPs. In 2015, Wei et al. first proposed the concept of converting liquid-phase colorimetric assay into enhanced surface-tethered electrochemical analysis [28]. Based on the strategy, Hg^{2+} was sensitively detected as the model analyte. As shown in Figure 4A, cysteamine-capped AgNPs were prepared and modified with thymine-1-acetic acid, in which thymine (T) could be specifically coordinated with Hg^{2+} by the formation of a T-Hg^{2+}-T bond. Meanwhile, the gold electrode was modified with thymine-1-acetic acid. Hg^{2+} was captured by thymine modified on the electrode surface, which allowed for the attachment of thymine-functionalized AgNPs (Ag-T) nanoprobes. Subsequently, surface-bound Ag-T nanoprobes could induce more Hg^{2+} and nanoprobes to be immobilized on the electrode, thus leading to the formation of Ag-T-based nanostructures. Finally, a strong and well-defined electrochemical signal was attained. The detection limit of the proposed electrochemical sensor was approximately two orders of magnitude lower than that of the AgNPs-based colorimetric assay of Hg^{2+}. Similarly, Cu^{2+} and Cr^{3+} were selectively detected through their coordination with 4-mercaptobenzoic acid and 3-mercaptopropanoic acid, respectively [56,57]. Recently, Gu et al. employed T-functionalized upconversion nanoparticles (UCNP) as sensing units (T-UCNP) to develop an ECL sensor for the analysis of Hg^{2+} [58]. After the target-induced aggregation, multiple UCNPs were assembled on the electrode surface, and an amplified ECL signal was generated.

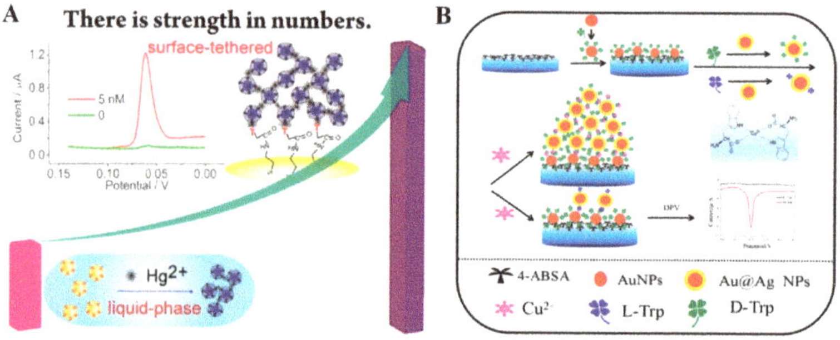

Figure 4. (**A**) Schematic representation of the electrochemical Hg^{2+} sensor based on Ag-T nanoprobes [28]. Copyright 2015 American Chemical Society. (**B**) Schematic representation of electroactive Au@Ag NP assembly driven signal amplification for ultrasensitive chiral discrimination of D-/L-Trp [59]. Copyright 2019 American Chemical Society.

Biomolecules that can specifically interact with metal ions can also been detected by this method. For example, Zhao et al. reported an electrochemical method for the chiral recognition of D-/L-tryptophan (Trp) based on the Cu^{2+}-assisted NPs aggregation [59]. As shown in Figure 4B, D-Trp functionalized AuNPs (D-Trp-AuNPs) were used to modify the glass carbon electrode. As a result of the high binding constant with NPs and D-Trp, Cu^{2+} could induce more electroactive Au@Ag NP to assemble on the D-Trp-AuNPs-modified electrode. In the presence of D-Trp, more Au@Ag NPs networks were formed on the electrode surface, and a strong DPV was observed. In addition, Wang et al. reported the detection of lipopolysaccharide based on the Cu^{2+}-induced formation of AuNPs aggregates as the signal labels on the electrode surface, in which anionic groups were coordinated with Cu^{2+} ions to induce the nanoparticle aggregation [60].

2.3. *Metal–Thiol and Boronate Ester Interactions*

Phenylboronic acid (PBA) and its derivates can bind with α-hydroxycarboxylate acids (such as citrate and tartrate) and o-diphenol/diol-containing species (such as catechol derivatives, nucleosides, and glycoproteins) via the formation of a covalent boronate ester bond [61]. Such interactions have been introduced into the design of various biosensors [62–64]. Capping reagents play an important role in the enhancement of the stability and solubility of nanomaterials [65,66]. Among them, trisodium citrate is the most frequently used reagent for the synthesis of negatively charged AgNPs and AuNPs. Unlike DNA, the ribose group in RNA contains an intact cis-diol structure in the pentose ring at the 3′-terminal, which can react with PBA to form a boronate ester covalent bond. Our group reported a label-free electrochemical method for the detection of miRNAs based on the in situ aggregation of AgNPs [67]. As shown in Figure 5, after the hybridization, the exposed cis-diol moiety in the ribose of the captured miRNA was derivatized by 4-mercaptophenylboronic acid (MPBA). Next, the thiol group of MPBA could grasp citrate-capped AgNPs via the formation of a Ag-S bond. Then, the surface-bound AgNPs could recruit more MPBA and AgNPs from solution through the formation of boronate ester and Ag-S bonds, finally resulting in the in situ generation of MPBA-AgNPs-based networks on the electrode surface. Based on the amplified electrochemical signal, miRNA-21 was sensitively determined with a detection limit down to 20 aM.

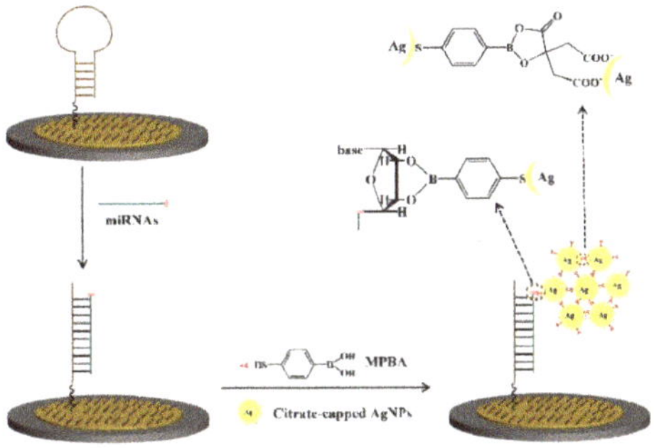

Figure 5. Schematic representation of the proposed electrochemical strategy for miRNAs detection based on MPBA-induced in situ formation of AgNPs aggregates as labels [67]. Copyright 2017 Elsevier B.V.

Tyrosinase can catalyze the oxidation of tyrosine residue in the substrate peptide and transform monophenol into o-diphenol, which could be recognized by MPBA. Based on the MPBA-AgNPs-based networks, our group developed two sensitive electrochemical strategies for the determination of protein kinase activity [68]. As shown in Figure 6, after the hydroxyl of tyrosine residue was phosphorylated in the presence of tyrosine kinase (Src) and 5-[-thio] triphosphate (ATP-S), the thiophosphate peptide could bind to AgNPs through the formation of a Ag-S bond. Eventually, under the MPBA-assisted in situ assembly of AgNPs, the nanoarchinectures were formed on the electrode. In another strategy, after the oxidation of tyrosine residue, MPBA reacted with the o-diphenol moiety, and then, AgNP was captured by MPBA through the Ag-S interaction, leading to the generation of AgNPs-based networks. However, once the tyrosine residue was phosphorylated, the oxidation and assembly process would be blocked, resulting in the decrease in the current intensity. Moreover, glycan on the surface of glycoprotein can also react with MPBA to induce the formation of MPBA-AgNPs-based networks on the electrode surface, thus allowing for the development of electrochemical glycoprotein aptasensors [69].

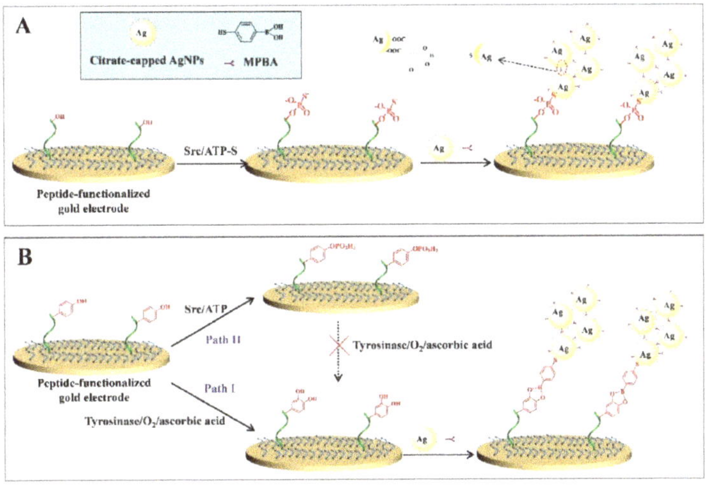

Figure 6. Schematic representation of the proposed electrochemical strategies for protein kinase detection based on the in situ formation of the AgNPs aggregates as labels. In the first design (**A**), ATP-S was used as the co-substrate. In the second design (**B**), ATP was used as the co-substrate, and tyrosinase was used to convert monophenol into o-diphenol [68]. Copyright 2017 Elsevier B.V.

Benzene-1,4-dithiol (BDT) with two thiol groups can be used as the connector for the assembly of AgNPs. Based on the BDT-induced in situ formation of AgNPs networks, Hou et al. constructed a modification-free amperometric biosensor for the detection of wild-type p53 protein [70]. As displayed in Figure 7, a dsDNA probe containing two consensus sites was employed to modify the gold electrode for the capture of wild-type p53 protein. After the binding between the probe and protein, the thiol and amine groups on the surface of p53 protein bind to AgNPs via the formation of Ag-S and Ag-N bonds. In the presence of BDT, more AgNPs were in situ assembled on the electrode surface to form the networks for signal amplification. This method has been successfully used to detect wild-type p53 protein in cell lysates with satisfactory results.

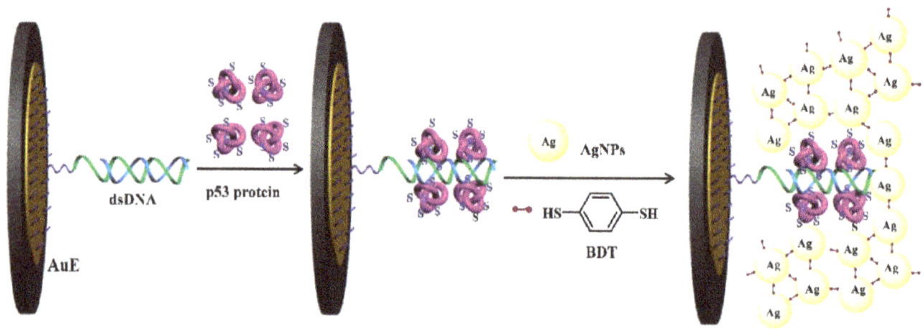

Figure 7. Schematic representation of the biosensor for the detection of wild-type p53 protein by the in situ formation of AgNPs networks for signal amplification [70]. Copyright 2020 Elsevier B.V.

2.4. Peptide-Induced Assembly of Nanoparticles

As a result of the diversity of structural units (amino acids), peptides can be synthesized with a specific sequence and used as the aptamer for the capture of the target. Moreover, a peptide with positive charges can induce the aggregation of negatively charged citrate-capped AuNPs and AgNPs via the electrostatic interactions. In this process, the peptide probe plays two roles (the target binder and the NPs aggregation inducer). Once the peptide was bound to the target, it would lose the ability to trigger the aggregation of NPs. Based on this fact, AuNPs and AgNPs-based liquid-phase colorimetric assays have been converted into surface-tethered electrochemical electrochemical assays [71]. For example, the tripeptide (Arg-Pro-Arg) with two positively charged arginine residues could lead to the AuNPs aggregation. Dipeptidyl peptidase-IV (DPP-IV) can induce the hydrolyzation of the peptide, thus preventing the aggregation of AuNPs. DPP-IV activity was determined by the colorimetric and electrochemical methods based on the peptide-induced AuNPs aggregation [72].

PrP(95−110) with an amino acid sequence of THSQWNKPSKPKTNMK was identified as the receptor of small amyloid-β (Aβ) oligomer (AβO) with high specificity and affinity [73–76]. Our group found that PrP(95−110) could induce the AuNPs or AgNPs aggregation with a color change [77,78]. For this view, we further developed an electrochemical platform for the detection of AβO based on the in situ formation of AgNPs networks for signal amplification [78]. As shown in Figure 8A, the adamantine (Ad)-modified PrP(95−110) could be attached on the surface of AgNPs to result in their aggregation. Based on the host–guest interaction, the formed Ad-PrP(95−110)-AgNPs networks were tethered on the β-CD (β-cyclodextrin)-modified electrode surface, thus producing an amplified electrochemical signal. However, in the presence of AβO, the binding of AβO to the peptide hindered the interaction of the peptide and AgNPs, thus leading to the reduced magnitude of aggregation on the electrode and decreasing the electrochemical signal from the oxidation of AgNPs.

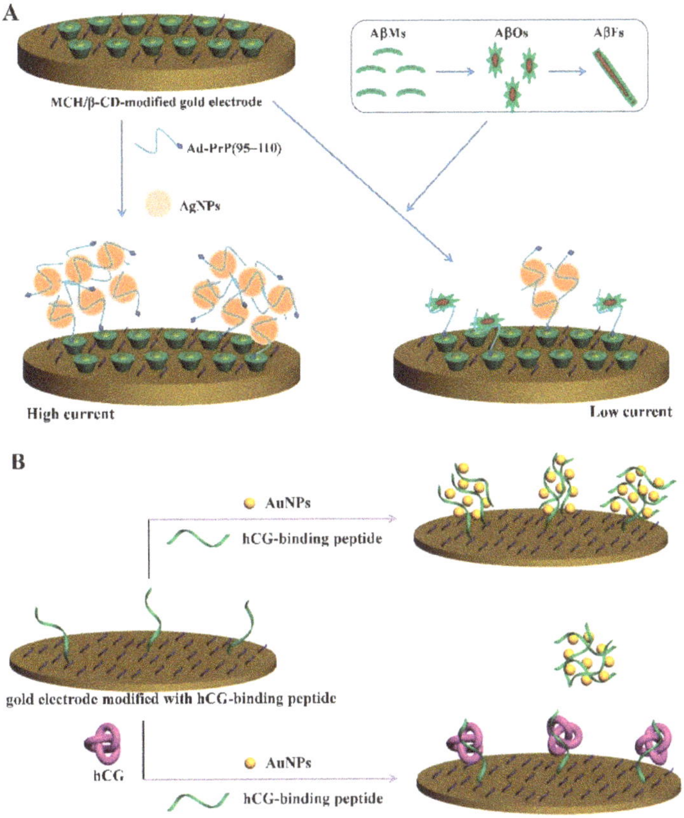

Figure 8. (**A**) Schematic representation of the electrochemical method for the selective detection of AβOs using AgNPs as the redox reporters and Ad-PrP(95−110) as the receptor [78]. Copyright 2016 American Chemical Society. (**B**) Schematic illustration of the electrochemical method for hCG detection using a peptide probe as the receptor of hCG and the inducer of AuNPs assembly [79]. Copyright 2017 Elsevier B.V.

As a hormone produced by placenta, human chorionic gonadotropin (hCG) is recognized as an important indicator for pregnancy and several cancers. Our group designed an electrochemical biosensor for hCG detection with a dual-functional peptide probe (PPLRINRHILTR) [79]. As shown in Figure 8B, the positively charged hCG-binding peptide used as the sensing unit can induce the aggregation of AuNPs via the electrostatic interactions and facilitate the in situ formation of AuNPs, which assemble on the electrode surface. The formed networks could significantly reduce the charge transfer resistance. However, in the presence of hCG, the stable complex of the peptide probe and hCG lost the coagulating ability toward AuNPs. The amount of AuNPs assemblies on the electrode was reduced, and the charge transfer resistance was intensified. This method based on the electrochemical impedance technique achieved the determination of hCG with a detection limit of 0.6 mIU/mL. By using AgNPs as the redox probes for a well-defined and amplified electrochemical signal, hCG could be sensitively measured by linear-sweep voltammetry (LSV) [80].

As one type of essential structural molecules, peptides with excellent self-assembly and molecular recognition ability can be self-assembled into various nanostructures. Moreover, through the careful encoding of peptide with a binding or reactive site, peptides can co-assemble with additional nanomaterials with unique optical and chemical properties into functional hybrid superstructures. Recently, Han et al. reported the co-assembly of peptides and carbon nanodots (CNDs) (Pep/CND) based on the $\pi-\pi$ stacking between tyrosine residues and CNDs (Figure 9A) [81]. The peptide and CNDs endowed Pep/CND co-assembly with the recognition capability and the catalytic activity, respectively. They further applied the Pep/CND co-assembly to construct an electrochemical method for the detection of transglutaminase 2 (TG2). As shown in Figure 9B, TG2 catalyzed the ligation of peptide P2 and peptide P1 on the electrode surface. Next, CNDs was bound to P2, subsequently triggering the co-assembly of a plenty of P3 and CNDs because the tyrosine was located at one terminal. The large amount of CNDs with catalytic ability could catalyze the redox reaction between H_2O_2 and 3,3′,5,5′-tetramethylbenzidine (TMB), resulting in the enhancement of electrochemical signal for the sensitive detection of TG2.

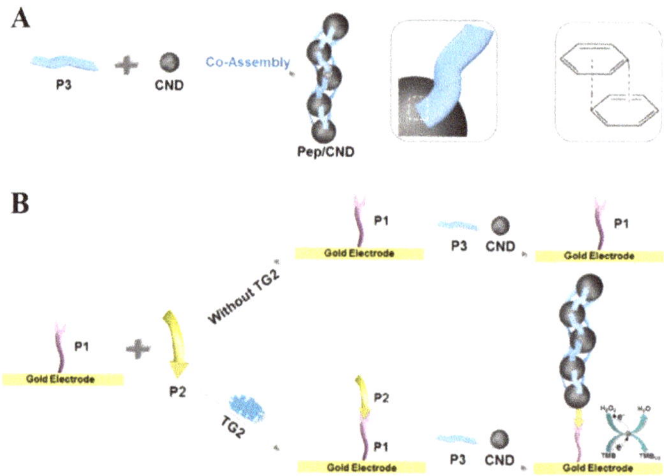

Figure 9. Schematic representation of (**A**) the co-assembly of P3 and CNDs and (**B**) the principle of the analysis of TG2 [81]. Copyright 2021 American Chemical Society.

2.5. SA–Biotin Interaction

SA is a tetrameric protein that can bind to four biotin molecules with high binding affinity in a wide pH range (K_d = 10^{-15} M) [82,83]. The specific and strong interaction is always utilized for the conjugation of antibodies or nucleic acids with enzymes or nanomaterials for signal output. It has been reported that aromatic phenylalanine (Phe) and its derivates can self-assemble into various nanostructures through the modulation of different parameters. Our group found that the biotinylated Phe (biotin-Phe) monomers can self-assemble into monodispersed biotin–Phe nanoparticle (biotin–FNP) by controlling the pH value [84]. Then, an impedimetric biosensor for the determination of caspase-3 activity and evaluation of cell apoptosis was based on the in situ assembly of biotin–FNP in the presence of SA [85]. As shown in Figure 10, tetrameric SA protein can be captured by the biotinylated DVED-containing peptide. Then, free biotin-binding sites of SA on the electrode surface allowed for the anchor of biotin–FNP and SA, finally leading to the formation of SA–biotin–FNP networks on the electrode surface. The direct electron transfer between $[Fe(CN)_6]^{3-/4-}$ and the electrode was seriously hindered. In the presence of caspase-3, the cleavage of peptide prevented the binding of SA and the follow-up formation of biotin–FNP networks on the electrode surface. The electron transfer resistance (R_{et}) was inversely proportional to the concentration and activity of caspase-3. The method was further employed to develop aptasensors for impedimetric detection of miRNAs and small molecules by competitive reactions [84,86].

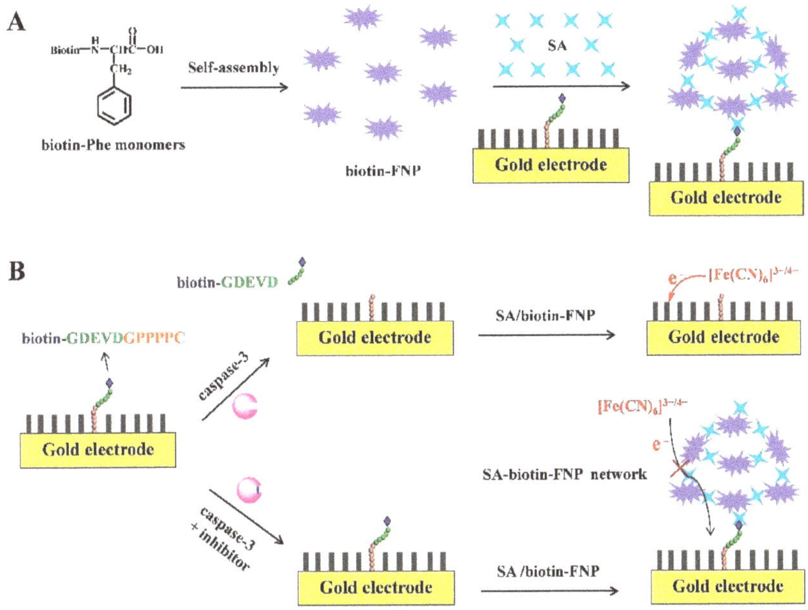

Figure 10. (**A**) Self-assembly of biotin–Phe monomers into biotin–FNP and the in situ formation of SA–biotin–FNP networks on the electrode surface. (**B**) Schematic representation of the biosensor for assay of caspase-3 activity via the signal amplification by SA-biotin–FNP networks [85]. Copyright 2020 Elsevier B.V.

3. In Situ Assembly of Small Molecules and Biomolecules

In DNA assembly-based electrochemical methods, the DNA assembly unit is usually conjugated with an electroactive molecule for signal output. Actually, the amount of electroactive molecule in the assembly product is low, and the insulation property of DNA and its assembly may decrease the detection performances. Recently, small molecules and biomacromolecules are proposed as the assembly units to form nanostructures on the electrode surface for improving the detection sensitivity. For example, inspired by the polymerization reaction, Hu et al. has reported the application of surface-initiated electrochemically mediated atom-transfer radical polymerization (SI-eATRP) as an amplification strategy for the electrochemical biosensing of different targets, including double-stranded DNA (dsDNA) and protein kinase activity [87,88]. Typically, they demonstrated the de novo growth of a polymers (dnGOPs)-based electrochemical biosensor for the detection of target DNA (tDNA) through SI-eATRP [89]. The principle of this method was illustrated in Figure 11A. Peptide nucleic acid (PNA) probes with a neutrally charged N-(2-aminoethyl)glycine units-composed backbone were immobilized on the electrode to specifically capture tDNA. After hybridization, phosphate groups with high density could bind to Zr^{4+} and α-bromophenylacetic acid (BPAA) through the phosphate-Zr^{4+}-carboxylate coordination chemistry. With the aid of a constant potential, the SI-eATRP was triggered, and numerous electroactive ferrocenylmethyl methacrylate (FMMA) monomers were polymerized into long polymeric chains on the electrode surface. Since the electrochemical response was greatly improved, tDNA was sensitively and selectively detected with a detection limit of 0.072 fM. However, the utilization of Cu^{2+} ions as catalysts may interfere with the next electrochemical experiments because of the non-specific interaction with nucleic acids and the electrochemical deposition of metal on the electrode. For this consideration, they further explored novel surface-initiated electrochemically controlled reversible-addition-fragmentation-chain-transfer (SI-eRAFT) polymerization without the use of transition metal ions for assays [90–92]. For instance, they reported a signal-amplified electrochemical sensing of thrombin activity by SI-eRAFT polymerization [93]. As shown in Figure 11B, after the thrombin-specific substrate peptide (Tb peptide) was cleaved, 4-cyano-4-(phenylcarbonothioylthio)pentanoic acid (CPAD) was bound to the peptide by the formation of the carboxylate–zirconium–carboxylate complexes. The initiation of SI-eRAFT resulted in the polymerization of a large number of FMMA on the electrode surface, leading to the significant increase in the current intensity. The results showed that the SI-eRAFT-based amplification strategy held great promise in the sensitive analysis of biomolecules. Recently, an in situ initiated ring-opening polymerization signal amplification strategy was also integrated with an electrochemical biosensor for the detection of CYFRA 21-1, which is a specific biomarker for non-small cell lung cancer [94]. In addition to the covalent bond for the polymerization of signal molecules, a peptide can self-assemble into various stable nanostructures via non-covalent interaction including hydrogen bonding, hydrophobic interaction, electrostatic interaction, non-specific Van der Waals and π–π stacking. Inspired by DNA assembly techniques, Huang et al. reported a signal amplification strategy based on the in situ self-assembly of peptides for the determination of AβO [95]. After being captured by the peptide CP4-PrP(95–110), the captured AβO could trigger the in situ self-assembly of the amphiphilic C16-GGG-PrP(95–110)-Fc peptide on the surface of the electrode under mild conditions. The accumulation of numerous Fc molecules generated a significantly amplified signal.

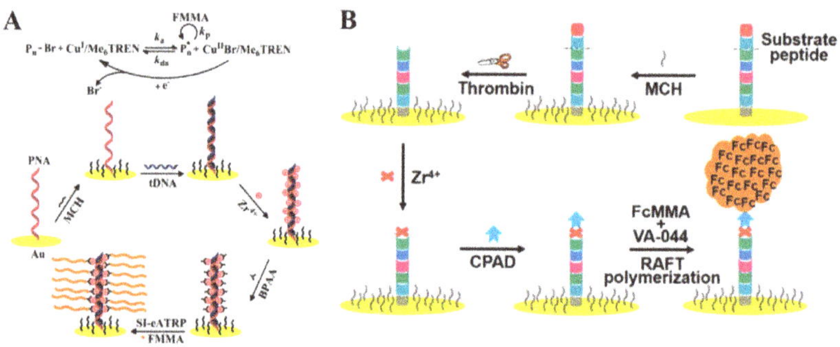

Figure 11. (**A**) Schematic representation of the principle of the dnGOPs-based electrochemical detection of DNA [89]. Copyright 2017 American Chemical Society. (**B**) Schematic representation of "signal-on" electrochemical biosensing of thrombin activity [93]. Copyright 2020 American Chemical Society.

Generally, the efficiency of enzyme-catalyzed hydrolysis or modification of substrate peptide or nucleic acid may be hampered due to the steric hindrance and low freedom of the substrate. Thus, it is useful to integrate the homogeneous assay with the surface-tethered electrochemical analysis, which can retain the high efficiency of the enzymatic reaction and the high selectivity of the interfacial analytical method. Our group reported an electrochemical caspase-3 biosensor by conversing a homogeneous assay into a surface-tethered electrochemical analysis based on the SA–biotin interaction [96]. As shown in Figure 12, SA molecules could co-assemble with the peptide substrates (biotin–GDEVDGK–biotin) to form (SA–biotin–GDEVDGK–biotin-SA)$_n$ aggregates on the SA-modified electrode surface. The in situ performed aggregates significantly blocked the electron transfer of $[Fe(CN)_6]^{3-/4-}$ and increased the R_{et}. However, after the homogeneous cleavage of the peptide by caspase-3, the amount of intact biotin–GDEVDGK–biotin decreased, and the product of biotin-labeled pieces further competed with the peptide substrate to bind SA, leading to the suppression of the in situ assembly of biotin–GDEVDGK–biotin and SA. DNA probes can be employed as the assembly units for the detection of miRNAs and enzyme activities. Our group also developed a DSN-based electrochemical biosensor for the analysis of miRNAs by integrating homogeneous enzymatic reaction with surface-tethered electrochemical analysis [97]. In the work, the biotinylated DNA (biotin–DNA–biotin) can trigger the in situ co-assembly of DNA and SA on the electrode. The insulating (SA–biotin–DNA–biotin)$_n$ assemblies could hamper the electron communication between $[Fe(CN)_6]^{3-/4-}$ and the electrode. However, when the biotinylated DNA was hybridized with the target miRNA, the biotin–DNA–biotin in the dsDNA would be hydrolyzed by DSN, resulting in the hybridization–enzymolysis cycle and the generation of abundant biotin–DNA fragments. The released fragments could further compete with biotin–DNA–biotin to bind SA and thus reduced the amount of (SA–biotin–DNA–biotin)$_n$ assemblies. In addition, our group also reported a (SA–biotin–DNA–biotin)$_n$ networks-based electrochemical biosensor for the detection of telomerase activity in cancer cells [98]. The primer was extended by telomerase to generate many (TTAGGG)n repeats on the electrode surface. The elongated primer could hybridize with its complementary sequence biotin–DNA–biotin, subsequently triggering the in situ co-assembly of biotin–DN–biotin and SA into (SA–biotin–DNA–biotin)$_n$ networks. Numerous phosphate groups in networks blocked the access of negatively charged $[Fe(CN)_6]^{3-/4-}$ to the electrode surface, resulting in the increase in R_{et}. Based on the signal-amplified strategy, telomerase extracted from two HeLa cells could be readily determined.

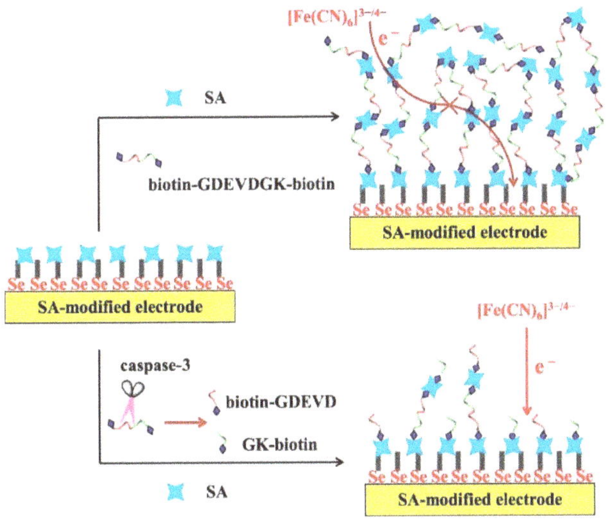

Figure 12. Schematic representation of the surface-tethered electrochemical analysis for caspase-3 detection [96]. Copyright 2021 American Chemical Society.

The applications and analytical performances of different assembly strategies are summarized in Table 1.

Table 1. Analytical performances of different electrochemical biosensors based on the in situ assembly of nanomaterials and molecules for the signal enhancement.

Interaction	Nanomaterials	Target	Detection Range	Detection Limit	Ref.
DNA hybridization	DNA-AuNPs	lysozyme	1 pM–1 nM	0.32 pM	[36]
	DNA-AgNPs	PDGF and thrombin	5 pg/mL–1000 ng/mL	1.6 pg/mL	[37]
	DNA-AuNPs	PKA	0.03–40 U/mL	0.03 U/mL	[38]
	DNA-AuNPs	miR-141	0.1 fM–10 nM	25.1 aM	[39]
	DNA-CeO_2	VEGF	1 fg/mL–0.1 ng/mL	0.27 fg/mL	[41]
	Thi-modified DNA-Fe_3O_4 NPs and Fc-CHO-modified DNA-Fe_3O_4 NPs	miR-141 and miR-21	1 fM–1 nM	0.44 fM for miR-141 and 0.46 fM for miR-21	[42]
	DNA-AgNPs	miR-17-5p	100 aM–100 pM	2 aM	[43]
DNA-based electrostatic interaction	Ag@Au CSNPs	Hg^{2+}	10 pM–2.5 nM	3.6 pM	[44]
	CTAB-capped AgNPs	PSA	0.1 pg/mL–75 ng/mL	0.033 pg/mL	[45]
	AuNPs	DNA	15 pM–1.0 nM	2.6 pM	[46]
DNA-based in situ metallization	AgNCs	DNA	0.2 fM–1 pM	0.16 fM	[48]
	AgNCs	miR-199a	1.0 fM–0.1 nM	0.64 fM	[49]
	AgNCs	methyltransferase	0.02–10 U/mL	0.0073 U/mL	[50]
	AgNPs	Pb^{2+}	1 pM–100 nM	0.24 pM	[51]
	AgNPs	Type b3a2	10 fM–10 nM	0.56 fM	[52]
	AgNCs	HAT	0.5–100 nM	0.49 nM	[53]
	CuNPs	folate receptor	0.01–100 ng/mL	3 pg/mL	[54]
	CuNCs	miR-21	10 pM–0.1 fM	10 aM	[55]
Metal ion–ligand coordination	thymine-modified AgNPs	Hg^{2+}	50 pM–50 nM	5 pM	[28]
	MBA-modified AgNPs	Cu^{2+}	0.1–100 nM	0.08 nM	[56]
	MPA-modified AgNPs	Cr^{3+}	200–5000 ppb	278 ppb	[57]
	Thymine-modified UCNPs	Hg^{2+}	10 pM–100 nM	0.4 pM	[58]
	Au@Ag NPs	D-tryptophan	5 pM–1 nM	1.21 pM	[59]
	L-cysteine-modified AuNPs	lipopolysaccharide	1.0–10 pg/mL	0.033 pg/mL	[60]
Metal–thiol and boronate ester interactions	citrate-capped AgNPs	tyrosinase	0.001–0.5 mU/mL	0.1 mU/mL	[61]
	citrate-capped AgNPs	thrombin	0.025–5 ng/mL	0.02 ng/mL	[61]
	citrate-capped AgNPs	H_2O_2	1 nM–0.6 μM	Not reported	[62]
	citrate-capped AgNPs	miR-21	0.1–50 fM	20 aM	[67]
	citrate-capped AgNPs	tyrosine kinase	0.1–25 ng/mL	0.1 ng/mL	[68]
	citrate-capped AgNPs	PSA	0.5–200 pg/mL	0.2 pg/mL	[69]
	citrate-capped AgNPs	wild-type p53	0.1–100 pM	0.1 pM	[70]
Peptide-induced assembly	citrate-capped AuNPs	PKA	0.01–1 U/mL	20 mU/mL	[71]
	citrate-capped AuNPs	DPP-IV	0.001–0.5 mU/mL	0.55 μU/mL	[72]
	citrate-capped AgNPs	AβO	0.01–200 nM	6 pM	[73]
	citrate-capped AgNPs	AβO	20 pM–100 nM	8 pM	[78]
	citrate-capped AuNPs	hCG	0.001–0.2 IU/mL	0.6 mIU/mL	[79]
	citrate-capped AgNPs	hCG	0.001–0.2 IU/mL	0.4 mIU/mL	[80]
	Carbon nanodots	transglutaminase 2	1 pg/mL–50 ng/mL	0.25 pg/mL	[81]
SA–biotin interaction	biotin-FNPs	aflatoxin B_1	0.05–3 pg/mL	Not reported	[84]
	biotin-FNPs	caspase-3	1–125 pg/mL	1 pg/mL	[85]
	biotin-FNPs	miR-21	0.1–250 fM	0.1 fM	[86]

Table 1. *Cont.*

Interaction	Nanomaterials	Target	Detection Range	Detection Limit	Ref.
In situ assembly of small molecules and biomolecules	Fc derivate	DNA	1.0 fM–1.0 nM	0.47 fM	[87]
	Fc derivate	PKA	0–140 mU/mL	1.63 mU/mL	[88]
	Fc derivate	DNA	0.1 fM–0.1 nM	0.072 fM	[89]
	Fc derivate	DNA	10 aM–10 pM	3.2 aM	[90]
	Fc derivate	PKA	0–140 mU/mL	1.02 mU/mL	[91]
	Fc derivate	trypsin	25–175 μU/mL	18.2 μU/mL	[92]
	Fc derivate	thrombin	10–250 μU/mL	2.7 μU/mL	[93]
	Fc derivate	CYFRA 21-1	1 pg/mL–1 μg/mL	9.08 fg/mL	[94]
	Fc-labeled peptide	AβO	0.005–5 μM	0.6 nM	[95]
	$(SA–biotin–peptide–biotin)_n$	caspase-3	0–50 pg/mL	0.2 pg/mL	[96]
	$(SA–biotin–DNA–biotin)_n$	miR-21	0.01–2.5 fM	10 aM	[97]
	$(SA–biotin–DNA–biotin)_n$	telomerase	20–5000 cells/mL	20 cells/mL	[98]

Abbreviation: AuNPs, gold nanoparticles; AgNPs, silver nanoparticles; PDGF, platelet-derived growth factor; PKA, protein kinase A; miR, microRNA; VEGF, vascular endothelial growth factor; Thi, thionine; Fc-CHO, ferrocene carboxaldehyde; CSNPs, core–shell nanoparticles; CTAB, cetyltrimethylammonium bromide; PSA, prostate specific antigen; AgNCs, silver nanoclusters; CuNPs, copper nanoparticles; CuNCs, copper nanoclusters; HAT, histone acetyltransferase; MBA, 4-mercaptobenzoic acid; MPA, 3-mercaptopropanoic acid; UCNPs, upconversion nanoparticles; DPP-IV, dipeptidyl peptidase-IV; AβO, amyloid-β oligomer; hCG, human chorionic gonadotropin; biotin–FNPs, biotin–phenylalanine-assembled nanoparticles; SA, streptavidin; Fc, ferrocene; CYFRA 21-1, cytokeratin19 fragment.

4. Conclusions

Liquid-phase aggregation-based assays, such as colorimetric and fluorescence assays, are simple and convenient. In contrast to the homogeneous analysis, heterogeneous interface assays are considered as more sensitive and accurate tools for biochemical detection. Nanomaterials in aggregation state may exhibit different optical and chemical properties compared to the monodispersed state. In this review, we summarize the advancements in the in situ assembly of nanomaterials for the signal output and amplification of electrochemical biosensors. The proposed strategies by converting liquid-phase aggregation-based assay into sensitive surface-tethered electrochemical analysis have exhibited more advantageous performances and have shown promising applications. It is expected that new assembly methods and units would be more effective and abundant to achieve highly sensitive and specific detection.

Author Contributions: Conceptualization, Z.S. and L.L.; writing—original draft preparation, Y.C.; writing—review and editing, N.X. and Y.H.; project administration, Z.S.; funding acquisition, Z.S. and L.L. All authors have read and agreed to the published version of the manuscript.

Funding: This research was funded by the Program for Innovative Research Team of Science and Technology in the University of Henan Province (21IRTSTHN005) and the National Natural Science Foundation of China (21905002).

Conflicts of Interest: The authors declare no conflict of interest.

References

1. Singh, P.; Parent, K.L.; Buttry, D.A. Electrochemical solid-state phase transformations of silver nanoparticles. *J. Am. Chem. Soc.* **2012**, *134*, 5610–5617. [CrossRef] [PubMed]
2. Yu, C.-X.; Xiong, F.; Liu, L.-L. Electrochemical biosensors with silver nanoparticles as signal labels. *Int. J. Electrochem. Sci.* **2020**, *15*, 3869. [CrossRef]
3. Xia, W.; Zhang, P.; Fu, W.; Hu, L.; Wang, Y. Aggregation/dispersion-mediated peroxidase-like activity of MoS_2 quantum dots for colorimetric pyrophosphate detection. *Chem. Commun.* **2019**, *55*, 2039–2042. [CrossRef] [PubMed]
4. Deng, S.Q.; Zou, H.Y.; Lan, J.; Huang, C.Z. Aggregation-induced superior peroxidase-like activity of $Cu_{2-x}Se$ nanoparticles for melamine detection. *Anal. Methods* **2016**, *8*, 7516–7521. [CrossRef]
5. Ni, P.; Dai, H.; Wang, Y.; Sun, Y.; Shi, Y.; Hu, J.; Li, Z. Visual detection of melamine based on the peroxidase-like activity enhancement of bare gold nanoparticles. *Biosens. Bioelectron.* **2014**, *60*, 286–291. [CrossRef]
6. Zhao, Y.; Huang, Y.; Zhu, H.; Zhu, Q.; Xia, Y. Three-in-one: Sensing, self-assembly, and cascade catalysis of cyclodextrin modified gold nanoparticles. *J. Am. Chem. Soc.* **2016**, *138*, 16645–16654. [CrossRef]

7. Chen, Y.; Zhang, W.; Zhao, Z.; Cai, Y.; Gong, J.; Kwok, R.T.K.; Lam, J.W.Y.; Sung, H.H.Y.; Williams, I.D.; Tang, B.Z. An easily accessible ionic aggregation-induced emission luminogen with hydrogen-bonding-switchable emission and wash-free imaging ability. *Angew. Chem. Int. Ed.* **2018**, *57*, 5011–5015. [CrossRef]
8. Feng, Q.; Li, Y.; Wang, L.; Li, C.; Wang, J.; Liu, Y.; Li, K.; Hou, H. Multiple-color aggregation-induced emission (AIE) molecules as chemodosimeters for pH sensing. *Chem. Commun.* **2016**, *52*, 3123–3126. [CrossRef] [PubMed]
9. Yang, H.; Liu, Y.; Guo, Z.; Lei, B.; Zhuang, J.; Zhang, X.; Liu, Z.; Hu, C. Hydrophobic carbon dots with blue dispersed emission and red aggregation-induced emission. *Nat. Commun.* **2019**, *10*, 1789–1800. [CrossRef]
10. Mou, M.; Wu, Y.; Niu, Q.; Wang, Y.; Yan, Z.; Liao, S. Aggregation-induced emission properties of hydrothermally synthesized Cu-In-S quantum dots. *Chem. Commun.* **2017**, *53*, 3357–3360. [CrossRef]
11. Rosi, N.L.; Mirkin, C.A. Nanostructures in biodiagnostics. *Chem. Rev.* **2005**, *105*, 1547–1562. [CrossRef]
12. Rao, H.; Xue, X.; Luo, M.; Liu, H.; Xue, Z. Recent advances in the development of colorimetric analysis and testing based on aggregation-induced nanozymes. *Chin. Chem. Lett.* **2021**, *32*, 25–32. [CrossRef]
13. Storhoff, J.J.; Elghanian, R.; Mucic, R.C.; Mirkin, C.A.; Letsinger, R.L. One-pot colorimetric differentiation of polynucleotides with single base imperfections using gold nanoparticle probes. *J. Am. Chem. Soc.* **1998**, *120*, 1959–1964. [CrossRef]
14. Perez, J.M.; Simeone, F.J.; Saeki, Y.; Josephson, L.; Weissleder, R. Viral-induced self-assembly of magnetic nanoparticles allows the detection of viral particles in biological media. *J. Am. Chem. Soc.* **2003**, *125*, 10192–10193. [CrossRef]
15. Liu, J.; Lu, Y. Fast colorimetric sensing of adenosine and cocaine based on a general sensor design involving aptamers and nanoparticles. *Angew. Chem. Int. Ed.* **2005**, *45*, 90–94. [CrossRef]
16. Liu, Z.D.; Zhu, H.Y.; Zhao, H.X.; Huang, C.Z. Highly selective colorimetric detection of spermine in biosamples on basis of the non-crosslinking aggregation of ssDNA-capped gold nanoparticles. *Talanta* **2013**, *106*, 255–260. [CrossRef]
17. Du, J.; Jiang, L.; Shao, Q.; Liu, X.; Marks, R.S.; Ma, J.; Chen, X. Colorimetric detection of mercury ions based on plasmonic nanoparticles. *Small* **2012**, *9*, 1467–1481. [CrossRef]
18. Li, Y.H.; Zhang, L.; Huang, J.; Liang, R.P.; Qiu, J.D. Fluorescent graphene quantum dots with a boronic acid appended bipyridinium salt to sense monosaccharides in aqueous solution. *Chem. Commun.* **2013**, *49*, 5180–5182. [CrossRef] [PubMed]
19. Wang, L.; Zeng, L.; Wang, Y.; Chen, T.; Chen, W.; Chen, G.; Li, C.; Chen, J. Electrochemical aptasensor based on multidirectional hybridization chain reaction for detection of tumorous exosomes. *Sens. Actuators B Chem.* **2021**, *332*, 129471–129480. [CrossRef]
20. Luo, C.; Tang, H.; Cheng, W.; Yan, L.; Zhang, D.; Ju, H.; Ding, S. A sensitive electrochemical DNA biosensor for specific detection of Enterobacteriaceae bacteria by Exonuclease III-assisted signal amplification. *Biosens. Bioelectron.* **2013**, *48*, 132–137. [CrossRef]
21. Li, M.; Yin, F.; Song, L.; Mao, X.; Li, F.; Fan, C.; Zuo, X.; Xia, Q. Nucleic acid tests for clinical translation. *Chem. Rev.* **2021**, *121*, 10469–10558. [CrossRef]
22. Manikandan, V.S.; Adhikari, B.; Chen, A. Nanomaterial based electrochemical sensors for the safety and quality control of food and beverages. *Analyst* **2018**, *143*, 4537–4554. [CrossRef]
23. Chen, A.; Chatterjee, S. Nanomaterials based electrochemical sensors for biomedical applications. *Chem. Soc. Rev.* **2013**, *42*, 5425–5438. [CrossRef]
24. Zhu, C.; Yang, G.; Li, H.; Du, D.; Lin, Y. Electrochemical sensors and biosensors based on nanomaterials and nanostructures. *Anal. Chem.* **2015**, *87*, 230–249. [CrossRef]
25. Sun, X.; Jian, Y.; Wang, H.; Ge, S.; Yan, M.; Yu, J. Ultrasensitive microfluidic paper-based electrochemical biosensor based on molecularly imprinted film and boronate affinity sandwich assay for glycoprotein detection. *ACS Appl. Mater. Interfaces* **2019**, *11*, 16198–16206. [CrossRef]
26. Chen, J.; Yu, C.; Gao, R.; Geng, Y.; Zhao, Y.; Niu, Y.; Zhang, L.; Yu, Y.; He, J. A palladium-platinum bimetal nanodendritic melamine network for signal amplification in voltammetric sensing of DNA. *Microchim. Acta* **2018**, *185*, 138–146. [CrossRef]
27. Li, H.; Sun, Z.; Zhong, W.; Hao, N.; Xu, D.; Chen, H.Y. Ultrasensitive electrochemical detection for DNA arrays based on silver nanoparticle aggregates. *Anal. Chem.* **2010**, *82*, 5477–5483. [CrossRef]
28. Wei, T.; Dong, T.; Wang, Z.; Bao, J.; Tu, W.; Dai, Z. Aggregation of individual sensing units for signal accumulation: Conversion of liquid-phase colorimetric assay into enhanced surface-tethered electrochemical analysis. *J. Am. Chem. Soc.* **2015**, *137*, 8880–8883. [CrossRef]
29. Li, H.; Qiang, W.; Vuki, M.; Xu, D.; Chen, H.-Y. Fluorescence enhancement of silver nanoparticle hybrid probes and ultrasensitive detection of IgE. *Anal. Chem.* **2011**, *83*, 8945–8952. [CrossRef]
30. Hyeon, J.E.; Kang, D.H.; Han, S.O. Signal amplification by a self-assembled biosensor system designed on the principle of dockerin–cohesin interactions in a cellulosome complex. *Analyst* **2014**, *139*, 4790–4793. [CrossRef]
31. Chen, C.; Liu, Y.; Zheng, Z.; Zhou, G.; Ji, X.; Wang, H.; He, Z. A new colorimetric platform for ultrasensitive detection of protein and cancer cells based on the assembly of nucleic acids and proteins. *Anal. Chim. Acta* **2015**, *880*, 1–7. [CrossRef]
32. Fu, C.; Jin, S.; Shi, W.; Oh, J.; Cao, H.; Jung, Y.M. Catalyzed deposition of signal reporter for highly sensitive surface-enhanced Raman spectroscopy immunoassay based on tyramine signal amplification strategy. *Anal. Chem.* **2018**, *90*, 13159–13162. [CrossRef]
33. Zahran, M.; Khalifa, Z.; Zahran, M.A.H.; Abdel Azzem, M. Recent advances in silver nanoparticle-based electrochemical sensors for determining organic pollutants in water: A review. *Mater. Adv.* **2021**, *2*, 7350–7365. [CrossRef]
34. Terracciano, M.; Rea, I.; De Stefano, L.; Rendina, I.; Oliviero, G.; Nici, F.; D'Errico, S.; Picciali, G.; Borbone, N. Synthesis of mixed-sequence oligonucleotides on mesoporous silicon: Chemical strategies and material stability. *Nanoscale Res. Lett.* **2014**, *9*, 317–325. [CrossRef] [PubMed]

35. Terracciano, M.; De Stefano, L.; Borbone, N.; Politi, J.; Oliviero, G.; Nici, F.; Casalino, M.; Piccialli, G.; Dardano, P.; Varra, M.; et al. Solid phase synthesis of a thrombin binding aptamer on macroporous silica for label free optical quantification of thrombin. *RSC Adv.* **2016**, *6*, 86762–86769. [CrossRef]
36. Cao, X.; Xu, J.; Xia, J.; Zhang, F.; Wang, Z. An electrochemical aptasensor based on the conversion of liquid-phase colorimetric assay into electrochemical analysis for sensitive detection of lysozyme. *Sens. Actuators B Chem.* **2018**, *255*, 2136–2142. [CrossRef]
37. Song, W.; Li, H.; Liang, H.; Qiang, W.; Xu, D. Disposable electrochemical aptasensor array by using in situ DNA hybridization inducing silver nanoparticles aggregate for signal amplification. *Anal. Chem.* **2014**, *86*, 2775–2783. [CrossRef]
38. Wang, Z.; Sun, N.; He, Y.; Liu, Y.; Li, J. DNA assembled gold nanoparticles polymeric network blocks modular highly sensitive electrochemical biosensors for protein kinase activity analysis and inhibition. *Anal. Chem.* **2014**, *86*, 6153–6159. [CrossRef] [PubMed]
39. Yu, S.; Wang, Y.; Jiang, L.P.; Bi, S.; Zhu, J.J. Cascade amplification-mediated in situ hot-spot assembly for microRNA detection and molecular logic gate operations. *Anal. Chem.* **2018**, *90*, 4544–4551. [CrossRef] [PubMed]
40. Jiang, J.; Wu, H.; Su, Y.; Liang, Y.; Shu, B.; Zhang, C. Electrochemical cloth-based DNA sensors (ECDSs): A new class of electrochemical gene sensors. *Anal. Chem.* **2020**, *92*, 7708–7716. [CrossRef]
41. Mei, C.; Pan, L.; Xu, W.; Xu, H.; Zhang, Y.; Li, Z.; Dong, B.; Ke, X.; McAlinden, C.; Yang, M.; et al. An ultrasensitive reusable aptasensor for noninvasive diabetic retinopathy diagnosis target on tear biomarker. *Sens. Actuators B Chem.* **2021**, *345*, 130398–130407. [CrossRef]
42. Yuan, Y.H.; Wu, Y.D.; Chi, B.Z.; Wen, S.H.; Liang, R.P.; Qiu, J.D. Simultaneously electrochemical detection of microRNAs based on multifunctional magnetic nanoparticles probe coupling with hybridization chain reaction. *Biosens. Bioelectron.* **2017**, *97*, 325–331. [CrossRef]
43. Miao, P.; Tang, Y.; Yin, J. MicroRNA detection based on analyte triggered nanoparticle localization on a tetrahedral DNA modified electrode followed by hybridization chain reaction dual amplification. *Chem. Commun.* **2015**, *51*, 15629–15632. [CrossRef]
44. Li, Z.; Miao, X.; Xing, K.; Peng, X.; Zhu, A.; Ling, L. Ultrasensitive electrochemical sensor for Hg(2+) by using hybridization chain reaction coupled with Ag@Au core-shell nanoparticles. *Biosens. Bioelectron.* **2016**, *80*, 339–343. [CrossRef]
45. Chen, X.; Wang, Y.; Zhang, J.; Zhang, Y. DNA concatemer-silver nanoparticles as a signal probe for electrochemical prostate-specific antigen detection. *Analyst* **2019**, *144*, 6313–6320. [CrossRef]
46. Li, Z.; Miao, X.; Xing, K.; Zhu, A.; Ling, L. Enhanced electrochemical recognition of double-stranded DNA by using hybridization chain reaction and positively charged gold nanoparticles. *Biosens. Bioelectron.* **2015**, *74*, 687–690. [CrossRef] [PubMed]
47. Dong, H.; Jin, S.; Ju, H.; Hao, K.; Xu, L.P.; Lu, H.; Zhang, X. Trace and label-free microRNA detection using oligonucleotide encapsulated silver nanoclusters as probes. *Anal. Chem.* **2012**, *84*, 8670–8674. [CrossRef]
48. Chen, Z.; Liu, Y.; Xin, C.; Zhao, J.; Liu, S. A cascade autocatalytic strand displacement amplification and hybridization chain reaction event for label-free and ultrasensitive electrochemical nucleic acid biosensing. *Biosens. Bioelectron.* **2018**, *113*, 1–8. [CrossRef]
49. Yang, C.; Shi, K.; Dou, B.; Xiang, Y.; Chai, Y.; Yuan, R. In situ DNA-templated synthesis of silver nanoclusters for ultrasensitive and label-free electrochemical detection of microRNA. *ACS Appl. Mater. Interfaces* **2015**, *7*, 1188–1193. [CrossRef]
50. Peng, X.; Zhu, J.; Wen, W.; Bao, T.; Zhang, X.; He, H.; Wang, S. Silver nanoclusters-assisted triple-amplified biosensor for ultrasensitive methyltransferase activity detection based on AuNPs/ERGO hybrids and hybridization chain reaction. *Biosens. Bioelectron.* **2018**, *118*, 174–180. [CrossRef]
51. Xie, X.; Chai, Y.; Yuan, Y.; Yuan, R. Dual triggers induced disassembly of DNA polymer decorated silver nanoparticle for ultrasensitive electrochemical Pb^{2+} detection. *Anal. Chim. Acta* **2018**, *1034*, 56–62. [CrossRef] [PubMed]
52. Feng, C.; Guo, J.; Li, G.; Ye, B.; Zou, L. Novel multiple strand displacement reaction coupled hybridization chain reaction for label-free and ultrasensitive electrochemical Type b3a2 biosensing. *Sens. Actuators B Chem.* **2021**, *326*, 128972–128978. [CrossRef]
53. Zou, Y.; Zhang, H.; Wang, Z.; Liu, Q.; Liu, Y. A novel ECL method for histone acetyltransferases (HATs) activity analysis by integrating HCR signal amplification and ECL silver clusters. *Talanta* **2019**, *198*, 39–44. [CrossRef]
54. Zhao, J.; Hu, S.; Cao, Y.; Zhang, B.; Li, G. Electrochemical detection of protein based on hybridization chain reaction-assisted formation of copper nanoparticles. *Biosens. Bioelectron.* **2015**, *66*, 327–331. [CrossRef]
55. Wang, Y.; Zhang, X.; Zhao, L.; Bao, T.; Wen, W.; Zhang, X.; Wang, S. Integrated amplified aptasensor with in-situ precise preparation of copper nanoclusters for ultrasensitive electrochemical detection of microRNA 21. *Biosens. Bioelectron.* **2017**, *98*, 386–391. [CrossRef] [PubMed]
56. Zhou, M.; Han, L.; Deng, D.; Zhang, Z.; He, H.; Zhang, L.; Luo, L. 4-mercaptobenzoic acid modified silver nanoparticles-enhanced electrochemical sensor for highly sensitive detection of Cu^{2+}. *Sens. Actuators B Chem.* **2019**, *291*, 164–169. [CrossRef]
57. Ejeta, S.Y.; Imae, T. Selective colorimetric and electrochemical detections of Cr(III) pollutant in water on 3-mercaptopropionic acid-functionalized gold plasmon nanoparticles. *Anal. Chim. Acta* **2021**, *1152*, 338272–338281. [CrossRef]
58. Gu, Y.; Jiang, Z.; Ren, D.; Shang, Y.; Hu, Y.; Yi, L. Electrochemiluminescence sensor based on the target recognition-induced aggregation of sensing units for Hg^{2+} determination. *Sens. Actuators B Chem.* **2021**, *337*, 129821–129826. [CrossRef]
59. Zhao, Y.; Cui, L.; Ke, W.; Zheng, F.; Li, X. Electroactive Au@Ag nanoparticle assembly driven signal amplification for ultrasensitive chiral recognition of d-/l-Trp. *ACS Sustain. Chem. Eng.* **2019**, *7*, 5157–5166. [CrossRef]
60. Wang, N.; Dai, H.; Sai, L.; Ma, H.; Lin, M. Copper ion-assisted gold nanoparticle aggregates for electrochemical signal amplification of lipopolysaccharide sensing. *Biosens. Bioelectron.* **2019**, *126*, 529–534. [CrossRef] [PubMed]

61. Xia, N.; Liu, L.; Chang, Y.; Hao, Y.; Wang, X. 4-mercaptophenylboronic acid-induced in situ formation of silver nanoparticle aggregates as labels on an electrode surface. *Electrochem. Commun.* **2017**, *74*, 28–32. [CrossRef]
62. Liu, L.; Sun, T.; Ren, H. Electrochemical detection of hydrogen peroxide by inhibiting the p-benzenediboronic acid-triggered assembly of citrate-capped Au/Ag nanoparticles on electrode surface. *Materials* **2017**, *10*, 40. [CrossRef] [PubMed]
63. Xia, N.; Wu, D.; Sun, T.; Wang, Y.; Ren, X.; Zhao, F.; Liu, L.; Yi, X. Magnetic bead-based electrochemical and colorimetric methods for the detection of poly(ADP-ribose) polymerase-1 with boronic acid derivatives as the signal probes. *Sens. Actuators B Chem.* **2021**, *327*, 128913–128920. [CrossRef]
64. Xia, N.; Wu, D.; Yu, H.; Sun, W.; Yi, X.; Liu, L. Magnetic bead-based electrochemical and colorimetric assays of circulating tumor cells with boronic acid derivatives as the recognition elements and signal probes. *Talanta* **2021**, *221*, 121640–121647. [CrossRef]
65. Shinde, S.; Kim, D.-Y.; Saratale, R.G.; Syed, A.; Ameen, F.; Ghodake, G. A spectral probe for detection of aluminum (III) ions using surface functionalized gold nanoparticles. *Nanomaterials* **2017**, *7*, 287. [CrossRef] [PubMed]
66. Ali, A.; Hussain, F.; Attacha, S.; Kalsoom, A.; Qureshi, W.A.; Shakeel, M.; Militky, J.; Tomkova, B.; Kremenakova, D. Development of novel antimicrobial and antiviral green synthesized silver nanocomposites for the visual detection of Fe^{3+} ions. *Nanomaterials* **2021**, *11*, 2076. [CrossRef] [PubMed]
67. Liu, L.; Chang, Y.; Xia, N.; Peng, P.; Zhang, L.; Jiang, M.; Zhang, J.; Liu, L. Simple, sensitive and label-free electrochemical detection of microRNAs based on the in situ formation of silver nanoparticles aggregates for signal amplification. *Biosens. Bioelectron.* **2017**, *94*, 235–242. [CrossRef]
68. Liu, L.; Cheng, C.; Chang, Y.; Ma, H.; Hao, Y. Two sensitive electrochemical strategies for the detection of protein kinase activity based on the 4-mercaptophenylboronic acid-induced in situ assembly of silver nanoparticles. *Sens. Actuators B Chem.* **2017**, *248*, 178–186. [CrossRef]
69. Xia, N.; Cheng, C.; Liu, L.; Peng, P.; Liu, C.; Chen, J. Electrochemical glycoprotein aptasensors based on the in-situ aggregation of silver nanoparticles induced by 4-mercaptophenylboronic acid. *Microchim. Acta* **2017**, *184*, 4393–4400. [CrossRef]
70. Hou, L.; Huang, Y.; Hou, W.; Yan, Y.; Liu, J.; Xia, N. Modification-free amperometric biosensor for the detection of wild-type p53 protein based on the in situ formation of silver nanoparticle networks for signal amplification. *Int. J. Biol. Macromol.* **2020**, *158*, 580–586. [CrossRef] [PubMed]
71. Sun, K.; Chang, Y.; Zhou, B.; Wang, X.; Liu, L. Gold nanoparticles-based electrochemical method for the detection of protein kinase with a peptide-like inhibitor as the bioreceptor. *Int. J. Nanomed.* **2017**, *12*, 1905–1915. [CrossRef]
72. Xia, N.; Wang, X.; Wang, X.; Zhou, B. Gold nanoparticle-based colorimetric and electrochemical methods for dipeptidyl peptidase-IV activity assay and inhibitor screening. *Materials* **2016**, *9*, 857. [CrossRef] [PubMed]
73. Xing, Y.; Feng, X.Z.; Zhang, L.; Hou, J.; Han, G.C.; Chen, Z. A sensitive and selective electrochemical biosensor for the determination of beta-amyloid oligomer by inhibiting the peptide-triggered in situ assembly of silver nanoparticles. *Int. J. Nanomed.* **2017**, *12*, 3171–3179. [CrossRef] [PubMed]
74. Balducci, C.; Beeg, M.; Stravalaci, M.; Bastone, A.; Sclip, A.; Biasini, E.; Tapella, L.; Colombo, L.; Manzoni, C.; Borsello, T.; et al. Synthetic amyloid-beta oligomers impair long-term memory independently of cellular prion protein. *Proc. Nat. Acad. Sci. USA* **2010**, *107*, 2295–2300. [CrossRef]
75. Laurén, J.; Gimbel, D.A.; Nygaard, H.B.; Gilbert, J.W.; Strittmatter, S.M. Cellular prion protein mediates impairment of synaptic plasticity by amyloid-beta oligomers. *Nature* **2009**, *457*, 1128–1132. [CrossRef] [PubMed]
76. Chen, S.; Yadav, S.P.; Surewicz, W.K. Interaction between human prion protein and amyloid-beta (Abeta) oligomers: Role of N-terminal residues. *J. Biol. Chem.* **2010**, *285*, 26377–26383. [CrossRef] [PubMed]
77. Xia, N.; Zhou, B.; Huang, N.; Jiang, M.; Zhang, J.; Liu, L. Visual and fluorescent assays for selective detection of beta-amyloid oligomers based on the inner filter effect of gold nanoparticles on the fluorescence of CdTe quantum dots. *Biosens. Bioelectron.* **2016**, *85*, 625–632. [CrossRef] [PubMed]
78. Xia, N.; Wang, X.; Zhou, B.; Wu, Y.; Mao, W.; Liu, L. Electrochemical detection of amyloid-beta oligomers based on the signal amplification of a network of silver nanoparticles. *ACS Appl. Mater. Interfaces* **2016**, *8*, 19303–19311. [CrossRef] [PubMed]
79. Xia, N.; Wang, X.; Yu, J.; Wu, Y.; Cheng, S.; Xing, Y.; Liu, L. Design of electrochemical biosensors with peptide probes as the receptors of targets and the inducers of gold nanoparticles assembly on electrode surface. *Sens. Actuators B Chem.* **2017**, *239*, 834–840. [CrossRef]
80. Xia, N.; Chen, Z.; Liu, Y.; Ren, H.; Liu, L. Peptide aptamer-based biosensor for the detection of human chorionic gonadotropin by converting silver nanoparticles-based colorimetric assay into sensitive electrochemical analysis. *Sens. Actuators B Chem.* **2017**, *243*, 784–791. [CrossRef]
81. Han, Y.; Zhang, Y.; Wu, S.; Jalalah, M.; Alsareii, S.A.; Yin, Y.; Harraz, F.A.; Li, G. Co-assembly of peptides and carbon nanodots: Sensitive analysis of transglutaminase 2. *ACS Appl. Mater. Interfaces* **2021**, *13*, 36919–36925. [CrossRef] [PubMed]
82. Chang, Y.; Ma, X.; Sun, T.; Liu, L.; Hao, Y. Electrochemical detection of kinase by converting homogeneous analysis into heterogeneous assay through avidin-biotin interaction. *Talanta* **2021**, *234*, 122649–122654. [CrossRef] [PubMed]
83. Huang, Y.; Sun, T.; Liu, L.; Xia, N.; Zhao, Y.; Yi, X. Surface plasmon resonance biosensor for the detection of miRNAs by combining the advantages of homogeneous reaction and heterogeneous detection. *Talanta* **2021**, *234*, 122622–122628. [CrossRef] [PubMed]
84. La, M.; Wu, D.; Gao, Y.; Xia, N.; Niu, Y.; Liu, L.; Yi, X. Competitive impedimetric aptasensors for detection of small molecule pollutants by the signal amplification of self-assembled biotin-phenylalanine nanoparticle networks. *Electrochem. Commun.* **2020**, *118*, 106791–106796. [CrossRef]

85. Xia, N.; Huang, Y.; Cui, Z.; Liu, S.; Deng, D.; Liu, L.; Wang, J. Impedimetric biosensor for assay of caspase-3 activity and evaluation of cell apoptosis using self-assembled biotin-phenylalanine network as signal enhancer. *Sens. Actuators B Chem.* **2020**, *320*, 128436–128442. [CrossRef]
86. La, M.; Zhang, Y.; Gao, Y.; Li, M.; Liu, L.; Chang, Y. Impedimetric detection of microRNAs by the signal amplification of streptavidin induced in situ formation of biotin phenylalanine nanoparticle networks. *J. Electrochem. Soc.* **2020**, *167*, 117505–117512. [CrossRef]
87. Hu, Q.; Wang, Q.; Kong, J.; Li, L.; Zhang, X. Electrochemically mediated in situ growth of electroactive polymers for highly sensitive detection of double-stranded DNA without sequence-preference. *Biosens. Bioelectron.* **2018**, *101*, 1–6. [CrossRef]
88. Hu, Q.; Wang, Q.; Jiang, C.; Zhang, J.; Kong, J.; Zhang, X. Electrochemically mediated polymerization for highly sensitive detection of protein kinase activity. *Biosens. Bioelectron.* **2018**, *110*, 52–57. [CrossRef] [PubMed]
89. Hu, Q.; Wang, Q.; Sun, G.; Kong, J.; Zhang, X. Electrochemically mediated surface-initiated de novo growth of polymers for amplified electrochemical detection of DNA. *Anal. Chem.* **2017**, *89*, 9253–9259. [CrossRef] [PubMed]
90. Hu, Q.; Han, D.; Gan, S.; Bao, Y.; Niu, L. Surface-initiated-reversible-addition-fragmentation-chain-transfer polymerization for electrochemical DNA biosensing. *Anal. Chem.* **2018**, *90*, 12207–12213. [CrossRef]
91. Hu, Q.; Kong, J.; Han, D.; Zhang, Y.; Bao, Y.; Zhang, X.; Niu, L. Electrochemically controlled RAFT polymerization for highly sensitive electrochemical biosensing of protein kinase activity. *Anal. Chem.* **2019**, *91*, 1936–1943. [CrossRef] [PubMed]
92. Hu, Q.; Su, L.; Chen, Z.; Huang, Y.; Qin, D.; Niu, L. Coenzyme-mediated electro-RAFT polymerization for amplified electrochemical interrogation of trypsin activity. *Anal. Chem.* **2021**, *93*, 9602–9608. [CrossRef]
93. Hu, Q.; Bao, Y.; Gan, S.; Zhang, Y.; Han, D.; Niu, L. Amplified electrochemical biosensing of thrombin activity by RAFT polymerization. *Anal. Chem.* **2020**, *92*, 3470–3476. [CrossRef] [PubMed]
94. Lu, J.; Hao, L.; Yang, F.; Liu, Y.; Yang, H.; Yan, S. Ultrasensitive electrochemical detection of CYFRA 21-1 via in-situ initiated ROP signal amplification strategy. *Anal. Chim. Acta* **2021**, *1180*, 338889–338895. [CrossRef] [PubMed]
95. Huang, Y.; Zhang, B.; Yuan, L.; Liu, L. A signal amplification strategy based on peptide self-assembly for the identification of amyloid-β oligomer. *Sens. Actuators B Chem.* **2021**, *335*, 129697–129707. [CrossRef]
96. Xia, N.; Sun, Z.; Ding, F.; Wang, Y.; Sun, W.; Liu, L. Protease biosensor by conversion of a homogeneous assay into a surface-tethered electrochemical analysis based on streptavidin-biotin interactions. *ACS Sens.* **2021**, *6*, 1166–1173. [CrossRef] [PubMed]
97. Liu, L.; Deng, D.; Wu, D.; Hou, W.; Wang, L.; Li, N.; Sun, Z. Duplex-specific nuclease-based electrochemical biosensor for the detection of microRNAs by conversion of homogeneous assay into surface-tethered electrochemical analysis. *Anal. Chim. Acta* **2021**, *1149*, 338199–338206. [CrossRef]
98. Liu, L.; Wu, D.; Zhen, S.; Lu, K.; Yi, X.; Sun, Z. Electrochemical detection of telomerase in cancer cells based on the in-situ formation of streptavidin-biotin-DNA-biotin networks for signal amplification. *Sens. Actuators B Chem.* **2021**, *334*, 129659–129665. [CrossRef]

nanomaterials

Review

Recent Advances in Electrochemical Sensing of Hydrogen Peroxide (H_2O_2) Released from Cancer Cells

Touqeer Ahmad [1], Ayesha Iqbal [2], Sobia Ahsan Halim [1], Jalal Uddin [3], Ajmal Khan [1,*], Sami El Deeb [1,4,*] and Ahmed Al-Harrasi [1,*]

1 Natural and Medical Sciences Research Center, University of Nizwa, P.O. Box 33, Birkat Al Mauz, Nizwa 616, Oman; touqeernano@hotmail.com (T.A.); sobia_halim@unizwa.edu.om (S.A.H.)
2 Division of Pharmacy Practice and Policy, School of Pharmacy, University of Nottingham, Nottingham NG7 2RD, UK; ayesharph@hotmail.com
3 Department of Pharmaceutical Chemistry, College of Pharmacy, King Khalid University, Abha 62529, Saudi Arabia; jalaluddinamin@gmail.com
4 Institute of Medicinal and Pharmaceutical Chemistry, Technische Universitaet Braunschweig, 38106 Braunschweig, Germany
* Correspondence: ajmalkhan@unizwa.edu.om (A.K.); s.eldeeb@tu-bs.de (S.E.D.); aharrasi@unizwa.edu.om (A.A.-H.)

Abstract: Cancer is by far the most common cause of death worldwide. There are more than 200 types of cancer known hitherto depending upon the origin and type. Early diagnosis of cancer provides better disease prognosis and the best chance for a cure. This fact prompts world-leading scientists and clinicians to develop techniques for the early detection of cancer. Thus, less morbidity and lower mortality rates are envisioned. The latest advancements in the diagnosis of cancer utilizing nanotechnology have manifested encouraging results. Cancerous cells are well known for their substantial amounts of hydrogen peroxide (H_2O_2). The common methods for the detection of H_2O_2 include colorimetry, titration, chromatography, spectrophotometry, fluorimetry, and chemiluminescence. These methods commonly lack selectivity, sensitivity, and reproducibility and have prolonged analytical time. New biosensors are reported to circumvent these obstacles. The production of detectable amounts of H_2O_2 by cancerous cells has promoted the use of bio- and electrochemical sensors because of their high sensitivity, selectivity, robustness, and miniaturized point-of-care cancer diagnostics. Thus, this review will emphasize the principles, analytical parameters, advantages, and disadvantages of the latest electrochemical biosensors in the detection of H_2O_2. It will provide a summary of the latest technological advancements of biosensors based on potentiometric, impedimetric, amperometric, and voltammetric H_2O_2 detection. Moreover, it will critically describe the classification of biosensors based on the material, nature, conjugation, and carbon-nanocomposite electrodes for rapid and effective detection of H_2O_2, which can be useful in the early detection of cancerous cells.

Keywords: analytical methods; biosensors; carbon materials; electrochemical sensing; H_2O_2; nanomaterial

Citation: Ahmad, T.; Iqbal, A.; Halim, S.A.; Uddin, J.; Khan, A.; El Deeb, S.; Al-Harrasi, A. Recent Advances in Electrochemical Sensing of Hydrogen Peroxide (H_2O_2) Released from Cancer Cells. *Nanomaterials* **2022**, *12*, 1475. https://doi.org/10.3390/nano12091475

Academic Editors: Dong Liu and Baiqing Yuan

Received: 16 February 2022
Accepted: 23 March 2022
Published: 26 April 2022

1. Introduction

Biosensors are simple devices that are small and are generally used in the field of medicine, pharmaceutical industries, environmental technology, and food industry. They are used for the measurement of many biological and chemical substances [1]. Owing to the advancement of science and technology, the research involved in biosensors has successfully made the biosensing devices small and efficient [2]. The use of the latest novel techniques and availability of a new biomaterial have made the biosensors efficient and have extended their use in multiple industries such as pharmaceutical, environmental, agriculture, and industrial laboratories [3]. Biosensors are of many types, and electrobiochemical biosensors have been commonly used for over 20 years in the field of diagnostics to detect biochemicals like glucose, lactate, cholesterol, urea, creatinine, DNA, antigens, antibodies, and cancer markers [4]. Electrobiochemical biosensors are also useful in the analysis of food materials and drinks and are extensively utilized in environmental and pharmaceutical laboratories [5,6]. Cancer is one of the most fatal diseases, and every year, more than 10 million new cases and 6 million deaths are reported worldwide [7,8]. Cancer is linked with high rates of morbidities and mortalities and more than 8.7 million deaths worldwide in 2015 [9]. The cancer incidence in high- and low-income countries is similar to the trend of increase in lower–middle-income countries because of the increase in risk factors associated with cancer [10]. In the United States of America (USA), cancer is the second leading cause of mortality, with heart disease being the first, and in a study published in 2017, it was estimated that more than 0.6 million people will die annually from cancer [8]. The survival rate of cancer patients drastically increases if the cancer is detected in earlier stages. The appearance of alarming systems in cancer patients is usually after cancer has spread in the body to multiple locations or has metastasized in different locations and organs, which characterizes an advanced stage of cancer. Most of the people are diagnosed with cancer at an advanced stage, which causes a high risk for mortality. To have a better disease prognosis, it is imperative that new research should focus on the early detection of cancerous cells in the body.

The use of biosensing devices, which have been designed to detect biochemicals, holds vast potential in the early diagnosis of cancers. Biosensors work by detecting a biological moiety or analyte and then converting it into an electrical signal, which can be detected and analyzed by the biosensor device. Many cancerous cells release specific chemicals called biomarkers, which can be detected using a biosensor device. The specific biomarker levels can also help in analyzing the effectiveness of anticancer therapy. The use of biosensor devices is a promising technique, which can help in early and accurate detection, imaging of cancerous cells, monitoring of angiogenesis, detection of proliferation, and tracking of metastatic changes and the efficacy of anticancer therapeutic regimens [11]. The latest research in biosensor devices using the latest techniques, such as nanotechnology-empowered diagnostics, can help in the identification of specific cancer biomarkers, which can help in the detection of cancer, disease progression, disease remission, and further proliferation. The biomarkers of cancer can be overexpressed proteins, surface antigens, active or inactive metabolites, miRNA, or the cancerous cells themselves. Many biosensors are excellent for use as an effective analytical device because of their capability to detect specific cancer biomarkers due to their highly sensitive, selective, robust, and miniaturized point-of-care cancer diagnostic capability [12]. Hydrogen peroxide (H_2O_2) is normally present inside the body and is vital in initiating and performing many important physiological processes. H_2O_2 is a by-product of respiratory chain and enzymes oxidases (glucose oxidase, cholesterol oxidase, glutamate oxidase, etc.) [13–17]. Hydrogen peroxide is a reactive oxygen species (ROS), which helps in regulating normal body functions such as cell growth, activation of the immune system, and programmed apoptotic changes [13–17]. The body system normally functions in homeostasis, and an increased level of H_2O_2 due to increased production can cause harm to the body. Increased levels of H_2O_2 can cause damage to normal cells [18], increase inflammatory responses [19], and cause cancer [20]. H_2O_2 regulates cancer cell characteristics, including invasion, proliferation, migration, apoptosis, and angiogenesis. Oxidative stress is associated with high levels of ROS, common in many types of cancerous cells. H_2O_2 has a specific role as a second messenger in pro-tumorigenic signaling pathways of cancerous cells [21,22]. GPX2 regulates cancer progression by regulating the hydrogen peroxide level in the cells, so when the level of H_2O_2 is downregulated to a normal level and the oxidative stress is relieved, it can help in dysregulating cancer cell homeostasis [23]. H_2O_2 has recently been a prime focus of research because of its high biological significance. When studied within living systems, it is noteworthy to check the concentration of H_2O_2 in mammals. The cellular compartment concentration has to be in the physiological range of 1 nM to 0.5 μM [24,25]. The latest research in the field of cancer diagnosis via biomarker-based techniques has been evaluated as successful because the process ensures high-precision, reliable, and sensitive data. The processes of biomarker-based cancer diagnostic are simple, which makes it a popular choice. Recent studies are focusing on profiling the cell functions with the efflux of endogenous H_2O_2 as potential biomarkers for diagnosing various cancers by measuring them using conventional biological assays [26]. However, before moving to the practical implications of using H_2O_2 as a potential biomarker for cancer diagnostics in living systems, detection of increased oxidative stress, prediction of neurodegenerative diseases, and detection of tumor growth inside living organisms, it is imperative to develop methods and techniques that can precisely detect and measure the level of H_2O_2 inside the cellular compartments [27].

The current problem with detecting H_2O_2 in cellular compartments is its low concentration in the body, as well as reactivity, which makes it difficult to separate its normal physiological concentration in a healthy organism from the concentration in a diseased or high-risk state. Therefore, scientists and researchers are focusing on developing sensors that can detect and quantify H_2O_2 in different systems and physiological conditions. Currently, many analytical techniques such as titrimetry, spectrophotometry, chemiluminescence, chromatography, fluorescence, and phosphorescence are measuring and determining the concentration of H_2O_2 in cellular compartments to develop consistent, precise, sensitive, fast, efficient, and low-cost methods. Currently, many methods investigated based on these analytical techniques have methodological disadvantages such as small sensitivity and selectivity, time-consuming and complex process, susceptibility to interference, and high-cost running instruments [28]. Alternatively, the advantages of "electrochemical sensors" are that they are highly sensitive and selective, reliable, quick, less costly, simple, and practical, and therefore, they are an optimum solution for exact and sensitive H_2O_2 detection [29]. The advancements of science and the latest research are looking for potential solutions to detect cancer at early stages and provide individualized therapeutic regimens. However, these techniques and methods still have many restrictions and limitations in the context of clinical examinations, histopathological analysis, imaging mammography, and chemotherapeutic adverse drug effects, as well as their high running costs [30,31]. In the detection of breast cancer, the patients are first exposed to a high amount of radiation in mammography because of its inadequate test sensitivity. Detection and disease progression are usually confirmed using a biopsy, an invasive procedure used to conduct histopathology of the disease [32]. These commonly used procedures are highly risky, uncomfortable, invasive, and costly. Therefore, the latest research is currently focused on developing noninvasive, inexpensive, highly selective screening, diagnostic, and therapeutic approaches to improve the disease prognosis. Several reviews regarding the importance of the carbon nanomaterial in the electrochemical sensing of H_2O_2 have been published. Out of those reviews, Yang et al. and Wang et al. conducted their reviews by comparing different nanomaterials for the creation of electrochemical biosensors and their applications for the detection of biomolecules [33,34]. Other materials have also been used in creating biosensors such as graphene electrochemical biosensors [35–37] and chemical sensors [38,39], which have been reviewed by Kuila and coworkers. Depending on the aim of the existing reviews, they provide an overview of biosensors that have been used to detect biological analytes. Similarly, Ping and coworkers [40] reviewed the strengths, advantages, and existing applications of 2D graphene-based aptasensors, whereas Chen and coworkers [13] focused their review on the carbon nanomaterial and transition metals in the electrocatalytic reduction of H_2O_2 in different samples. Similarly, Zhang et al. aimed to review the role of carbon materials in improving the sensitivity of H_2O_2 biosensors [41]. Regardless of these existing review articles, a comprehensive overview on the carbon-based nanomaterials and their composite with metal NPs, metal oxides, and biomolecules for the electrochemical detection of H_2O_2 secreted from cancerous cells is still missing in the literature.

This review highlights the recent development in the application of carbon nanomaterials and metal nanoparticles for H_2O_2 detection. A subdivision of the sensors has been made depending on the nanomaterial used: (i) metal nanoparticles, (ii) graphene modified with metal or metal oxide nanoparticles to form "graphene nanocomposites," (iii) enzyme-loaded graphene-like 2D nanomaterials, and (iv) carbon nanotubes modified with metal or metal oxide nanoparticles. Finally, the current potential and challenges of using carbon nanomaterials for H_2O_2 detection are outlined. Relative electrochemical properties such as limit of detection (LOD), sensitivity, and stability are reported for each sensor, and a critical comparison between the results has been carried out by summarizing the strengths and weaknesses of the various sensors found in the review.

2. Classical Methods for H_2O_2 Detection

2.1. Electrochemical Systems

Electrobiochemical biosensors are built on the principle that they have a biorecognition component providing an electroactive constituent after reacting with an analyte that is transformed into an electrical signal that is measured using a physiochemical transducer as shown in Figure 1. The electrochemical transducers help in the detection and monitoring of the changes in the electrical current and potential. The commonly used easily detected biorecognition elements are enzymes. Antibodies, complete cells, and DNA can also be conventionally used for the construction of electrochemical biosensors as biorecognizable elements [1,42] by loading on a metal surface or carbon electrodes [43]. The most used electrochemical transducers are amperometric, potentiometric, conductometric, and impedimetric. The advantage of using electrochemical biosensors in analytical techniques is that they can be effectively used to reduce the size of the device. Additional advantages are that these sensors are low cost, highly sensitive, stable, and reproducible; show a linear response; can detect turbid samples; and are efficient due to using low sample volumes and chemical consumption [5,6,44,45].

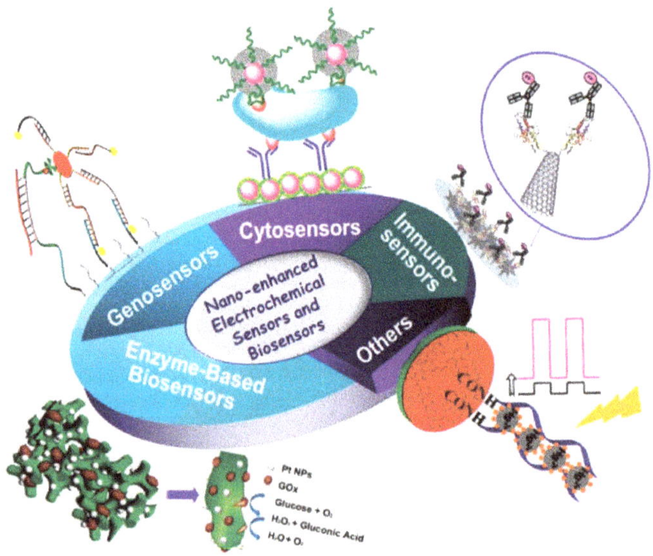

Figure 1. Distinct strategies for the electrochemical detection of H_2O_2 including Cyto-biosensors, Immuno-biosensors, Enzymatic and non-enzymatic biosensors. Figure reproduced with permission from [46]. Copyright 2014. American Chemical Society (Washington, DC, USA).

2.2. *Potentiometric*

Biosensors act on the principle of measuring the electric potential generated by an electrode in the absence of a substantial amount of current via a reference and functional electrode that has been adjusted to sense selective analytes, with membranes that cover the electrode surface. When a target analyte interacts with the membrane covering the electrode surface, a subsequent change in electric potential is detected and measured by the electrode [47]. Potentiometric biosensors work by measuring the electric potential at zero current, which helps in differentiating the reference and functional electrodes. Potentiometric sensors measure the generated electric potential of ion-selective electrodes during biological reactions with target-specific ions. In potentiometric biosensors, the enzyme is attached on the surface of the electrode by glutaraldehyde crosslinking or an adsorption process. The probe of pH meter is covered by the membrane, where biological reaction either produces or absorbs hydrogen ions. The alteration in hydrogen ions causes a change in pH, which is a measure of the concentration of the analyte [48]. For instance, two example of potentiometric biosensors depending upon the type of electrode used are as follows: (a) Potentiometric biosensors use a Nafion membrane/Pt electrode for H_2O_2 determination, with an additional advantage of a perm-selective barrier [49]. Ascorbate and redox-active species reduce the overall electrode response, which further potentiates coupling between the redox potential on the Pt electrode and Donnan potential and increases sensitivity in detecting H_2O_2. The present potentiometric biosensor has a sensitivity of 125.1 $\pm$ 5.9 mV/decade, linear range of 10–1000 μM, and LOD of 10 μM [49]. (b) Zheng et al. used an MnO_2/CPE to detect H_2O_2 [50]. The biosensor shows sensitivity of 19.4–121 mV/decade, with a wide linear range of 0.3–363 μM and LOD of 0.12 μM. The analytical parameters of the MnO_2 doped/CPE biosensor were superior to those of the up-to-date potentiometric biosensors, i.e., Nafion membrane/Pt electrode [49], because of the enhanced electrode surface area, linear range, and LOD, except for the sensitivity. The coupling between the redox potential on the Pt electrode and Donnan potential made the Nafion membrane/Pt electrode superior to the MnO_2 doped/CPE in terms of sensitivity [50].

2.3. Amperometric Biosensor

Zhao et al. developed an amperometric biosensor by immobilizing HRP on a silica sol–gel matrix on CPE for the determination of extracellular H_2O_2 excreted from breast cancer cells. The amperometric biosensor detected H_2O_2 via a sequence-specific peptide immobilized on the electrode surface and explicitly bound with horseradish peroxidase (HRP) in an auspicious orientation. The composed biosensors showed a linear range from 1.0×10^{-7} to 1.0×10^{-4} M, with a detection limit (LOD) of 3.0×10^{-8} M [51]. Zhao et al. showed a linear calibration of H_2O_2, i.e., 2×10^{-5} to 2.6×10^{-3} M, under optimum conditions [52]. In another strategy, a hydrogen peroxide biosensor was fabricated by coating a sol–gel–horseradish peroxidase LSPR layer onto a Nafion-methylene green modified electrode to develop a probe for H_2O_2 detection. The developed electrode exhibited sensitivity of 13.5 μA mM^{-1}, with a detection limit of 1.0×10^{-7} M and response to 95% of the steady-state current in <20 s [53]. To detect H_2O_2, Tripathi et al. developed a novel biosensor by entrapping HRP in a new Ormosil composite doped with ferrocene monocarboxylic acid–bovine serum albumin conjugate and multiwalled carbon nanotubes (MWNTs). The developed biosensor showed a linear range of 0.02–4.0 mM, with a LOD of 5.0 μM (S/N = 3) [54]. In another study, a titania sol–gel matrix entrapping hemoglobin (Hb) was used as a peroxidase mimetic to sense H_2O_2 with a linear range from 5.0×10^{-7} to 5.4×10^{-5} M and a detection limit of 1.2×10^{-7} M [55]. Povedano et al. used His-Tag–Zinc finger commercial (His–Tag–ZFP) protein. The His–Tag–ZFP prefers to bind with RNA hybrids over ssRNAs, ssDNAs, and dsDNAs. These were further conjugated with streptavidin–HRP (Strep–HRP) in order to detect H_2O_2 with a LOD of 0.91 nM [56]. Reduced graphene oxide wrapped $ZnMn_2O_4$ microspheres ($ZnMn_2O_4$@rGO)-modified glassy carbon electrode (GCE) ($ZnMn_2O_4$@rGO/GCE) was used to make amperometric biosensors to detect H_2O_2. The resultant electrode showed a linear detection with a wide concentration range of 0.03–6000 μM and was used to detect H_2O_2 released from human breast carcinoma cells (MCF-7) as low as 0.012 μM [57]. Dong et al. reported a high-performance sensor using high-index facets of Au–Pd nanocubes loaded on large surface reduced grapheme oxide (rGO), and GCEs were modified by physical adsorption of both nanocomposites. The synthesized biosensor with three-dimensional nanocomposites possessed high sensitivity toward H_2O_2, with a minimum LOD of 4 nM, wide linear range from 0.005 μM to 3.5 mM, and swift response time [58]. Later, Jia and his coworker developed a nonenzymatic biosensor composed of poly(diallyldimethylammonium chloride) (PDDA)-capped rGO nanosheets loaded with a trimetallic AuPtAg nanoalloy. This biosensor detects H_2O_2 released from carcinoma cells with a LOD of 1.2 nM and a wide linear range from 0.05 μM to 5.5 mM [59]. In another study, hierarchical $Mo_2C@MoS_2$ consisting of interlayer-expanded MoS_2 nanosheets wrapped on Mo_2C nanorods was built as a highly sensitive, bifunctional electrochemical biosensor to detect H_2O_2 produced by cancerous cells, with sensitivity of 1080 μA mM^{-1} cm^{-2} and LOD of 0.2 μM [60]. Thiruppatthi et al. reported a simple stimulus responding aminophenol, pre-anodized screen-printed carbon electrode (SPCE*/AP) that could detect NADH and H_2O_2. The electrode was built by adsorbing aminophenol on the surface of the electrode, prepared from aminophenylboronic acid via boronic acid deprotection with H_2O_2. The resulting biosensor displayed linear ranges from 50 to 500 μM and from 200 μM to 2 mM, with detection limits (S/N = 3) of 4.2 and 28.9 μM for NADH and H_2O_2, respectively [61]. Maji et al. used cetyltrimethylammonium bromide-loaded gold nanorods (AuNRs) immobilized on a GC electrode to construct an amperometric biosensor (AuNRs/GC for the electrocatalytic detection of H_2O_2 under localized surface Plasmon resonance (LSPR) excitation (808 nm, 2 W cm^{-2}). This biosensor showed an exaggerated improvement in its biosensing properties (~2–4-fold), with a wide linear range from 5.0 μM to 5.0 mM, LOD of 1.8 μM, and sensitivity of 1.6 μA mM^{-1} cm^{-2} [62]. In another study, self-supported MoS_2 nanosheet arrays were built, and they showed highly potent electrocatalytic performance, with a LOD of 1.0 μM (S/N = 3) and high sensitivity of 5.3 mA mM^{-1} cm^{-2}. This biosensor with self-supported MoS_2 nanosheet arrays successfully detected trace amounts of H_2O_2 released from live A549 cancer cells [63].

2.4. Calorimetric Biosensors

A new area of nanotechnology and its integration with biosensors has introduced the concept of calorimetric biosensors for cancerous cell diagnosis and detection. Li et al. used a microfluidic paper-based analytical device (μ-PAD) for the synchronous sensitive and visual detection of H_2O_2 released from cancer cells. μ-PAD construction was done using a layer-by-layer modification of concanavalin A, graphene quantum dots (GQDs)-labeled flower-like Au@Pd alloy nanoparticles (NPs) probe, and vertical alignment of cancerous cells on the surface of ZnO [64]. In the study by Zhang et al. porous platinum NPs on graphene oxide (Pt-NPs/GO) were used in building a calorimetric biosensor. The resultant nanocomposite functioned as a peroxidase mimetic, which could catalyze peroxidase substrate reaction in the presence of H_2O_2. Building on this phenomenon, Pt-NPs/GO acts as a signal transducer in developing a colorimetric assay for cancerous cell detection [65]. Additionally, porous, alloy-structured PtPd nanorods (PtPd PNRs) were used as a peroxidase mimetic for H_2O_2 detection. The PtPd PNRs were found to have a detection limit of 8.6 nM and a linear range from 20 nM to 50 mM and were used as a signal transducer to develop an innovative detection method for studying the flux of H_2O_2 released from cells [66]. Folate and iron-substituted polyoxometalate [$(FeOH_2)_2SiW_{10}O_{36}$] provided a novel method for the detection of H_2O_2 with good sensitivity, with a linear range of 1.34×10^{-7} to 6.7×10^{-5} mol L^{-1}, and low detection limit (1×10^{-7} mol L^{-1}) and swift response toward H_2O_2. Ye et al. showed a new analysis method based on calorimetric analysis. The colorimetric biosensing strategy was based on iodide-responsive Cu–Au nanoparticles (Cu–Au NPs) combined with the iodide-catalyzed H_2O_2–3,3,5,5-tetramethylbenzidine (TMB) reaction system. The bimetallic Cu–Au NPs absorbed iodide, thus indirectly inducing the colorimetric signal variation of the H_2O_2–TMB system. The results demonstrated economically effective, simple, label-free visualization of H_2O_2 from cancerous cells with high selectivity and sensitivity. The resultant calorimetric biosensor operates with a linear range from 50 to 500 cells/mL and a LOD of 5 cells in 100 μL [67]. Calorimetric biosensors can be designed using single wall nano tube (SWNT), which is subsequently embedded within a collagen matrix. When there is an angiogenic stimulation of human umbilical vein endothelial cells (HUVECs), H_2O_2 molecules are released, which can be detected using this SWNT sensor. The constructed calorimetric biosensor shows calibration from 12.5 to 400 nM and can measure H_2O_2 at a nanomolar concentration in HUVEC from humans, with 1 s temporal and 300 nm spatial resolutions [68]. Other biosensors based on the principle of calorimetry include a biosensor with an electrode made of a 2D hybrid material (RGO–PMS@AuNPs). This biosensor displayed remarkable electrochemical performance and possessed high sensitivity and high selectivity in detecting H_2O_2 in 0.1 M phosphate-buffered saline as compared to enzymatic biosensors. The developed biosensor has an additional advantage over other sensors because it is nontoxic and can detect H_2O_2 without any intrusion by common interfering agents, with high sensitivity of 39.2 $\mu A\ mM^{-1}\ cm^{-2}$, broad detection ranges from 0.5 µM to 50 mM, and a LOD of 60 nM. The sensor has high efficiency and can detect H_2O_2 in trace amounts, i.e., as low as nanomolar, secreted from living HeLa and HepG2 tumor cells [69].

2.5. Chemiluminescence Material for the Detection of H_2O_2

For the early diagnosis and detection of cancerous cells, it is important that molecules that indicate changes or biomarkers should be efficiently imaged and sensed. These parameters are important especially in studies that are evaluating the clinical mechanisms and designing effective chemotherapeutic agents [70,71]. The diagnostic and therapeutic methods for multiple detections are slow and need repetitive sampling, which results in low sensitivity and accuracy, because of heterogeneous sampling for separate detections [72]. The multiple fluorescence (FL) technique has promising results when used in in situ concurrent detection of multiple biomolecules. This technique has certain limitations such as weak compatibility with different biological systems, toxicity to living cells, and necessity for specialized synthesis and preparation [70]. Additionally, the FL signals generated using this technique faced interference, changes from background effects, and photobleaching while operating. Thus, it is highly desirable that in situ sequential detection of multiple biomolecules using within a complex biological sample is greatly desirable, the FL technique should be researched in cancer diagnostics without the current limitations [73]. The chemiluminescence (CL) technique is based on the principle that light is generated because of the energy released during a chemical reaction due to the de-excitation of the high energy moieties to the ground state or through energy transfer to luminophore molecules as shown in Figure 2 [74–76]. CL methods have gained popularity because these techniques are highly sensitive, are free of interference, phototoxicity, and photobleaching, and show no changes from background effects. The combination of CL methods with enzymes and analytes such as firefly luciferase (FFLuc) for 5′-triphosphate disodium salt (ATP) [77,78] and HRP [79,80] may result in a highly sensitive and competent method for H_2O_2 detection. In the recent advancement in the field of NPs, the multifunctional NPs with shell-like structures in their core have promising results in simultaneous diagnosis and treatment in living systems [81–84]. These multifunctional NPs were synthesized in the study by Ren et al. where dual functioning NPs were developed by combining HRPSiO_2@FFLuc NPs with the enzyme-based core–shell structures, where the enzymes HRP and FFLuc were the main components of the core and shell of the NPs. They used the dual functioning NPs for the simultaneous in situ sequential detections and imaging of two biomolecules, namely, ATP and H_2O_2, in the same biological system. The surroundings of tumor cells or tissues are slightly acidic, and SiO_2 is sensitive to an acidic environment, which causes the breakage of the SiO_2 layer/component and exposes FFLuc and HRP (outside) and the SiO_2 core (inside) to catalytic reactions. This results in the emission of two separate but simultaneous chemiluminescence signals for the sequential detection of ATP and H_2O_2, which avoids the signal interference between each other [73]. In another study by Lee et al., a novel contrasting agent was successfully synthesized, which was highly sensitive and specific and could image H_2O_2 in living systems [85]. The authors used peroxalate NPs to image H_2O_2 by inducing a chemiluminescent reaction using three components: H_2O_2, peroxalate esters, and fluorescent dyes. The peroxalate NPs were coated with peroxalate esters (hydrophobic polymer in its matrix). These NPs image H_2O_2 via a dual step process. Firstly, H_2O_2 diffusion occurs in the NPs, which then causes a reaction with the peroxalate ester groups and generates dioxetanedione, creating high energy inside the NPs [86,87], which subsequently then chemically excites the encapsulated fluorescent dyes, via a chemically initiated electron-exchange luminescence mechanism [88,89], leading to CL from the NPs and allows imaging of H_2O_2. Additionally, Lee et al. developed a method to synthesize peroxalate micelles, with a composition of amphiphilic peroxalate-based copolymers, rubrene (fluorescent dye), and a "stealth" polyethylene glycol (PEG) molecule to evade macrophage phagocytosis, which could successfully detect H_2O_2 through CL. These peroxalate-loaded micelles detected H_2O_2 within nanomolar concentrations (>50 nM) and were highly sensitive in detecting H_2O_2 in low physiological concentrations inside living systems [90].

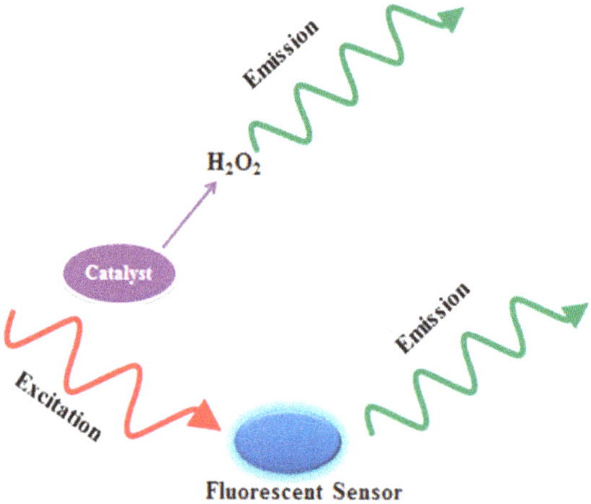

Figure 2. Mechanism of the chemiluminescent material for the detection of H_2O_2 released in cancer cells. Excitation and de-excitation of chemiluminescence materials can be seen during chemical reaction.

Another study found that using peroxyoxalate chemiluminescent (POCL) NPs, H_2O_2 could be detected in trace amounts within living systems (in vivo) using optimized CL techniques in the near-infrared (NIR) wavelength. The detection of H_2O_2 using NIR is efficient in living systems because the penetration power of these NIR rays is higher because of the reduced photon-limiting interferences (scattering and absorption) happening within biological mediums [91–93]. CL using luminol was synthesized using o-benzyl alcohol-decorated block poly(carbonate)s copolymer, viz., PMPC–ONA, giving the resultant micelles a high H_2O_2 detection ability. In these micelles, luminol, fluorophore, and hemin were wrapped, forming an L/H/S@PMPC–ONA nanoprobe. These micelles work based on the principle that in the presence of H_2O_2 in the system, H_2O_2 diffuses within NPs, reacts with the hemin, and generates high energy reactive oxygen. The high energy reactive oxygen then chemically excites the luminol, activating the CL to expose nitrosobenzaldehyde recognition sites. This process destabilizes the micelles and releases the fluorescent indicator (fluorophore), which helps in imaging H_2O_2 [94]. Lee et al. additionally synthesized a nanoprobe using multiple molecule integration, i.e., dye/peroxalate NPs, which exhibited more enhanced and controlled CL, and hence displayed widespread applications in biomedical imaging of H_2O_2. This new enhanced nanoprobe was synthesized using nanoscopic coaggregation of a dye, which exhibited the aggregation-enhanced fluorescence phenomenon with a peroxalate, which had a high response to H_2O_2, which converted the energy generated from the chemical reaction to electronic excitation [95]. Additionally, Lee et al. successfully detected and imaged H_2O_2 via CL resonance energy transfer in the NIR wavelength using quantum dots functionalized with a luminol derivative [96]. Geng et al. devised a method to detect H_2O_2 via aggregation-induced emission fluorogen using 2,3-bis(4-(phenyl(4-(1,2,2-triphenylvinyl)-phenyl)amino)phenyl)-fumaronitrile

(TPETPAFN), resulting in dye-encapsulated NPs [97]. A polyoxometalate (POM), vanadomolybdophosphoric heteropoly acid ($H_5PMo10V_2O_{40}$, PMoV2), shows similar activity like peroxidases and functions by catalyzing the luminol/H_2O_2 reaction to generate CL. This phenomenon was shown in the study by Jia et al. where the study results showed an enzyme-free luminol/H_2O_2/$PMoV_2$ CL system, which could be utilized for its high sensitivity in detecting H_2O_2. This enzyme-free luminol/H_2O_2/$PMoV_2$ CL system exhibited good linear dependence with respect to H_2O_2 concentration within a wide range of up to 5 to 5000 nM (LOD) [98].

2.6. Titrimetry

The titrimetry technique can be used to analyze an unknown amount of H_2O_2 in a known sample concentration. The titrimetric technique uses iodometry, permanganate, and cerium (IV) in an acidic medium. In the study by Klassen et al., the concentration of H_2O_2 was assessed at 300 μM using the I_3^- method after the calibration with permanganate. ε max measurement was made at 351 nm as 25,800 M^{-1} cm^{-1} using the calibration plot of the I_3^- method titrated against potassium dichromate ($KMnO_4$) [99]. In the study by Murty et al., the concentration of H_2O_2 was measured potentiometrically in an acidic medium using 8–11 M phosphoric acid [47]. Kieber and Helz synthesized a method for the detection of H_2O_2 by modifying the iodometric titration method using water matrices, where iodine was liberated as follows:

$$H_2O_2 + 2H^+ + 2I^- \rightarrow I_2 + 2H_2O$$

$$I_2 + 2H_2O + C_6H_5AsO = 2I^- + 2H^+ + C_6H_5AsO(OH)_2$$

The I_2 produced was consumed by adding an excess of phenylarsine oxide. The end result was declared by titrating the remaining amount of phenylarsine oxide with iodine [100] when the intense blue color of the starch–iodine complex disappeared. The LOD was 0.02 μM. In another study, a two-step absorbance, microtiter plate method was developed by titrating an acidified H_2O_2 solution with standard cerium (IV) sulfate. In the second step, cerium (IV) sulfate was converted into cerium (III) sulfate, and potassium iodide was converted into iodine [101]. This process is commonly used and possesses additional advantages over the other methods because of its simplicity and low running costs, but its limitation is its inaccuracy at lower concentrations. Additionally, the other limitations of the method are that it consumes more time and requires skilled personnel to perform the calibration of the instrument.

2.7. Spectroscopy

One of the most common, convenient, and extensively used methods for determining and measuring H_2O_2 is spectroscopy. This method is based on the principle that colored compounds are formed with respect to absorbance measurements comparative method of methyl blue and toluidine. A method comparing the reaction of methyl blue and toluidine blue with iodine solution was introduced for determination of H_2O_2 based on the following reaction:

$$H_2O_2 + 2KI + 2HCl \rightarrow I_2(aq) + 2KCl + 2H_2O.$$

In the comparison, methyl blue when reacted with iodine gave a single-peak visible spectrum with a higher extinction coefficient (=49,100 M^{-1} cm^{-1}) [102]. In another study by Matsubara et al., a method using a mixture of titanium IV and 2,4((5-bromopyridyl)azo)5-(*N*-propyl-*N*-sulfopropyl amino) phenol disodium for determining H_2O_2 [103] was demonstrated. Molar absorptivity was found to be 5.7104 M^{-1} cm^{-1} at 539 nm. In the study by Clapp et al., the measurement of H_2O_2 was done using an aqueous solution with titanium (IV) sulfate. This method yielded a yellow peroxotitanium species at a wavelength of 407 nm [104]. An in vitro method for the detection of H_2O_2 was developed using the

1,10-phenanthroline method. The advantages of this method are its short processing time, increased sensitivity, and high reproducibility [105]. In another study, catalytic decomposition of H_2O_2 was demonstrated by monomeric molybdenum (VI) by mixing hydroquinone, ammonium molybdate, and anilinium sulfate with varying H_2O_2 concentrations and determining the absorbance at 550 nm [106]. Zhang and Wong demonstrated a method for the estimation of the concentration of H_2O_2 in marine water at acidic pH of 4 in the presence of HRP at 592 nm using leuco crystal violet oxidation. The LOD for H_2O_2 was found to be 20 nM with ±1% accuracy [107]. In the study by Huang et al., a fast, reproducible, and reliable method for the detection and measurement of H_2O_2 was demonstrated. This method used 4AAP-DEA-βCD-hemin, and the LOD was 8.4 × 10^{-5}, with a molar absorption coefficient of 1.65 × 10_4 mol/L/cm [108]. Zhang et al. showed the determination of H_2O_2 in pulp bleaching effluents. The study shows that H_2O_2, in the presence of sulfuric acid solution, chemically reacted with vanadium pentoxide and formed a peroxovanadate complex that is reddish-brown [109].

2.8. Colorimetry

The method of determining H_2O_2 using iodide and starch was first developed in 1943 by Eisenberg. The H_2O_2 samples were treated with a titanium sulfate reagent, and the changes in color were quantified with the presence of H_2O_2. The chemical reaction of H_2O_2 with the titanium sulfate reagent is shown as follows:

$$Ti^{+4} + H_2O_2 + 2H_2O = H_2TiO_4 + 4H^+.$$

The formation of a yellow compound called pertitanic acid determined the H_2O_2 concentration within a range of 0.2–3.0 mg/100 mL [110]. Another study showed a more sensitive method using colorimetry. The study showed that the oxidation of iodide takes place in the presence of $(NH_4)_2MoO_4$ (ammonium molybdate), which helps in determining the concentration of H_2O_2 even in micromolar quantities. The study determined the molar absorptivity of the starch–iodine complex (intense blue) at a value of 39.45 $mmol^{-1}$ cm per liter at a wavelength of 570 nm [111]. The colorimetric method based on enzymes using plant extracts was developed by Fernando et al., where a sharp pink quinoneimine dye was formed. The pink dye formation took place when H_2O_2 reacted with phenol, 4-aminoantipyrine, and HRP in 0.4 M phosphate buffer with pH of 7.0 [112]. The assay results were considered optimum when the assay conditions were maintained at pH 7.0, temperature of 37 °C, 0.7 mM H_2O_2 concentration, and 1 U/mL enzyme concentration within 30 min. The optimum assay resulted in a limit of quantitation and LOD of 411 and 136 mM, respectively. Another simple method to detect the H_2O_2 released by cells within a tissue culture was based on the principle that phenol red oxidizes in the presence of H_2O_2. The study results showed a direct linear relationship between the concentration of H_2O_2 and absorbance, which had a range of 1 to 60 nmol/mL. The absorbance was measured at 520 nm [113]. Another fast and reliable method for determining H_2O_2 was developed using a colorimetry technique. In the method, 4-nitrophenylboronic acid was utilized for determining the concentration of H_2O_2 in an aqueous medium, where nitrophenylboronic acid reacted with H_2O_2 and produced 4-nitrophenol. The LOD was found to be ~1.0 μM [114]. Nitinaivinij et al. used the principle of colorimetry and demonstrated the determination of H_2O_2 in a very low concentration. The method utilized the technique of chromaticity analysis of silver nanoprisms (AgNPrs). The AgNPrs decomposed in the presence of H_2O_2, producing yellow color, and showed the H_2O_2 concentration at 1.57 mM with high accuracy and sensitivity [115]. The advantage of this method is that the determination of H_2O_2 can be carried out using a simple apparatus, but this method could give false-positive readings, and the results were not applicable to determine H_2O_2 within turbid samples.

2.9. Chromatography

Chromatographic techniques are commonly used for separation. High-performance liquid chromatography (HPLC) is an analytical technique used for the detection and separation of different moieties. In the study by Takahashi, separation of H_2O_2 was achieved using an electrochemical detector and a cation-exchange resin gel column of sulfonated styrene-divinylbenzene copolymer. This method was found to have a linearity of 0.9984. The LOD was measured at 0.2 pmol [116]. In another method by Wada et al., H_2O_2 separation was achieved using an octadecylsilyl column, and the LOD was measured at 1.1 μM [117]. In another study, H_2O_2 was determined using gas chromatography in the presence of oxidized butyric acid, and its absorbance was found at a wavelength of 517 nm [118]. Another method in a study used a ligand exchange-type column for the separation of H_2O_2. The column was packed using a sulfonated polystyrene/divinylbenzene cation-exchange [119]. Steinberg et al. used the principle of reverse-phase chromatographic techniques in HPLC to determine H_2O_2. Iodovanillic acid was formed and was detected using UV absorption at 280 nm with a LOD of ~0.1 μM [120]. The advantages of chromatographic techniques in H_2O_2 determination are that these methods are relatively simple, have low operational costs, and use a wide range of stationary phases and columns. The limitations of this technique are its costly overall equipment, its long operational time, interferences, and the necessity for a specialized operator to run the machine.

2.10. Fluorescence

Another common method to detect H_2O_2 that has wide applications is based on the principle of fluorescent signal detection. In fluorescence sensors, the excitation of electrons is achieved from an external photon source, in contrast to CL, where light is generated via a chemical reaction [121]. Many fluorescent probes have been constructed using different materials. The probes include naphthofluorescein disulfonate [122], homovanillic acid [123], peroxyfluor-1 [124], peroxyresorufin-1 [124], single-walled carbon nanotubes [125], peroxyxanthone-1 [124], and phosphine-based fluorescent reagents [126]. In one study, the fluorescent biosensors helped in the detection of intracellular H_2O_2 in mice peritoneal macrophages [122]. In the study by Miller et al., three fluorescent probes that were detectable via confocal and two-photon spectroscopic methods from the peroxysensor family were successfully developed. Each fluorescent probe emitted at a different wavelength from the other, which allowed these probes to be used in various applications with respect to specific emitting wavelengths [124]. Recently, intracellular H_2O_2 concentration can be measured using HyPer, a genetically encoded H_2O_2 biosensor (Figure 3) [127]. HyPer is a chimeric protein [128] composed from the permuted yellow fluorescent protein (cpYFP) and H_2O_2-sensitive domain of the bacterial transcription factor OxyR, which is responsible for sensing H_2O_2 [129]. In the study by Belousov et al., an H_2O_2 sensor named HyPer was developed and studied. The HyPer sensor was successful in detecting an increase in H_2O_2 levels in HeLa cells during Apo2L/TRAIL protein-induced apoptosis (programmed cell death). This sensor also detected increased levels of H_2O_2 in cells taken from rat adrenal medulla (PC-12) that had been previously exposed to nerve growth factor [128]. The HyPer family includes five probes: HyPer [128], HyPer2 [130], HyPer3 [131], HyPer7 [132], and HyPerRed [133]. GEFIs of this family consist of a circularly permuted fluorescent protein (cpYFP for the numbered HyPers or cpmApple for HyPerRed) integrated via short peptide linkers into the bacterial transcription factor OxyR lacking a DNA-binding domain. Upon oxidation by H_2O_2, OxyR forms an intramolecular disulfide bond [134] that elicits conformational rearrangements. These rearrangements are then transmitted into the chromophore center of a fluorescent moiety of a GEFI, causing fluorescence alterations that can be subsequently detected. HyPer and its improved derivates, HyPer2 and HyPer3, contain cpYFP. cpYFP has two excitation peaks at 420 and 500 nm and a single emission peak at 516 nm. When the OxyR domain is oxidized by H_2O_2, the intensity of fluorescence excited at approximately 420 nm (F_{420}) decreases, whereas the intensity of fluorescence excited at approximately 500 nm (F_{500}) increases proportionally. A sensor readout is generated as a F_{500}/F_{420} ratio [128].

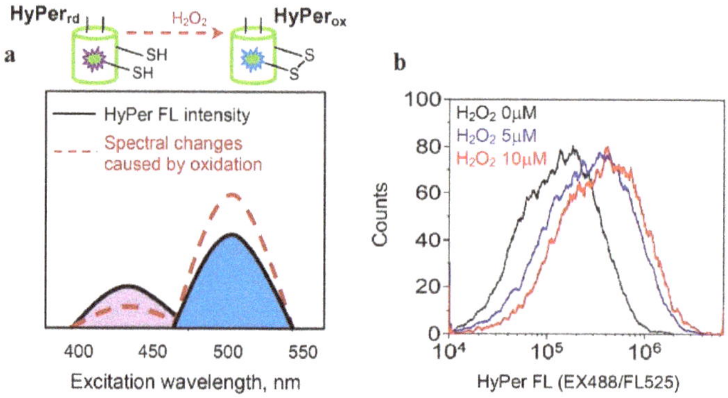

Figure 3. Analysis of HyPer fluorescence in K562 cells exposed to extracellular H_2O_2. (**a**) Scheme demonstrating the changes in the excitation spectrum of HyPer upon oxidation. (**b**) Flow cytometry histograms of K562 cells measured after two-minute exposure to different concentrations of H_2O_2. Reproduced with permission from [127]. Copyright 2019, Science Direct (Amsterdam, The Netherlands).

Another study by Xu et al. showed a specific probe called Mito-H_2O_2, which is used to detect mitochondrial-associated H_2O_2 levels in HeLa cells. The study further showed that Mito-H_2O_2 was an effective, sensitive, and quick mitochondrial-targeted sensor [135]. Xiao et al. also developed another fluorescent probe called ER-H_2O_2 specifically for targeting the endoplasmic reticulum, which was equally effective, sensitive, and quick in the detection of H_2O_2. Xiao et al. induced apoptosis in both the organelles using L-buthionine sulfoximine, and both of these probes were tested for H_2O_2 specificity and selectivity [136]. Shen et al. developed a microfluidic method, which had droplets in combination with gold nanoclusters. This method was demonstrated to have high sensitivity for the detection of H_2O_2 secreted by a single cell. When a single cell was isolated using a microdroplet (with a volume of 4.2 nL), it can secrete H_2O_2, which causes florescent changes in HRP-gold nanoclusters with high specificity and high sensitivity of 200–400 attomole. The high throughput performance (~103 single-cell encapsulated microdroplets per minute) of the resultant microfluidic device makes it a powerful tool to investigate cell-to-cell heterogeneity in releasing H_2O_2 at the large scale, promising revelation of new knowledge to understand the biological role of H_2O_2 in tumor cells [137]. Moreover, Wang et al. fabricated a Ce6@Lum-AuNPs nanoprobe using green syntheses methods. They successfully loaded luminol-gold NPs with the fluorescent receptor Chlorin e6 (Ce6). The resultant fluorescent Ce6@Lum-AuNPs proved successful towards fluorescent bioimaging of cancerous cells [138].

3. Recent Advances

3.1. Current Approaches in the Construction of Biosensors

Over the past 200 years, the use of enzymes was common because of specific substrate sensitivity. However, enzymes are highly unstable and sensitive and are prone to denaturation caused by environmental changes such as pH and temperature. Therefore, recent studies have focused on using an artificial pseudo-catalyst instead of enzymes to overcome the drawbacks [139–141]. Denaturation of enzymes is common in enzyme groups such as peroxidases, catalases, monoamine oxidase, choline oxidase, uricase, and ascorbate oxidase. Peroxidases, also known as heme proteins, constitute the prosthetic group, i.e., ferriprotoporphyrin, and are usually found to have a molecular weight of 30 to 150 kDa [142,143]. Peroxidases are oxidoreductases and are produced by many animals, plants, and microorganisms. Peroxidases reduce H_2O_2 and help in the oxidation of aromatic amines, phenols, and organic and inorganic substrates [144] and are extensively utilized in biochemistry, enzyme immunoassays, wastewater treatment plants containing phenol compounds, synthesis of aromatic compounds, and removing H_2O_2 from food materials [143]. The application of peroxidase enzymes is extensive, and they are commonly used in analytical techniques for the detection of glucose [145], cholesterol [146], uric acid [147], H_2O_2 [148], alcohols [149], and phenols [144]. Peroxidase enzymes are also used in the pharmaceutical industry for the construction of biosensors for the detection of different drugs in the body. As previously mentioned, enzymes are prone to degradation; hence, the latest research involves the replacement of enzymes with pseudo-catalysts, i.e., inorganic/organic. These materials are chosen because of their low cost, stability, and convenience [150–152].

3.2. Electrochemical Sensing of H_2O_2 via Metal Nanoparticles

Nanotechnology has advanced, and there are many types of NPs available nowadays. NPs can be classified based on the nanomaterial used to synthesize them. Commonly available nanomaterials include metal NPs [153,154], carbon nanomaterials [155,156], and metallic oxide nanostructures [157]. Nowadays, NPs are used in manufacturing H_2O_2 electrochemical sensors, exhibiting distinctive electrical and catalytic properties toward the reduction or oxidization of H_2O_2 and having a broad range of stability based on the nanomaterial used in them. However, until now, most of the studies reporting cost-effective H_2O_2 detecting electrochemical sensors have a detection limit of the sub-micromolar level [38,158]. To detect H_2O_2 in cellular matrices, the electrobiochemical sensors should be sensitive enough to sense H_2O_2 concentration in nanomoles. Currently, those biosensors that are highly sensitive and have optimum H_2O_2 detection limits have been developed using HRP and metal nanoparticles [159,160], which decreases the long-term operational stability and increases the operational costs. In the study by Wang et al., real-time electrochemical detection of H_2O_2 via small MoS_2 NPs in Raw 264.7 cancerous cells was performed. The resulting device had a detection limit lower than 2.5 nM and a wide linear range of up to five orders of magnitude [161]. In vivo monitoring of H_2O_2 secreted from living cells is essential in understanding cellular signaling pathways. The release of H_2O_2 from living cells is very low because the selective detection of H_2O_2 at a low level is challenging. To overcome this difficulty of detecting endogenous H_2O_2 from live cells, Dou et al. synthesized three hybrid metal nanoflower sensors for the detection and monitoring of H_2O_2 concealed from living MCF-7 cancerous cells. The three-hybrid metal Au–Pd–Pt nanoflower-decorated MoS_2 nanosheet-modified sensors were developed using simple wet chemistry. The three-hybrid metal nanoflower sensors (Au–Pd–Pt/MoS_2) show a synergetic increase in the electrocatalytic reduction of H_2O_2 with an ultrasmall detection limit as low as the sub-nanomolar level. Immobilization of aminin glycoproteins on the nanocomposite surface will result in an increase of its biocompatibility, which, in turn, enhances composite adherence to cells. This property of nanocomposites can be effectively used in future applications directed toward monitoring the secretion of H_2O_2 from living cells and cellular apparatus and may be utilized in developing highly efficient and sensitive cancer diagnostics sensors [162]. Sun et al. synthesized a dumbbell-shaped Pt_xPd_{100-x}–Fe_3O_4 NP composite, which could effectively determine the secretion of H_2O_2 from Raw 264.7 cells with a detection limit of 5 nM [160]. Chang et al. developed a sensitive fluorescent assay to determine H_2O_2 with a wide linear range of 1 to 100 μM and detection limit of 0.8 μM. A fluorescent biosensor based on the inner filter effect (IFE) was manufactured using poly (vinyl pyrrolidone)-protected gold nanoparticles (PVP–AuNPs) and fluorescent BSA-protected gold nanoclusters (BSA–AuNCs). The BSA–AuNCs acted as an IFE fluorophore pair. The high extinction coefficient of PVP–AuNPs served as a dominant absorber and influenced the emission of the fluorophore in the BSA–AuNCs assay. The surface Plasmon resonance (SPR) of PVP–AuNPs was significantly enhanced with an increase in H_2O_2 concentration. The increased H_2O_2 then caused the significant induction of the fluorescent quenching effect of BSA–AuNCs [163]. Cui et al. showed a fast, simple, and reagent-free method for H_2O_2 detection. The study used luminol-reduced Au NPs for the determination of H_2O_2. The resulting biosensor had the electrochemiluminescence application in effectively determining the concentration of H_2O_2 within limits of 3×10^{-7}–1.0×10^{-3} mol L^{-1} with a low detection limit of 1.0×10^{-7} mol L^{-1} ($S/N = 3$) [164].

Liu et al. synthesized porphyrin functionalized ceria (Por-Ceria) uniform nanoparticles as a calorimetric probe for H_2O_2 detection [165]. A nickel phosphide nanosheet array on a titanium mesh (Ni_2P NA/TM) possesses superior analytical performance with a rapid retort time of <5 s. Manufactured biosensors showed high selectivity and stability, with a wide linear range of 0.001–20 mM, ultrasmall LOD of 0.2 μM ($S/N = 3$), and high sensitivity of 690.7 μA mM^{-1} cm^{-2} [166]. Small (10–30 nm) platinum nanoparticles (Pt-NPs) were prepared via protein-directed one-pot reduction. The resultant BSA/Pt-NPs composite shows colorimetric determination of H_2O_2 with a linear range from 50 μM to 3.0 mM, LOD of 7.9 μM, and visually detected lowest concentration of 200 μM [167].

Ultrathin silver nanosheets that can detect H_2O_2 with a LOD of 0.17 μM, linear range of 5–6000 μM, and fast response time <2 s were synthesized by Ma et al. The synthesized biosensors showed real-time determination of H_2O_2 released from living HeLa and SH-SY5Y cells, with high sensitivity of 320.3 μA mM^{-1} cm^{-2} [168].

The synergistic combination of p-type semiconductive channels of layered double hydroxides (LDHs) exhibited multifunctional properties, a distinctive morphology, and abundant surface active sites. The Fe_3O_4@CuAl NSs modified electrode exhibited excellent electrocatalytic activity toward H_2O_2 reduction. The projected biosensor revealed prominent electrochemical sensing of H_2O_2 with an extensive linear range of eight orders of magnitude and a low detection limit of 1 nM ($S/N = 3$) [169]. Copper(I) phosphide nanowires on 3D porous copper foam (Cu_3P NWs/CF) were fabricated via electrochemical anodized $Cu(OH)_2$ NWs to manufacture noble metal-free electrocatalysts. The Cu_3P NWs/CF-based sensor exhibited first-rate electrocatalytic reduction of H_2O_2 with a detection limit as low as that achieved by noble metal-free electrocatalysts, i.e., 2 nM. The developed sensor assured sensitive and consistent determination of H_2O_2 excretion from living tumorigenic cells [170]. Xiong et al. developed a nickel phosphide nanosheet array on a titanium mesh (Ni_2P NA/TM) using an economical and effective metal toward electrocatalytic H_2O_2 reduction. Ni_2P NA/TM, being a nonenzymatic H_2O_2 sensor, presented superior analytical performance, with a swift response time <5 s and wide linear range of 0.001–20 mM. The resultant electrode exhibited high sensitivity of 690.7 μA mM^{-1} cm^{-2} and ultrasmall detection limit of 0.2 μM ($S/N = 3$) [166]. A Prussian blue nanocube-decorated molybdenum disulfide (MoS_2-PBNCs) nanocomposite was designed for the electrochemical sensing of H_2O_2. Interestingly, a sensor for label-free sensing of carcinoembryonic antigen (CEA) can be constructed by using MoS_2-PBNCs nanocomposites. The electrochemical response of the MoS_2-based immunosensor was linear, with a CEA concentration range from 0.005 to 10 ng mL^{-1} and minimum recognition limit of 0.54 pg mL^{-1} [171].

3.3. H_2O_2 Detection Using Enzymatic Biosensors

Various analytical techniques, i.e., chemiluminescence [95], fluorescence [172], and electrochemistry [80,160], have been employed for the analysis of H_2O_2 at the cellular level. Among them, electrochemical sensors are an area of high interest and provide fast, economically effective, and real-time determination via a simple mechanism with ultrahigh sensitivity and selectivity. Electrochemical detection is considered a powerful tool for the determination of other electroactive metabolites such as glucose [173], dopamine [174], and O_2 [175] secreted from live cells. The high selectivity and sensitivity of enzymes made them valuable for the electrochemical biosensing of H_2O_2. Horseradish peroxidase (HRP) enzymes draw considerable attention for the construction of electrochemical biosensors because of their efficient catalysis of H_2O_2 [176,177]. Wang et al. reported a highly sensitive sequence-selective DNA sensor composed of an HRP-labeled probe. The proposed biosensor successfully detected the K-ras gene, which is associated with colorectal cancer. Thiol (–SH) modified capture probe adsorbed chemically on the gold electrode via self-assembly and exhibiting a detection limit of 5.85×10^{-12} mol L^{-1}, hybridization of nucleic acid (target DNA:K-ras gene), and a HRP labeled oligonucleotide detection probe can be achieved using the sandwich way method. Wang et al. developed an extremely sensitive sequence-selective DNA sensor on an HRP-labeled probe to detect the specific K-ras gene which is associated with colorectal cancer. At first, the capture probe modified with –SH was chemically adsorbed on the gold electrode by self assembly. Then, a complementary nucleic acid (target DNA:K-ras gene) was hybridized with an HRP labeled oligonucleotide detection probe in a sandwich way with a detection limit of 5.85×10^{-12} mol L^{-1} [178]. Bruno et al. developed horseradish peroxidase conjugated gold nano biosensors for detection of H_2O_2 released by prostate cancerous cells. The proposed biosensor can detect hydrogen peroxide (H_2O_2) in a wide linear range from 2 to 100 μM with a low detection limit of 0.01 μM [179]. A Cyt c loaded nanostructured TiO_2 film was successfully prepared by Luo, which exhibits natural enzymatic activity toward H_2O_2, redox formal potential ($E^{0'}$) of 108.0 ± 1.9 mV versus Ag | AgCl, and an heterogeneous electron transfer rate constant (ks) of 13.8 ± 2.1 s^{-1} [157]. To stabilize the enzyme model, Zhou et al. used an enzyme cytochrome c (Cyt c), to facilitate the transfer of electrons between the redox enzyme and electrode. Cyt c was immobilized stably into the molecular hydrogel to maintain its innate bioactivity toward H_2O_2. The use of Cyt c is a consistent methodology to regulate H_2O_2 at an optimized potential with high selectivity over other ROS, oxygen, metal ions, and ascorbic acid. The in vivo sensing of H_2O_2 from living cells, a small molecular hydrogel provides long-lasting stability and good reproducibility [180].

3.4. *Carbon-Based Material for H_2O_2 Sensing*

The success of graphene has boosted great research in the synthesis and characterization of graphene-like 2D materials, single and few-atom-thick layers of van der Waals materials, which show fascinating and useful properties. The single atom layer of C is the most transparent, strongest, and thinnest material and exhibits electrical conductance much better than Cu, with the ability to endure a current density that is six orders of magnitude [36,181,182]. The structure of some of the carbon-based material is shown in Figure 4. Recently, graphene has attracted great interest in the development of biosensors, i.e., optical and electrochemical, with improved performance owing to its integration with different nanomaterials (metals and metal oxides) and quantum dots [183–185]. Researchers have developed great interest in the emerging class of carbon-based 2D materials (graphene) because of their distinctive properties with applications in sensing and biosensing, electronics, catalysis, composites, and coatings. The excellent optical and electrical properties of carbon-based 2D materials made their use emergent in sensing and biosensing and showed real-time application in the field of biochemistry and nanomedicines [186,187]. Graphene-like 2D layered nanomaterials boron nitride (BN), transition metal dichalcogenides, graphite–carbon nitride (gC_3N_4), graphenes, and transition metal oxides have been investigated broadly [188,189]. Boron nitride nanosheets contain alternate nitrogen and boron atoms in a honeycomb lattice structure with extensive band gap, and BN is an insulator [190]. Instead of various transduction techniques, electrochemical methods are well known for analytical biomarker detection via graphene 2D-based sensors [191].

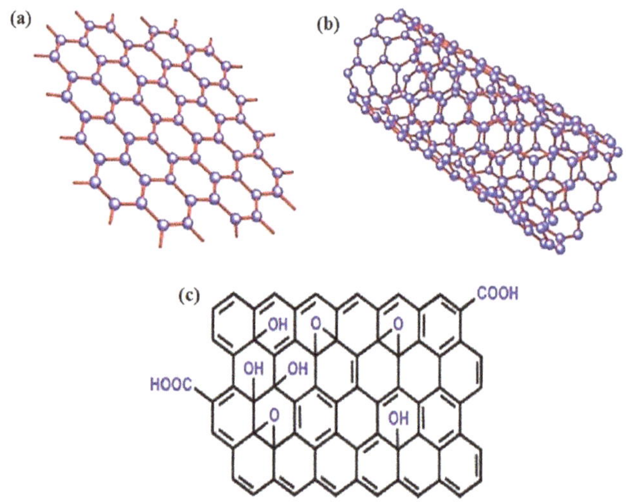

Figure 4. Different types of carbon-based materials, i.e., (**a**) Graphene, (**b**) Carbon nanotube, and (**c**) reduced grapheme oxide, used in electrochemical sensing of H_2O_2.

3.4.1. Graphene-Based Metal-Free Electrocatalysts

The application of carbon materials in analytical and industrial electrochemistry is well known owing to their low cost and electrocatalytic potential in a number of redox reactions [192]. Recently, groups of researchers showed that surface functionalization of graphene materials results in diverse behavior, which made them benevolent in sensing in contrast to intrinsic graphene. Zhou et al. showed the chemical reduction of graphene oxide into chemically reduced graphene oxide (CR-GO) via hydrazine, and the resultant GCE constructed from the obtained CR-GO showed excellent sensing capability for H_2O_2 detection. The synthesized electrochemical biosensor exhibited a lower detection limit of 0.05 μM and wide linear range from 0.05 to 1500 μM, which precedes the use of functionalized carbon materials in electrochemical sensing [155].

In another work, Takahashi et al. reported rGO modified GCE via electrodeposition. The electrochemical studies showed an enhancement in the sensing performance of the rGO modified electrode that was considerably better than the original electrode for hydrogen peroxide detection. Some studies showed a high electron density on the defective sites (edges) of modified graphene oxide, which made it a potential candidate for the electrocatalytic reduction of H_2O_2 [193]. The synthesis of novel quality graphene is important in exploiting graphene application for electrochemical sensing. Chemical and physical (thermal method) reduction of GO (hydrophilic GO to hydrophobic graphene) is the most effective method to manufacture graphene on a large scale. During chemical and physical reduction, exfoliated graphene becomes disorderly aggregated, which results in the decrease in their disperse behavior in water and limits their practical applications [194]. Later on, some researchers fixed this problem using various dispersants, i.e., sodium dodecyl sulfate, cetyltrimethyl ammonium bromide (CTAB), and DNA. These dispersants enhanced the disperse behavior and stability of graphene in an aqueous environment. Lv et al. simply introduced DNA molecules on the graphene surface using the self-assembly method and formed graphene–DNA hybrids (GN/DNA). DNA–graphene was found to show a physical interaction, i.e., π–π stacking via aromatic rings of graphene and *N*-containing functional moiety in DNA, which results in a strong interaction between graphene and DNA. Stacking DNA on the graphene surface not only enhanced graphene dispersion in aqueous media but also imparted an electron-rich character to graphene by forming a GN/DNA composite. Comparative studies showed that the GN/DNA modified electrode exhibited higher sensitivity, wide detection range, and swift response time in contrast to the GN-modified electrode for the electrochemical sensing of H_2O_2 [195]. Woo et al. fabricated a multiwalled carbon nanotube–graphene composite (MWCNT–graphene) via a direct in situ chemical reduction of graphene oxide and pre-treated MWCNT mixture. The prepared component showed a uniform network of ultrathin graphene sheets stuck between nanotube bundles. Structural analysis showed that the morphology of graphene present between nanotube bundles was comparatively higher than pure graphene, which showed wrinkled and aggregated morphology. The electrochemical sensor constructed from the resultant MWCNT–graphene exhibited a wide detection range from 20 μM to 2.1 mM and low detection limit of 9.4 μM. Synergic increase in the electrochemical performance of the MWCNT–graphene composite is attributed to high electrical conductivity of MWCNTs [196]. Recently, metal-free electrocatalysts, heteroatom-doped graphene, play a crucial role in H_2O_2 detection. The electronic properties of graphene can be altered drastically by doping graphene with N, S, and B, which play a crucial role in operating the electronic properties. Wang and coworkers used the nitrogen plasma treatment strategy to produce N-doped graphene from reduced graphene oxide as a starting material. Spectral studies of N-graphene showed that the nitrogen atom was substituted into graphene sheets with three different nitrogens, including graphitic N, pyridinic N, and pyrrolic N. The concentration of nitrogen in graphene sheets was optimized by monitoring the plasma exposure time, and the resultant N-doped graphene showed improved electrocatalytic performance as compared to pristine graphene in electrochemical sensing [197].

Wu et al. reported the synthesis of N-doped graphene using hydrazine as a nitrogen source, with a 4.5% N/C atomic ratio, and reducing agent. Structural studies of N-doped graphene were made via XPS measurements. Structural analyses showed 28% pyridinic N, 49% pyrrolic N, 19% graphitic N, and 4% oxidized N [198]. Increased sensitivity, a wide linear range, and a low detection limit were achieved using N-doped graphene as compared to pristine graphene. In addition to N, Yeh et al. successfully synthesized boron-doped graphene nanosheets (BGNs) using B_2O_3 and graphene nanosheets through an atmospheric-pressure carbothermal reaction. Boron doping on the graphene surface created defects in nearby sites and uneven charge separation, which, in turn, facilitated the charge transfer to neighbor atoms. The resultant BGN-doped graphene showed a wide linearity range from 1.0 to 20.0 mM, detection limit of 3.8 μM, and much higher sensitivity (266.7 $\mu A\ mM^{-1}\ cm^{-2}$) compared with undoped GNs [199]. Recently, the electrochemical performance of the detection of H_2O_2 was further improved using co-doped graphene with two elements. Yang et al. synthesized N and B co-doped graphene (NB-G) using a microwave-activated chemical–thermal treatment strategy. In this strategy, they first developed N-graphene using GO and cyanamide as a precursor, followed by microwave treatment. The boron atom was doped on N-modified graphene via the pyrolysis of the N-G and B_2O_3 mixture at 900 °C for 0.5 h in an Ar atmosphere to obtain BN-G [200]. Electrochemical studies of NB-G were made using ferric/ferrous coupling of $K_3[Fe(CN)_6]/K_4[Fe(CN)_6]$. The prepared electrode exhibited outstanding electrocatalytic reduction of H_2O_2 and a rapid response time, with a linear range from 0.5 μM to 5 mM and detection limit as low as 0.05 μM. The excellent electrochemical performance of the NB-G electrode is attributed to the novel structural network, with high charge transfer and large surface area, and the synergistic effect between the two heteroatoms of B and N [200]. Table 1 shows electrochemical performance of non-enzymatic metal free H_2O_2 sensors based on graphene.

Table 1. Non-enzymatic metal free H_2O_2 electrochemical sensors based on graphene.

Carbon Material	Sensitivity $\mu A\ mM^{-1}\ cm^{-2}$	Liner Range (μM)	Detection Limit (μM)	Ref.
CR-GO	-	0.05–1500	0.05	[155]
Graphene-MWCNT	32.91	20–2100	9.4	[196]
rGO/nPPy	47.69	0.1–4	0.034	[201]
IL-GR-s-PANI	280.0	0.5–2000	0.06	[202]
rGO/Tyrosine	69.07	100–2100	80	[203]
Poly(o-Phenylenediamine)/GO	16.2	2.5–25	0.84	[204]
BGNs	266.7	1000–20,000	3.8	[199]
NB-G	-	0.5–5000	0.05	[200]
GSnano/CS	18.78	5.22–10,430	2.6	[205]
GN-HN-SWCNT	0.015	0.2–400	0.05	[206]
H-GNs/PEDOT	235	0.1–10	0.08	[207]
NS-GQD/G	-	0.4–33	0.026	[208]
Functionalized 3D Graphene	169.7	0.4–660	0.08	[209]
rGO/GO hybrid MEA	-	0.18–9.6	-	[210]
3D-G/GCE	-	0.2–41,200	0.17	[211]

3.4.2. Carbon Composite with Enzymes for H_2O_2 Detection

Noble metals, nonnoble metal oxide, and sulfide-modified graphene composites are used to immobilize HRP for the construction of enzymatic H_2O_2 biosensors [212,213]. Song at al. [214] reported MoS_2–graphene (MoS_2-Gr)-based biocompatible biosensors for the ultrasensitive detection of H_2O_2. MoS_2-Gr nanosheets were prepared using GO and $NaMoO_4$ as precursors using the solvothermal method, and a change in solution color from reddish-brown (GO) to black confirmed the dispersion of dark flower-like MoS_2 nanoparticles on the Gr surface. Structural analyses were made using XRD results, which confirmed the formation of MoS_2-Gr composites. Electrostatic interaction arose between negatively charged MoS_2-Gr nanosheets and positively charged HRP and resulted in the formation of the HRP-MoS_2-Gr composite. The appearance of the peak in the UV–Vis spectra at 402 nm confirmed the immobilization of HRP on MoS_2-Gr, whereas no peak was noticed in case of MoS_2-Gr nanosheets. The HRP-MoS_2-Gr fabricated biosensor showed excellent stability and enhanced electrocatalytic performance for H_2O_2 detection. The resultant biosensors exhibited a low detection limit of 0.049 μM and broad linear range from 0.2 μM to 1.103 mM.

Later, Yu et al. immobilized horseradish peroxidase (HRP) on Au-decorated graphene oxide. The fabricated biosensors showed a fast response with remarkable performance, such as low detection limit (7.5×10^{-9} M) and real-time measurement of cellular H_2O_2 in living cells [27]. Liu et al. used horseradish peroxidase (HRP) immobilized on 3D porous graphene (PGN) to develop a real-time biosensor for the detection of H_2O_2 from living cells. Nanoporous graphene plays a significant role in the excess absorption of HRP, accelerates the diffusion rate, and shows excellent electrochemical performance toward H_2O_2 with a LOD of 0.0267 nM and wide linear range of seven orders of magnitude [215]. Enzymatic biosensors suffer from two major problems, namely, enzymatic loss and inactivation, which greatly affect biosensor performance. Fan and his coworker overcame this problem by encapsulating horseradish peroxidase on biomimetic graphene capsules (GRCAPS) using $CaCO_3$ as a porous sacrificial template to mimic the existence form of bioenzymes in organisms as shown in Figure 5. As a result, the synthesized biosensor showed a low detection limit of 3.3 mmol L^{-1} and wide linear range of 0.01–12 mmol L^{-1} [216]. Wu et al. used another strategy to construct horseradish peroxidase–attapulgite nanohybrids on glassy carbon to fabricated biosensors. The prepared biosensor showed a rapid response, high sensitivity, and a low detection limit with a wide linear range for the detection of hydrogen peroxide released from RAW 264.7 macrophage cells [217]. Table 2 shows electrochemical performance of enzymatic H_2O_2 biosensor loaded on graphene.

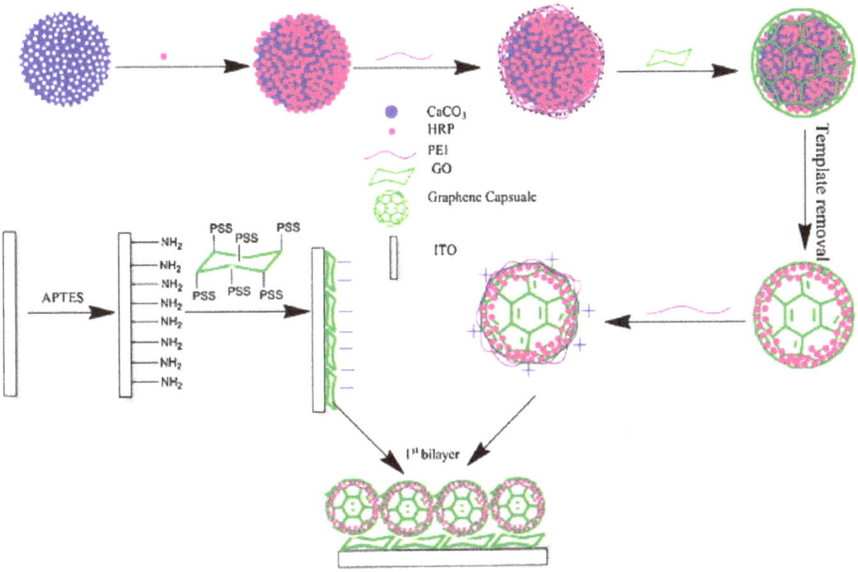

Figure 5. Mechanism of the synthesis of the graphene enzyme composite for the electrochemical sensing of H_2O_2. Reproduced with permission from [216]. Copyright 2015, Science Direct.

Table 2. Graphene-based enzymatic biosensors for H_2O_2 detection in Cancerous cells.

Graphene-Based Materials	Sensitivity $\mu A\ mM^{-1}\ cm^{-2}$	Linear Range (μM)	Detection Limit (μM)	Ref.
GE/Fe_3O_4/Hb GCE	0.3837	100–1700	6.00	[218]
rGO-CMC/Hb	-	0.083–13.94	0.08	[219]
Hb/AuNPs/ZnO/Gr	-	6.0–1130	0.8	[220]
HRP/graphene	-	0.33–14.0	0.11	[221]
Hb/Au/GR-CS	3.47×10^5	2.0–935	0.35	[222]
Hb/Au NPs-Gr	-	0.1–70	0.03	[223]
HRP/P-L-His-rGO	2.6×10^5	0.2–5000	0.05	[224]
GS-PSS/GRCAPS	-	10–12,000	3.3	[216]
HRP/AuNP/ThGP	0.086	0.5–1800	0.01	[225]
PANI/HRP/GE-CNT/AuPt NPs	370	0.5–100	0.17	[226]
Au/graphene/HRP/CS	-	5.0–5130	1.7	[227]
MP11/DMPG-AuNPs/PDDA-G	243.7	20–280	2.6	[228]
(HRP-Pd)/f-graphene	92.82	25–3500	0.05	[229]
HRP-f-graphene-Ag	143.5	25–19,350	5.0	[230]
HRP/CeO_2-rGO	4.65	0.1–500	0.021	[212]
HRP-MoS_2-Gr	679.7	0.2–1103	0.049	[214]
Catalase/AuNPs/graphene-NH_2	13.4	0.3–600	0.05	[231]
Cyt c/GO-MWCNT/Au NP	0.533	1×10^{-5}–1.4×10^{-4}	27.7×10^{-6}	[232]
RGO-MWCNT-Pt/Mb	1.990	1×10^{-5}–1.9×10^{-4}	16×10^{-6}	[233]
PPY-He-RGO	-	0.1–10	0.13	[41]
HRP/PGN/GCE	-	8.0×10^{-11}–6.64×10^{-7}	2.6×10^{-5}	[215]
PGR/catalase/GCE	-	1.0×10^{-7}–7.7×10^{-6}	1.5×10^{-3}	[234]
FeS_x/graphene	-	-	5×10^{-4}	[235]
F-MoS_2-FePt NCs	-	8–300	2.24	[236]

3.4.3. Graphene Composite with Metal Nanoparticles for H_2O_2 Detection

Dai et al. prepared heterogeneous Co_3O_4 dodecahedrons that contain carbon, and encapsulated Au nanoparticles (Au@C-Co_3O_4) were proposed via the pyrolysis of Au nanoparticle-encapsulated zeolitic imidazolate framework-67 (Au@ZIF-67). A remarkable increase in electrocatalytic performance with ultrahigh sensitivity of 7553 μA mM^{-1} cm^{-2} and with a detection limit of 19 nM was observed using the electrode fabricated from the porous Au@C-

Co_3O_4 even with the 0.85% Au content in the composite. The synthesized biosensors were applicable for monitoring H_2O_2 concentration, which will be helpful in identifying cancerous cells [237]. A metal organic framework consisting of porphyrinand iron metal decorated on well-ordered mesoporous carbon (OMC) for hydrogen peroxide (H_2O_2) secreted from viable cells. Porphyrinic iron metal-organic framework (pFeMOF)-decorated ordered mesoporous carbon (OMC) was developed to detect hydrogen peroxide (H_2O_2) released from viable cells. Increased stability and electrical conduction were noticed with the introduction of OMC. Electrocatalytic reduction of H_2O_2 was observed at two different linear ranges, i.e., from 70.5 to 1830.5 μM and from 0.5 to 70.5 μM, with high sensitivity of 67.54 μA mM^{-1} at a low concentration and 22.29 μA mM^{-1} at a high concentration and with a detection limit (LOD) as low as 0.45 μM [238]. A nonenzymatic H_2O_2 electrochemical sensor was developed by immobilizing 2D ultrathin MnO_2 nanosheets onto glassy carbon electrodes (GCE) with a Nafion film. The amperometric study showed an excellent increase in electrocatalytic reduction of H_2O_2 with an extreme low detection limit (5 nM), wide linear range (25 nM^{-2} μM and 10–454 μM), and high sensitivity of 3261 mA M^{-1} cm^{-2} via the immobilization of the MnO_2 nanosheets. The constructed biosensors were efficaciously employed for real-time monitoring of H_2O_2 released from SP2/0 cells in trace amounts [239].

The functionalized hollow-structured nanospheres (HNSs) centered on Pd nanoparticles (NPs) adorned double shell-structured N-doped graphene quantum dots (N-GQDs)@N-doped carbon (NC) HNSs, with ultrafine Pd NPs and "nanozyme" N-GQDs as dual signal-amplifying nanoprobes, act as an exceedingly effective electrochemical sensor for the detection of H_2O_2 released from cancer cells. The hybrid HNS material-based synthesized electrochemical biosensors demonstrate excellent performance, which involves an ultrasmall detection limit as low as nanomolar and a rapid response time. The extra sensitivity, selectivity, and reproducibility of the synthesized biosensors make them valuable for real-time tracking of H_2O_2 released from different living cancer cells in a normal state and treated with chemotherapy and radiotherapy [26]. Heteroatom-doped graphene (N and B) exhibits multidimensional electron transport pathways, which make their use valuable in electrocatalytic sensing of H_2O_2 with excellent stability and response time. Tables 3 and 4 show electrochemical performance of graphene-supported non-noble metal and noble metal nanoparticles.

Table 3. Graphene-supported non-Noble metal nanoparticles for electrochemical detection of H_2O_2.

Graphene-Based Materials	Sensitivity μA mM^{-1} cm^{-2}	Linear Range (μM)	Detection Limit (μM)	Ref.
Nafion/EGO/Co_3O_4	560	1–100	0.3	[240]
CoHCFNPs/GR	0.0007	0.6–379.5	0.1	[41]
VS_2 NPs/GCE	41.96 2	0.5–2.5	0.224	[241]
CoO_xNPs/ERGO	148.6	5–1000	0.2	[242]
CoTPP/RGO	0.0013	0.1–4600	0.02	[243]
rGO/CoPc-COOH	14.5	100–12,000	60	[244]
(PDDA-G/Fe_3O_4)n	61.2	20–6250	2.5	[245]
Fe_3O_4/GO-PAMAM	1.385	20–1000	2.0	[246]
CoS/RGO	2.519	0.1 to 2542.4	0.042	[247]
rGO-Fe_2O_3	0.085	50–9000	6.0	[248]
Fe_3O_4/rGO	387.6	1–20,000	0.17	[249]
Ni_2P NA/TM	690.7	0.001–20	0.2	[166]
Fe_3O_4/RGO	22.27	0.5–3000	0.18	[41]
Fe_3O_4/RGO	0.0168	4.0–1000	2.0	[250]
PB/TiO_2-GR	480.97	0.04–2000	0.0086	[251]
RGO/Fe_3O_4	688.0	100–6000	3.2	[252]
Cu-MOF-GN	57.73	10–11,180	2.0	[33]
PFECS/rGO	117.142	10–190	1.253	[253]
FeTSPc-GR-Nafion	36.93	0.2–5000	0.08	[254]
RGO/ZnO	13.49	0.02–22.48	0.02	[255]
Cu_2O/N-graphene	26.67	5.0–3570	0.8	[256]
Cu_2O-rGO	0.0207	30–12,800	21.7	[257]
CuS/RGO	0.035	5–1500	0.27	[80]
Cu_2O/GNs	-	300–7800	20.8	[258]
GO/MnO_2	38.2	5.0–600	0.8	[259]
MnO_2-ERGO	59.0	100–45,400	10	[260]
MnO_2 nanosheet/graphene	-	10–900	2.0	[261]
CoFe/NGR	435.7	-	0.28	[262]
Co@NCS	-	0.5–7500	0.08	[263]

The resultant N and B co-doped graphene (NB-G)-based electrochemical sensor showed a linear response from 0.5 μM to 5 mM with a LOD of 0.05 μM ($S/N = 3$). This increase in sensitivity with an ultralow detection limit to microlevel attributed to the NB-G constructive structure and special effects arose from the co-doping of N and boron in graphene [200]. $CuFe_2O_4$ nanoparticle-doped reduced graphene oxide based on a CPE was used as a voltammetric sensor for hydrogen peroxide (H_2O_2) sensing. The synthesized sensors showed a rapid amperometric response in less than 2 s and wide linear range of 2 to 200 μM with a low detection limit of 0.52 μM under optimum conditions (pH 5) [264]. Xi et al. synthesized N and S dual-doped graphene (NSG) co-doped carbocatalyst via one-pot syntheses. The NSG-modified electrode showed superior catalytic activity toward sensing, including a linear range up to 1.7 mM. $ZnMn_2O_4$-wrapped reduced graphene oxide microspheres ($ZnMn_2O_4$@rGO) act as an excellent electrocatalyst for H_2O_2 reduction. The $ZnMn_2O_4$@rGO-pasted glassy carbon electrode ($ZnMn_2O_4$@rGO/GCE) displayed a linear detection range of 0.03–6000 μM with a detection limit of 0.012 μM. The resultant biosensor showed promising results in physiology and diagnostics and was applicable in the determination of H_2O_2 secreted from human breast cancer cells (MCF-7) [57]. An AuNPs-NH_2/Cu-MOF composite was prepared via ammoniation of Au NPs, anchored with a Cu-based metal organic framework (Cu-MOF). The synthesized AuNPs-NH_2/Cu-MOF composite was further modified with a GCE to prepare an AuNPs-NH_2/Cu-MOF/GCE electrode. The synthesized AuNPs-NH_2/Cu-MOF/GCE composites possessed high sensitivity and selectivity, and they can be used as an electrochemical enzyme-free sensor for the quantitative detection of H_2O_2. Instead of quantitative H_2O_2 detection, the synthesized electrochemical sensor showed a wide linear response toward H_2O_2 concentrations ranging 5–850 μM with a LOD down to 1.2 μM [265]. Wang et al. improved the sensitivity of the electrode using hemin-capped biomineralized gold nanoparticles (Hem@AuNPs)-doped reduced graphene oxide (rGO), followed by coating with chitosan (CS). The resultant electrode from the prepared nanohybrids showed excellent electrocatalytic reduction of H_2O_2 with superior sensitivity, stability, and response time of few seconds. The most important feature of the synthesized electrode from the resultant nanohybrid is its lower detection limit of 9.3 nM and linear range of five orders of magnitude. Such characteristics enable this biosensor to detect H_2O_2 releasing from living Hela cells accurately and make this biosensor valuable for ultrasmall detection of H_2O_2 from living HeLa cells precisely [266]. Sun and his coworkers designed a novel nonenzymatic hydrogen peroxide sensor using intermetallic PtPb nanoplates (PtPb/G) supported on graphene with enhanced electrochemical performance for H_2O_2 detection in neutral solution and H_2O_2 released from the cells. The nanocomposite exhibited excellent electrocatalytic activity for the electrochemical reduction of H_2O_2 in half-cell test and with wide linear detection range of 2 nM to 2.5 mM and ultralow detection limit of 2 nM. An experiment further showed that the sensitivity of intermetallic PtPb nanoplates is 12.7 times higher than that of a commercial Pt/C electrode for the detection of H_2O_2 released from Raw 264.7 cells [267]. A graphene/Nafion/azure/I/Au nanoparticle composites modified glass carbon electrode (graphene/Nafion/AzI/AuNPs/GCE) was used for the construction of a nonenzymatic H_2O_2 sensor. The performance of the synthesized biosensors was recorded under optimum conditions, i.e., pH of 4.0 and potential of −0.2 V, upon the addition of H_2O_2. A stable current was obtained in less than 3 s, with a detection limit of 10 μM ($S/N = 3$) and a linear range of 30 μM to 5 mM [268]. Ju et al. reported a green and simple strategy for the in situ growth of surfactant-free Au nanoparticles (Au NPs) on nitrogen-doped graphene quantum dots (Au NPs–N-GQDs). The reported strategy showed the in situ formation of the Au NPs–N-GQDs hybrid by simple mixing of N-GQDs and $HAuCl_4 \cdot 4H_2O$ without any reductant and surfactant. The prepared nanocomposite (Au NPs–N-GQDs) exhibited a low detection limit of 0.12 μM and sensitivity of 186.22 μA/mM cm^2 for the electrochemical detection of hydrogen peroxide (H_2O_2) [79]. Another research group developed a microelectrode with high sensitivity, a wide linear range, and good selectivity for the detection of H_2O_2 released from female cancer cells. The synthesized hierarchical nanohybrid microelectrode was composed of 3D porous graphene enfolded activated carbon fiber (ACF). This technique, i.e., green ionic liquid (IL), plays a crucial role in the simultaneous superficial and effective electrodeposition and electrochemical reduction of GO nanosheets on ACF to form a 3D porous ionic liquid functionalized electrochemically reduced GO (ERGO)-wrapped ACF (IL–ERGO/ACF) [269].

Table 4. Graphene supported noble metal nanoparticles for electrochemical detection of H_2O_2.

Graphene Based Material	Sensitivity $\mu A\ mM^{-1}\ cm^{-2}$	Linear Range (μM)	Detection Limit (μM)	Ref.
Au-PEI/GO	460.0	0.5–1680	0.2	[270]
AgNPs-MWCNT-rGO	0.833	100–100,000	0.9	[271]
RGO-Au-PTBO	63.39	5.0–1077.1	0.2	[272]
Ag-MnOOH-GO	59.14	0.5–17,800	0.2	[273]
Au NPs@POM-G	58.87	5.0–18,000	1.54	[274]
AgNPs-TWEEN-GO	0.7459	20–23,100	8.7	[41]
GO-ATP-Pd	504.85	0.1–10,000	0.016	[275]
GN/Au-NPs	-	0.5–500	0.22	[276]
GN-Pt	0.01	2–710	0.5	[277]
Ag NWs-graphene	12.37	10.0–34,300	1.0	[278]
GR-AuNRs	389.2	30–5000	10	[279]
Au@C-Co_3O_4	7553	-	0.019	[237]
Au NPs-N-GQDs	186.22	0.25–13,327	0.12	[79]
AuNPs-NH2/Cu-MOF/GCE	1.71	5–850	1.2	[265]
GO/Au@Pt@Au	-	0.05–17,500	0.02	[280]
NG-hAuPd	5095.5	0.1–20	0.02	[281]
PDA-RGO/Ag NP	0.0111	0.5–8000	2.07	[282]
AgNPs/GN	-	100–100,000	0.5	[283]
Ag/SG	-	100–136,500	0.14	[284]
Pt/PG	341.14	1–1477	0.5	[285]
PDDA-RGO/MnO_2/AuNPs	1132.8	5.0–500	0.6	[286]
AgNP/rGO	-	100–80,000	7.1	[287]
AgNPs-GO	-	10–20,000	0.5	[288]
RGO-AuNP	5.3	250–22,500	6.2	[289]
GNPs/SGS	27.7	20–15,000	0.2	[290]
AgNPs-CNT-rGO	-	10–10,000	1.0	[291]
PpyNFs/AgNPs-rGO	0.7367	100–5000	1.099	[292]
polystyrene@RGO-Pt	0.0675	0.5–8000	0.1	[41]
Graphene/Nafion/Azl/AuNPs	-	30–5000	10	[268]
Pt/GN	0.0204	2.5–6650	0.8	[41]
RGO-AuNPs (B)	9.5	25–41,500	5.0	[273]
PtAu/G-CNTs	313.4	2.0–8561	0.6	[293]
PtAuNPs-CTAB-GR	0.1654	0.005–4.8	0.0017	[294]
PtAu/RGO	4.105	0.015–8.73	0.008	[295]
Pt/graphene-CNT paper	1.41	0–25.0	0.01	[296]
pFeMOF/OMC	67.54	70.5–1830.5	0.45	[238]
Pd-PEI/GO	-	0.5–459	0.2	[297]
Pd-NPs/GN	0.019	0.001–2000	0.0002	[298]
PdNPGNs	2.75	0.1–1000	0.05	[299]
RGO-PMS@AuNPs	39.2	0.5–50,000	0.06	[69]
2Au1Ag-PDA/CFME	12966	0–55	0.12	[300]
TiO_2NTs/r-GO/AgNPs	1152	15,500–50,000	2.2	[301]
PtPb/G	-	2–2.5	0.02	[267]
3DGA-AuNPs/cytc/GCE	351.57	-		[302]
PdPt NCs@SGN/GCE	-	1–300	0.3	[303]
AuNFs/Fe_3O_4@ZIF-8-MoS_2	-	5–120	0.9	[304]

3.4.4. Graphene-Loaded Biomolecules for Selective Detection of H_2O_2

Recently, graphene-based heme protein electrodes have gained wide attention for H_2O_2 detection. These graphene-based materials offer an appropriate microenvironment to maintain the redox bioactivity of proteins and make the transfer of electrons feasible between redox proteins (active centers) and the principal electrode [232]. A mixture of a strong acid and an oxidizing agent is used for the synthesis of graphene oxide from graphite [289]. GO serves as a precursor of graphene and as a sensing element. Several proteins, including cytochrome c, horseradish peroxidase (HRP), and myoglobin (Mb), were incubated. Zuo et al. [227] fabricated a heme proteins-modified GO electrode from GO suspension. Immobilization of protein on a GO sheet is associated with strong hydrophobic and electrostatic interactions between proteins and GO. The innate characteristics of the proteins remain unaltered in the presence of GO, which offers an appropriate microenvironment for the immobilization of protein with an intact structure. Studies revealed that the protein-based GO modified electrodes have an advantage over the featureless voltammograms because of the emittance of redox peaks from proteins on these electrodes, which stipulate an efficient electrical wiring of the redox centers of proteins to the surface of the electrode in the presence of GO. Importantly, the proteins retained their intrinsic peroxidase activity upon forming mixtures with GO and the catalytic properties provide a high sensing performance for H_2O_2 detection with low detection limit and wide detection range. Furthermore, Mani and coworkers improved the performance of Mb-based H_2O_2 biosensors using an RGO-MWCNT-Pt/Mb electrode [233]. The RGO-MWCNT-Pt composite was prepared using the wet chemical method, which provides good affinity and a large surface area for the accumulation of excess Mb. The Pt nanoparticles in the RGO-MWCNT-Pt/Mb composite showed excellent electrocatalytic activity and efficiently prohibited the accumulation and restacking of graphene sheets and CNTs. The resultant electrode (RGO-MWCNT-Pt/Mb) showed an excellent wide linear range from 10 pM to 0.19 nM with a detection limit of 6 pM and much higher sensitivity (1.99 $\mu A\ pM^{-1}\ cm^{-2}$) compared to other biosensors.

Additionally, HRP-fabricated H_2O_2 electrochemical biosensors were prepared using nano-graphene for the direct electron transfer from HRP to the substrates (electrode) [305]. These HRP-anchored graphene-based materials determine H_2O_2 with higher selectivity and sensitivity [224,306]. Zhang et al. reported immobilization of HRP and lysozyme enzymes on graphene oxide sheets in phosphate buffer solution by incubating GO with enzymes at 4 °C. The immobilized enzyme molecules were studied in situ using AFM, which clearly disclosed HRP molecules (bright spots) on the surface of GO. Strong hydrogen bonding and electrostatic interaction play a key role in loading enzymes (HRP and lysozyme) on graphene oxide, which was much higher than that on previously reported studies and was found to be the optimum solid substrate for the immobilization of the enzyme [307]. Moreover, Fan and coworkers applied the same method to generate graphene-poly (sodium 4-styrenesulfonate)/biomimetic graphene capsules (GS-PSS/GRCAPS) nanocomposites for direct electrochemical sensing of H_2O_2. Initially, porous $CaCO_3$ was used as a support for HRP encapsulation in GRCAPS. Afterward, a GS-PSS/GRCAPS composite was synthesized via layer-by-layer electrostatic self-assembly, in which negatively charged GS-PSS electrostatically interact with positively charged PEI@GRCAPS [261]. GRCAPS was revealed to mimic the existing enzymes in living cells and provide a satisfactory microenvironment for HRP to realize direct electron transfer at the modified electrode. The resultant electrochemical biosensor exhibited long-term stability, low detection limit, extensive linear range, and an excellent anti-interference ability. A nonenzymatic and highly electrocatalytic H_2O_2 biosensor was proposed using a novel electrode composed of hemin-capped biomineralized gold nanoparticles (Hem@AuNPs), rGO, and chitosan (CS). The excellent rGO conductivity and outstanding electrocatalytic performance of Hem@AuNPs make them suitable for developing ultrasensitive biosensors for real-time determination of H_2O_2. Taking advantages of the peroxidase-like activities of nanohybrids, the resultant electrode demonstrated a highly selective and outstanding electrochemical performance toward H_2O_2 with fast response, improved sensitivity, and stability. More significantly, the lower determination limit of 9.3 nM and wider linear ranges of five orders of magnitude enable this biosensor to accurately detect H_2O_2 released from living HeLa cells [266]. Jiao et al. reported nonenzymatic biosensors for dynamic, most significant ROS. Intracellular nonenzymatic monitoring of H_2O_2 was achieved via loading of AuPtAg nanoalloy on rGO capped with poly (diallyldimethylammonium chloride). The constructed biosensor showed rapid and precise measurement of H_2O_2 released from cancerous cells. The precise and accurate detection of H_2O_2 is due to the remarkable rGO and PDDA conductivity with outstanding synergistic electrocatalytic performance of ternary alloys. The remarkable electrochemical performance of the resultant biosensor, with a low detection limit (1.2 nM) and wide linear range (from 0.05 μM to 5.5 mM), is due to the peroxidase-like activity of the AuPtAg nanoalloy [59]. In another study, a cytochrome c (Cyt c)-immobilized Au nanoparticle-loaded 3D graphene aerogel (3DGA) was synthesized for the detection of H_2O_2. Morphological and surface study of the 3DGA-AuNPs revealed efficacious formation of 3D-networked assembly, which helps in enhancing conductivity and effective enzyme immobilization. The large surface of 3DGA and biocompatibility of AuNPs help in enabling direct electron transfer between the electrode and Cyt c. The as-prepared 3DGA-AuNPs/Cyt c/GCE exhibited a pair of well-defined redox peaks of a $Fe^{III/II}$ redox couple of Cyt c and revealed excellent electrocatalytic potential toward H_2O_2 with high sensitivity of 351.57 $\mu A\ mM^{-1}\ cm^{-2}$ [302].

3.5. *Carbon Nanotubes (CNTs)*

CNTs, an allotropic form of carbon, are composed of a graphene sheet packed in a tube constituting a cylinder (single-walled CNTs (SWCNTs)) or concentric and closed tubules (multiwalled CNTs (MWCNTs)) [308–310]. The combination of CNTs in biosensors offers numerous advantages, including increased surface area, smooth charge transfer, stacking of various biomolecules, and improved conductivity of the resulting platform as shown in Table 5 [310–312].

3.5.1. H_2O_2 Electrochemical Sensors Based on the Association of CNTs and Hemoproteins

Direct electrochemical assignment of proteins in biosensors is an area of high interest. However, direct electron transfer between proteins and electrodes is faced with major problems, i.e., the distance between the redox center and electrode and protein denaturation. Different methods, including polymer adsorption, covalent binding, and layer-by-layer film assembly, are well known for the deposition of protein molecules on the electrode surface. The excellent electrocatalytic properties of CNTs make them valuable in loading biomolecules and for use as biosensors. CNTs function as nanowires and boost the electron transfer from the protein's redox center to the electrode. The heme-containing proteins (Hb, Mb, Cyt c, and HRP) are the most common analytes for protein detection [313]. Heme proteins are the center of several biological redox reactions. Therefore, several studies reported the efficacy of these proteins as a biosensor for H_2O_2, nitrite, or hydrogen sulfide detection. Yang et al. have developed a method to directly bind hemoglobin to a vertically aligned CNT surface. They modified the nanotubes so they could use diazonium chemistry to directly bind hemoglobin. In amperometric detection of H_2O_2, an Hb-ACNT electrode exhibits a wide concentration range (40 μM to 3 mM), LOD of 5.4 M, high sensitivity, and long-term stability [314]. This aligned NT forest shows accumulation of Hb on a large area rather than immobilization in unsystematic tangled webs of CNT. Furthermore, Esplandiu et al. immobilized Mb to detect H_2O_2, studied direct electron transfer kinetic, and showed that vertically aligned NT forests possess better kinetics compared to the epoxy incorporated SWCNT/Mb sensor [315]. In addition, their LOD was 50 nM for H_2O_2, superior to other random and aligned NT methods. The release of H_2O_2 from living HepG2 cancer cells was studied by Zhang and coworkers, who constructed an enzyme-based biosensor with a LOD of 0.23 μM using SWCNTs as a robust scaffold for Hb immobilization. The constructed biosensor was also used for the quantification of H_2O_2 released from HepG2 cells via in situ biosynthesis of ZnO quantum dots, which was further confirmed by fluorescence staining [316]. Wang et al. applied a simple dispersion method to coat a GCEs with SWCNT and heme proteins in the presence of cetylrimethylammonium bromide (CTAB) [317]. CTAB-suspended NTs facilitate the immobilization of Mb, Cyt c, and HRP on the electrode surface. Redox chemistry of heme was studied in the presence of SWCNTs. The developed electrode was precisely used for nitrite and H_2O_2 detection, which gave rise to a new-fangled peak in cyclic voltammograms with decreased in heme reduction peak. The results indicate that the electrode exhibited a response time of only 4 s, LOD of 3.6 M, and less sensitivity for H_2O_2 detection. Several researchers have also demonstrated the efficacy of the immobilization of heme proteins on polymers. For instance, Hb was immobilized on polyelectrolyte surfactant polymers [318], where Hb retained a secondary structure, thus reducing the effect of protein denaturation in polymers. Moreover, the addition of SWCNT to the nanocomposite enhanced the reaction kinetics, and H_2O_2 was sensed with a LOD of 0.8 M. Likewise, MWCNT and Mb were immobilized on the collagen polymers [319], where H_2O_2 was measured with a linear range from 0.6 to 39 M. Nagaraju et al. used self-assembled monolayers of 4-aminothiophenol on gold electrodes with immobilized Cyt c for H_2O_2 detection [320]. Three orders of magnitude of faster electron transfer kinetics were observed with SWCNT in these monolayers as compared to the non-SWCNT monolayer. The results confirmed that NTs increased the direct electron transfer. However, large step changes (3.8 mM) in H_2O_2 were used, and no LOD was calculated. Thus, the sensitivity was not very significant. A layer-by-layer approach was also implemented to immobilize proteins rather than simultaneous deposition of all components, e.g., chitosan-stabilized NTs were placed on GCE, followed by the accumulation of gold NPs on chitosan, and subsequently, Hb was bound to the gold surface [321]. This method is beneficial in retaining the bioactivity of Hb and increases the amount of enzyme activity. The method showed a LOD of 0.2 M for H_2O_2 detection. The heme-based biosensors showed rapid and fast detection of changes in H_2O_2. In general, HRP is the most widely used in biosensing as compared to other heme-containing proteins, which shows the best results compared to others. The lowest limits of detection for H_2O_2 were achieved using an aligned NT geometry, which supports the accumulation of the heme protein and which, in turn, leads to fast electron transfer from proteins to electrodes. Future studies are required to address the reproducibility of electrode fabrication, practical geometries, and uses for real samples. SWCNTs, HRP, and 1-butyl-3-methylimidazolium tetrafluoroborate ($BMIM \cdot BF_4$) were combined to construct a cellular H_2O_2 sensor. At a working potential of −0.35V, HRP-BMIM·BF4/SWCNTs/CFUME showed a dynamic range of ~10.2 μM, with a low detection limit of 0.13 μM (S/N = 3) and high sensitivity of 4.25 A/M cm^2. Due to its small dimension and low working potential, HRP-BMIM·BF_4/SWCNTs/CFUME allowed direct amperometric real-time monitoring of H_2O_2 in HeLa cells treated with camptothecin (an anticancer drug) without complex data processing and extra surface coatings to prevent interference. Thus, its testing evidently demonstrated a significantly high level of H_2O_2 in HeLa cells under camptothecin stress [322]. A schematic presentation of enzyme loaded CNTs for the detection of H_2O_2 in living cells is shown in Figure 6.

Table 5. CNTs-supported metal nanoparticles biosensors for electrochemical sensing of H_2O_2 secreted from cancerous cells.

CNTs H_2O_2 Biosensors	Sensitivity $\mu A\ mM^{-1}\ cm^{-2}$	Linear Range (μM)	Detection Limit (μM)	Ref.
ZnO/COOH-MWNTs	-	1–21	–	[323]
GCE/MWCNTs-CDs	0.039	3.5–300	0.25	[324]
$((APy)_6[H_2W_{12}O_{40}])$/(SWCNT-COOH)	-	-	0.4	[325]
GCE/CNTs-PAMAM DENs-PtNCs	987.5	3–400	0.8	[326]

Table 5. *Cont.*

CNTs H_2O_2 Biosensors	Sensitivity $\mu A\ mM^{-1}\ cm^{-2}$	Linear Range (μM)	Detection Limit (μM)	Ref.
GCE/C_{60}-MWCNTs CS-IL/MB/CuNP	0.0243	2–4	0.055	[327]
3D PB NPs/G-CNTs	0.11343	1–3161	0.095	[328]
CF@N-CNTAs–AuNPs	142	1–4300	0.05	[329]
CDs/MWCNTs/GCE	–	–	-	[324]
OECT/PET/CE-CNTs/PtNPs	-	0.5–100	0.2	[330]
ZNBs/fMWCNTs	-	0.049–22	0.035	[331]
ZnONPs/MWCNTs	-	1000–200,000	-	[332]
N-CNTs	30	-	0.5	[333]
(GC) (BG-CNPs/GC)	-	-		[334]
GCE/rGONRs/MnO_2	0.0142	0.25–2245	0.071	[335]

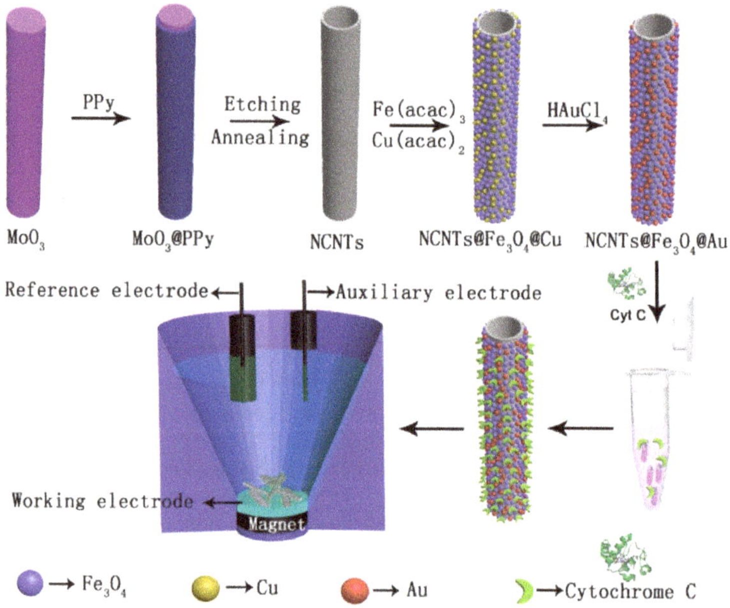

Figure 6. Enzyme-loaded CNTs for the detection of H_2O_2 in living cells. Reproduced with permission from [336]. Copyright 2019, Elsevier Ltd (Amsterdam, The Netherlands).

3.5.2. H_2O_2 Electrochemical Sensors Based on the Association of Metallic Nanoparticles and CNTs

In the last few years, the connotation of metal nanoparticles with CNTs has been considered a valuable alternate for the development of highly sensitive and selective sensors for H_2O_2 detection shown in Figure 7 [41,337,338]. Zhang et al. developed a remarkable stretchy nanohybrid microelectrode using carbon fibers [329]. The constructed microelectrode reproduced a remarkable analytical signal at 0.300 V, with an ultrasmall LOD as low as nanomolar. Rapid and ultrasmall sensing of H_2O_2 excreted from MCF-7 and MDA-MB-231 cells was achieved because of the synergistic catalytic activity of the N-CNTA-decorated AuNPs. Table 6 showed non-enzymatic CNTs biosensors for H_2O_2 sensing.

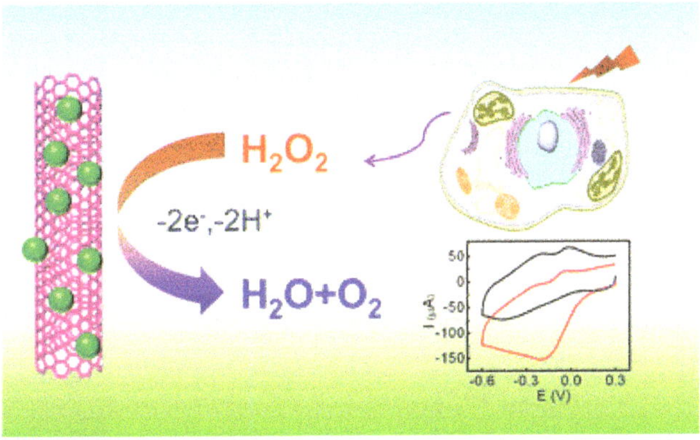

Figure 7. Mechanism of metal nanoparticle loaded CNTs for real-time analyses of H_2O_2 secreted from live cells. Reproduced with permission from [324], Copyright 2016 Spinger Ltd (Berlin/Heidelberg, Germany).

Table 6. Ultra-sensitive electrochemical biosensors for detection of H_2O_2.

H_2O_2 Biosensors	Sensitivity μA mM^{-1} cm^{-2}	Linear Range (μM)	Detection Limit (μM)	Ref.
GN-HN-SWCNT	0.015	0.2–400	0.05	[206]
Cyt c/GO-MWCNT/Au NP	0.533	1×10^{-5}–1.4×10^{-4}	27.7×10^{-6}	[232]
HRP/P-L-His-rGO	2.6×10^5	0.2–5000	0.05	[224]
HRP/AuNP/ThGP	0.086	0.5–1800	0.01	[225]
RGO-MWCNT-Pt/Mb	1.990	1×10^{-5}–1.9×10^{-4}	16×10^{-6}	[233]
CoTPP/RGO	0.0013	0.1–4600	0.02	[243]
GN-Pt	0.01	2–710	0.5	[277]
PDA-RGO/Ag NP	0.0111	0.5–8000	2.07	[282]
Pt/GN	0.0204	2.5–6650	0.8	[41]
PtAuNPs-CTAB-GR	0.1654	0.005–4.8	0.0017	[294]
Pd-NPs/GN	0.019	0.001–2000	0.0002	[298]
GCE/MWCNTs-CDs	0.039	3.5–300	0.25	[324]
GCE/C_{60}-MWCNTs CS-IL/MB/CuNP	0.0243	2–4	0.055	[327]
GCE/rGONRs/MnO_2	0.0142	0.25–2245	0.071	[335]

Bai et al. developed an electrode with high electrocatalytic activity using carbon dots (CDs) and oxidized MWCNTs modified GCEs [324]. The excellent biosensing capability of the MWCNT/CD/GCEs composite is associated with the large surface area and electron acceptor characteristics of MWCNTs and with the excellent donor capacity of the CDs. The analytical performance of the sensor is highly dependent on the MWCNT:CD ratio, so 10:1 was selected for electrode manufacturing. The developed biosensors successfully quantified the H_2O_2 secreted from HeLa cells with a linear range of 3.5×10^{-6} and 3×10^{-4} M and a LOD of 0.25 mM.

Liu and Ding used a Pt-encapsulated poly(amidoamine) dendrimer with amine terminations (G6-NH_2 PAMAM dendrimer) covalently attached to a carboxylated CNT composite [326]. Elaborated architectures showed a rapid, reproducible, and steady response at 0.150 V, with a linear range between 3 and 400 mM. The biosensor successfully detected H_2O_2 in MCF-7 cells with an LOD of 0.8 mM. Liu at al. modified GCEs by dendrimer-encapsulated Pt nanoclusters and carbon nanotubes (Pt-DENs/CNTs) to detect extracellular H_2O_2 excreted from live cells. Those Pt-DENs/CNTs nanocomposites were characterized by UV-Vis spectra, SEM, energy-dispersive X-ray spectroscopy, and TEM. The nonenzymatic sensor displayed exceptional catalytic activity in H_2O_2 reduction. The effective nonenzymatic sensing capability of H_2O_2 reduction revealed that the Pt-DENs/CNTs sensor has potential application in screening H_2O_2 in cellular processes [326].

Real-time monitoring of H_2O_2 secreted from living cells is important to understand the occurrence of diseases and searching of new therapeutic strategies. Zhao et al. successfully synthesized three-dimensional carbon nanotubes spaced graphene aerogel decorated with Prussian blue nanoparticles (3D PB NPs/G-CNTs) by one-step mild temperature treatment, in which the PB NPs acquire intrinsic peroxidase-like activity. The 3D porous structure of G-CNTs with large surface and high electrical conductivity can efficiently enhance catalytic performance and help in real-time detection of H_2O_2 released from living cells. The composite exhibited good catalytic performance toward H_2O_2 reduction with sensitivity of 134.3 μA mM^{-1} cm^{-2}, LOD of 95 nM, and wide linear range of 1–3161 μM [328].

Zhang et al. integrated Fe_3O_4 and Cu nanoparticles (NPs) into the NCNTs to produce N-doped carbon@Fe_3O_4-Cu nanotubes (NCNTs@Fe_3O_4@Cu) through a one-pot high-temperature decomposition. Then, Au NPs were assembled on the magnetic NCNTs to obtain an NCNTs@Fe_3O_4@Au composite by galvanic replacement with Cu NPs. The resultant composite provided a suitable platform for the immobilization of the enzyme to fabricate biosensors for H_2O_2 monitoring. After the cytochrome c (Cyt c) was accumulated by the NCNTs@Fe_3O_4@Au composite, the Cyt c/NCNTs@Fe_3O_4@Au gathered to the surface of the electrode with an external magnet [336]. Tabrizi et al. developed a flow injection amperometric sandwich-type aptasensor for the detection of human leukemic lymphoblasts (CCRF-CEM). An amperometric biosensor was synthesized by decorating nanogold on poly(3,4-ethylenedioxythiophene) (PEDOT-Au_{nano}). PEDOT-Au_{nano} acts as a nano-platform for immobilizing a thiolated sgc8c aptamer and MWCNTs loaded PdNPs/3,4,9,10-perylene tetracarboxylic acid (MWCNTs-Pd_{nano}/PTCA) to assemble a catalytic labeled aptamer. In this strategy, the CCRF-CEM cancer cells were sandwiched between the immobilized sgc8c aptamer on PEDOT-Au_{nano} (electrode) and sgc8c aptamer MWCNTs-Pd_{nano}/PTCA/aptamer (catalytic site). The resultant sandwich-type aptasensor determined the CCRF-CEM cancer cell concentration using 0.1 mM H_2O_2 (electroactive component). The MWCNTs-Pd_{nano} nanocomposites enhanced the electrocatalytic reduction of H_2O_2, which further boost sensor sensitivity toward CCRF-CEM cancer cells. The proposed sandwich-type aptasensor displayed outstanding analytical performance for real-time determination of CCRF-CEM cancer cells with high selectivity, ranging from 1.0×10^1 to 5.0×10^5 cells mL^{-1} with an LOD of 8 cells mL^{-1} [339].

3.5.3. In Vivo Sensing of H_2O_2 Release from Carcinoma Cells

Considering the significance of cellular H_2O_2 in cell pharmacology and pathophysiology, accurate and reliable in vivo detection of cellular H_2O_2 is sorely needed. Sensing H_2O_2 at cellular level is constrained by several factors, including small cell size, low concentration of cellular H_2O_2, and interferences in the culture medium [215,340]. Such in situ monitoring of the cellular release of H_2O_2 provides a new in vitro drug screening platform for personalized medicine and cancer therapy. Enzyme-based electrochemical sensing is efficient for continuous in situ monitoring of H_2O_2 because of its high sensitivity, rapid response, and selectivity [27,267]. In vivo monitoring of H_2O_2 secreted from living cells is essential in understanding cellular signaling pathways. The release of H_2O_2 from living cells is very low because the selective detection of H_2O_2 at a low level is challenging.

Chen et al. used a flow-through mode sensing strategy based on cell-in-lumen configuration for ultra-small detection of H_2O_2 secreted from the H1299 carcinoma cell. The current strategy involved the growth of cells on the inner surface of a porous hollow fiber (PHF), while a sensing layer comprised of multi-walled carbon nanotubes, gold nanoparticles (AuNPs), and enzymes accumulated on the outer surface of the PHF. The porous structure of the resultant electrode proved beneficial in the exchange of H_2O_2 from cell to sensing layer in a short time span. The resultant electrode exhibits ultra-small sensitivity to detect H_2O_2 at the nanomolar level having a detection limit of 6 nM with a wide linear range of 0.01–5 [341]. Ye et al. fabricated a PdPt NCs@SGN/GCE non enzymatic electrochemical biosensor comprised of Pd-Pt nanocages and SnO_2/graphene nanosheets. The resultant electrode displayed excellent catalytic activity toward H_2O_2 in situ secreted from human cervical cancer cells (Hela cells) with high selectivity and sensitivity, a low detection limit of 0.3 mM, and a large linear range from 1 mM to 300 mM [303]. Fe_3O_4 quantum dot was decorated on three-dimensional graphene nanocomposites (Fe_3O_4/3DG NCs) for real time in-situ monitoring of H_2O_2 released from living cancer cells. The fabricated electrochemical sensor mimics peroxidase-like activity with high sensitivity of 274.15 mA M^{-1} cm^{-2}, a low detection limit (78 nM), fast response (2.8 s), and outstanding reproducibility [342].

A nonenzymatic electrochemical sensor was constructed by immobilizing 2D ultrathin MnO_2 nanosheets onto glassy carbon electrodes (GCE) with a Nafion film for real-time monitoring of H_2O_2 released from SP2/0 cells in trace amounts. The amperometric study showed an excellent increase in electrocatalytic reduction of H_2O_2 with an extreme low detection limit (5 nM), wide linear range (25 nM^{-2} μM and 10–454 μM), and high sensitivity of 3261 mA M^{-1} cm^{-2} via the immobilization of the MnO_2 nanosheets [239]. Xi et al. synthesized N and S dual-doped graphene (NSG) co-doped carbocatalyst via one-pot syntheses. The NSG-modified electrode showed superior catalytic activity toward sensing, including a linear range up to 1.7 mM. The prepared electrode showed high sensitivity of 0.266 mA cm^{-2} mM^{-1} with a detection limit as low as 1 μM (S/N = 3), with good discernment, reproducibility, stability, and biocompatibility with real-time determination of H_2O_2 secreted from live cancerous cells [343]. Later on, Zhao et al. used a well-controlled strategy for the syntheses of the graphene fiber microelectrode via MnO_2 nanowire (MnO_2-NWs) assembly (MnO_2-NWs@Au-NPs). The prepared microelectrode showed proficient catalytic performance toward the redox reaction of H_2O_2. The nanohybrid microelectrode showed in vivo real-time detection of H_2O_2 released from human breast cancer cells [344]. Recently, Chen and his co-worker established high-index facets of an Au-Pd nanocubes loaded rGO composite. The resultant electrode comprised of three-dimensional nanocomposites showed a detection limit of 4 nM, a wide linear range from 0.005 μM to 3.5 mM, and real time monitoring of endogenous H_2O_2 in human serum samples released from a living breast cancer cell [345].

3.6. *MXenes Materials*

So far, various nanomaterials have been reported and utilized for the development of incrementally efficient biosensors. Among the most recently reported nanomaterials available for biosensors, MXenes have attracted much attention for their huge potential in biosensor development because of their unique characteristics [346]. MXenes are two-dimensional inorganic compounds with a thickness of a few atomic layers and they are composed of transition metal carbides, nitrides, or carbonitrides such as titanium carbide (Ti_3C_2) and titanium carbonitride (Ti_2CN), which confers them with exceptional characteristics, including high conductivity and superior fluorescent, optical, and plasmonic properties [347,348]. Moreover, the biocompatible property of MXenes enables their biomedical application [349,350]. Since they were first reported in 2011, MXenes have been used to develop various types of advanced biosensors, including electrochemical, fluorescent/optical, and surface-enhanced Raman spectroscopy (SERS) biosensors, by augmenting MXene characteristics to make them suitable for specific types of biosensors or by combining them with other nanomaterials [351,352]. Recent studies on the development of highly effective MXene biosensors show that this novel nanomaterial is the most ideal candidate for biosensor development at present. So far, no considerable development was seen in MXenes-based biosensors for detection of H_2O_2 released from a cancer cell. However, we foresee MXenes as having outstanding potential for detection of H_2O_2 at an ultra-low level with durable stability and long working hours.

4. Conclusions and Future Perspectives

Carbon nanomaterials have gained prodigious attention over the last two decades because of their higher applicability in electrochemical sensors. This review shed light on the application of carbon nanomaterials and their composite with metal, metal oxides, and biomolecules for the fabrication of electrochemical sensors for real-time monitoring of hydrogen peroxide. Initially, we discussed the recent advancement in the development of heme protein biosensors with carbon nanomaterials as immobilization matrix and their application in the detection of H_2O_2. Subsequently, the synthesis and application of graphene-supported nano-catalysts (metal-free, noble metals, and nonnoble metals) was discussed in detail for the construction of nonenzymatic H_2O_2 electrochemical sensors. Despite the extensive advancement in the design and application of carbon nanomaterials for the electrocatalytic determination of H_2O_2, it is crucially important to develop new techniques and methods for the synthesis of carbon-based electrocatalysts with a novel structure and extraordinary activity. Some of the most highly ultra-sensitive biosensors for detecting H_2O_2 at an ultra-low level are displayed in Table 6. Furthermore, the comprehensive understanding and exploration of the structure–property relationship of carbon nanomaterials and their extensive use in H_2O_2 sensors require more efforts and research. Particularly, its excellent electrical conductivity, electron mobility, small band gap, and ultrahigh surface area make it widely applicable in biosensors. These advantages of graphene would bestow good conductivity to the capsule film, further facilitate fast electron transfer between enzyme and basal electrode, and enhance the sensitivity and detection limit of biosensors. We prophesy excellent biosensing potential of new MXenes materials and carbon-based material for detection of H_2O_2 released from cancer cells at an ultra-low level with remarkable stability and selectivity.

Author Contributions: Conceptualization, A.K., A.A.-H. and S.E.D.; methodology, T.A. and A.I.; formal analysis, S.A.H. and A.K.; investigation, A.K. and S.E.D.; resources, A.K. and A.A.-H.; data curation, T.A., A.I. and S.A.H.; writing—original draft preparation, T.A., A.I. and A.K.; writing—review and editing, S.A.H., J.U., A.K. and A.A.-H.; supervision, A.K., A.A.-H. and S.E.D.; project administration, A.K., A.A.-H. and S.E.D.; funding acquisition, S.E.D. All authors have read and agreed to the published version of the manuscript.

Funding: The project was supported by grant from the Oman Research Council (TRC) through the funded project (BFP/RGP/HSS/19/198).

Informed Consent Statement: Not applicable.

Data Availability Statement: All data are available from the corresponding author upon request.

Acknowledgments: The authors would like to thank the University of Nizwa for the generous support of this project. The authors also extend their appreciation to the Deanship of scientific research at King Khalid University for funding this work through the research groups program under Grant No. RGP.1/259/42.

Conflicts of Interest: The authors declare no conflict of interest.

References

1. Martinkova, P.; Kostelnik, A.; Válek, T.; Pohanka, M. Main streams in the construction of biosensors and their applications. *Int. J. Electrochem. Sci.* **2017**, *12*, 7386–7403. [CrossRef]
2. Li, P.; Lee, G.-H.; Kim, S.Y.; Kwon, S.Y.; Kim, H.-R.; Park, S. From diagnosis to treatment: Recent advances in patient-friendly biosensors and implantable devices. *ACS Nano* **2021**, *15*, 1960–2004. [CrossRef] [PubMed]
3. Huang, Y.; Xu, J.; Liu, J.; Wang, X.; Chen, B. Disease-related detection with electrochemical biosensors: A review. *Sensors* **2017**, *17*, 2375. [CrossRef]
4. Monošík, R.; Stred'anský, M.; Šturdík, E. Application of electrochemical biosensors in clinical diagnosis. *J. Clin. Lab. Anal.* **2012**, *26*, 22–34. [CrossRef] [PubMed]
5. Monosik, R.; Stredansky, M.; Tkac, J.; Sturdik, E. Application of enzyme biosensors in analysis of food and beverages. *Food Anal. Methods* **2012**, *5*, 40–53. [CrossRef]
6. Faridbod, F.; Gupta, V.K.; Zamani, H.A. Electrochemical sensors and biosensors. *Int. J. Electrochem.* **2011**, *2011*. [CrossRef]
7. World Health Organization. *The World Health Report: 2004: Changing History*; World Health Organization: Geneva, Switzerland, 2004.
8. Siegel, R.L.; Miller, K.D.; Fedewa, S.A.; Ahnen, D.J.; Meester, R.G.; Barzi, A.; Jemal, A. Colorectal cancer statistics, 2017. *CA Cancer J. Clin.* **2017**, *67*, 177–193. [CrossRef] [PubMed]
9. Fitzmaurice, C.; Allen, C.; Barber, R.M.; Barregard, L.; Bhutta, Z.A.; Brenner, H.; Dicker, D.J.; Chimed-Orchir, O.; Dandona, R.; Dandona, L. Global, regional, and national cancer incidence, mortality, years of life lost, years lived with disability, and disability-adjusted life-years for 32 cancer groups, 1990 to 2015: A systematic analysis for the global burden of disease study. *JAMA Oncol.* **2017**, *3*, 524–548. [PubMed]
10. Steward, B.; Kleihues, P. *Colorectal Cancer*; World Cancer Report; IACR Press: Lyon, France, 2003.

11. Iannazzo, D.; Espro, C.; Celesti, C.; Ferlazzo, A.; Neri, G. Smart biosensors for cancer diagnosis based on graphene quantum dots. *Cancers* **2021**, *13*, 3194. [CrossRef]
12. Mahato, K.; Kumar, A.; Maurya, P.K.; Chandra, P. Shifting paradigm of cancer diagnoses in clinically relevant samples based on miniaturized electrochemical nanobiosensors and microfluidic devices. *Biosens. Bioelectron.* **2018**, *100*, 411–428. [CrossRef]
13. Chen, W.; Cai, S.; Ren, Q.-Q.; Wen, W.; Zhao, Y.-D. Recent advances in electrochemical sensing for hydrogen peroxide: A review. *Analyst* **2012**, *137*, 49–58. [CrossRef] [PubMed]
14. Wang, X.; Martindale, J.L.; Liu, Y.; Holbrook, N.J. The cellular response to oxidative stress: Influences of mitogen-activated protein kinase signalling pathways on cell survival. *Biochem. J.* **1998**, *333*, 291–300. [CrossRef] [PubMed]
15. Schreck, R.; Rieber, P.; Baeuerle, P.A. Reactive oxygen intermediates as apparently widely used messengers in the activation of the NF-kappa B transcription factor and HIV-1. *EMBO J.* **1991**, *10*, 2247–2258. [CrossRef] [PubMed]
16. Abe, J.-I.; Berk, B.C. Fyn and JAK2 mediate Ras activation by reactive oxygen species. *J. Biol. Chem.* **1999**, *274*, 21003–21010. [CrossRef] [PubMed]
17. Elias, H.; Vayssié, S. Reactive peroxo compounds generated in situ from hydrogen peroxide: Kinetics and catalytic application in oxidation processes. *Peroxide Chem. Mech. Prep. Asp. Oxyg. Transf.* **2000**, 128–138. [CrossRef]
18. Imlay, J.A.; Linn, S. Mutagenesis and stress responses induced in Escherichia coli by hydrogen peroxide. *J. Bacteriol.* **1987**, *169*, 2967–2976. [CrossRef] [PubMed]
19. Mittal, M.; Siddiqui, M.R.; Tran, K.; Reddy, S.P.; Malik, A.B. Reactive oxygen species in inflammation and tissue injury. *Antioxid. Redox Signal.* **2014**, *20*, 1126–1167. [CrossRef] [PubMed]
20. Sen, S.; Chakraborty, R.; Sridhar, C.; Reddy, Y.; De, B. Free radicals, antioxidants, diseases and phytomedicines: Current status and future prospect. *Int. J. Pharm. Sci. Rev. Res.* **2010**, *3*, 91–100.
21. Nogueira, V.; Hay, N. Molecular pathways: Reactive oxygen species homeostasis in cancer cells and implications for cancer therapy. *Clin. Cancer Res.* **2013**, *19*, 4309–4314. [CrossRef]
22. Martinez-Outschoorn, U.E.; Balliet, R.M.; Lin, Z.; Whitaker-Menezes, D.; Howell, A.; Sotgia, F.; Lisanti, M.P. Hereditary ovarian cancer and two-compartment tumor metabolism: Epithelial loss of BRCA1 induces hydrogen peroxide production, driving oxidative stress and NFκB activation in the tumor stroma. *Cell Cycle* **2012**, *11*, 4152–4166. [CrossRef]
23. Brigelius-Flohe, R.; Kipp, A. Glutathione peroxidases in different stages of carcinogenesis. *Biochim. Biophys. Acta (BBA)-Gen. Subj.* **2009**, *1790*, 1555–1568. [CrossRef] [PubMed]
24. Weinstain, R.; Savariar, E.N.; Felsen, C.N.; Tsien, R.Y. In vivo targeting of hydrogen peroxide by activatable cell-penetrating peptides. *J. Am. Chem. Soc.* **2014**, *136*, 874–877. [CrossRef]
25. Zhu, L.; Zhang, Y.; Xu, P.; Wen, W.; Li, X.; Xu, J. PtW/MoS_2 hybrid nanocomposite for electrochemical sensing of H_2O_2 released from living cells. *Biosens. Bioelectron.* **2016**, *80*, 601–606. [CrossRef] [PubMed]
26. Xi, J.; Xie, C.; Zhang, Y.; Wang, L.; Xiao, J.; Duan, X.; Ren, J.; Xiao, F.; Wang, S. Pd nanoparticles decorated N-doped graphene quantum dots@ N-doped carbon hollow nanospheres with high electrochemical sensing performance in cancer detection. *ACS Appl. Mater. Interfaces* **2016**, *8*, 22563–22573. [CrossRef] [PubMed]
27. Yu, C.; Wang, L.; Li, W.; Zhu, C.; Bao, N.; Gu, H. Detection of cellular H_2O_2 in living cells based on horseradish peroxidase at the interface of Au nanoparticles decorated graphene oxide. *Sens. Actuators B Chem.* **2015**, *211*, 17–24. [CrossRef]
28. Razmi, H.; Mohammad-Rezaei, R.; Heidari, H. Self-assembled Prussian blue nanoparticles based electrochemical sensor for high sensitive determination of H_2O_2 in acidic media. *Electroanalysis* **2009**, *21*, 2355–2362. [CrossRef]
29. Liu, Y.; Wang, D.; Xu, L.; Hou, H.; You, T. A novel and simple route to prepare a Pt nanoparticle-loaded carbon nanofiber electrode for hydrogen peroxide sensing. *Biosens. Bioelectron.* **2011**, *26*, 4585–4590. [CrossRef]
30. Liu, C.-J.; Yu, S.-L.; Liu, Y.-P.; Dai, X.-J.; Wu, Y.; Li, R.-J.; Tao, J.-C. Synthesis, cytotoxic activity evaluation and HQSAR study of novel isosteviol derivatives as potential anticancer agents. *Eur. J. Med. Chem.* **2016**, *115*, 26–40. [CrossRef]
31. Cardoso, A.R.; Moreira, F.T.; Fernandes, R.; Sales, M.G.F. Novel and simple electrochemical biosensor monitoring attomolar levels of miRNA-155 in breast cancer. *Biosens. Bioelectron.* **2016**, *80*, 621–630. [CrossRef] [PubMed]
32. Yahalom, G.; Weiss, D.; Novikov, I.; Bevers, T.B.; Radvanyi, L.G.; Liu, M.; Piura, B.; Iacobelli, S.; Sandri, M.T.; Cassano, E. An antibody-based blood test utilizing a panel of biomarkers as a new method for improved breast cancer diagnosis. *Biomark. Cancer* **2013**, *5*, 71–80. [CrossRef] [PubMed]
33. Yang, C.; Denno, M.E.; Pyakurel, P.; Venton, B.J. Recent trends in carbon nanomaterial-based electrochemical sensors for biomolecules: A review. *Anal. Chim. Acta* **2015**, *887*, 17–37. [CrossRef]
34. Wang, Z.; Dai, Z. Carbon nanomaterial-based electrochemical biosensors: An overview. *Nanoscale* **2015**, *7*, 6420–6431. [CrossRef] [PubMed]
35. Vashist, S.K.; Luong, J.H. Recent advances in electrochemical biosensing schemes using graphene and graphene-based nanocomposites. *Carbon* **2015**, *84*, 519–550. [CrossRef]
36. Lawal, A.T. Synthesis and utilisation of graphene for fabrication of electrochemical sensors. *Talanta* **2015**, *131*, 424–443. [CrossRef] [PubMed]
37. Kuila, T.; Bose, S.; Khanra, P.; Mishra, A.K.; Kim, N.H.; Lee, J.H. Recent advances in graphene-based biosensors. *Biosens. Bioelectron.* **2011**, *26*, 4637–4648. [CrossRef]
38. Liu, Y.; Dong, X.; Chen, P. Biological and chemical sensors based on graphene materials. *Chem. Soc. Rev.* **2012**, *41*, 2283–2307. [CrossRef]

39. Wu, S.; He, Q.; Tan, C.; Wang, Y.; Zhang, H. Graphene-based electrochemical sensors. *Small* **2013**, *9*, 1160–1172. [CrossRef]
40. Ping, J.; Wang, Y.; Fan, K.; Wu, J.; Ying, Y. Direct electrochemical reduction of graphene oxide on ionic liquid doped screen-printed electrode and its electrochemical biosensing application. *Biosens. Bioelectron.* **2011**, *28*, 204–209. [CrossRef] [PubMed]
41. Zhang, R.; Chen, W. Recent advances in graphene-based nanomaterials for fabricating electrochemical hydrogen peroxide sensors. *Biosens. Bioelectron.* **2017**, *89*, 249–268. [CrossRef]
42. Pohanka, M.; Skládal, P. Electrochemical biosensors—Principles and applications. *J. Appl. Biomed.* **2008**, *6*, 57–64. [CrossRef]
43. Vigneshvar, S.; Sudhakumari, C.; Senthilkumaran, B.; Prakash, H. Recent advances in biosensor technology for potential applications–An overview. *Front. Bioeng. Biotechnol.* **2016**, *4*, 11. [CrossRef]
44. Lazcka, O.; del Campo, F.J.; Munoz, F.X. Pathogen detection: A perspective of traditional methods and biosensors. *Biosens. Bioelectron.* **2007**, *22*, 1205–1217. [CrossRef] [PubMed]
45. He, F. Development of Capillary-Driven Microfludic Biosensors for Food Safety and Quality Assurance. Ph.D. Thesis, University of Massachusetts Amherst, Amherst, MA, USA, 2014.
46. Zhu, C.; Yang, G.; Li, H.; Du, D.; Lin, Y. Electrochemical sensors and biosensors based on nanomaterials and nanostructures. *Anal. Chem.* **2015**, *87*, 230–249. [CrossRef]
47. Pundir, C.S.; Deswal, R.; Narwal, V. Quantitative analysis of hydrogen peroxide with special emphasis on biosensors. *Bioprocess Biosyst. Eng.* **2018**, *41*, 313–329. [CrossRef]
48. Yunus, S.; Jonas, A.M.; Lakard, B. Potentiometric biosensors. In *Encyclopedia of Biophysics*; Roberts, G.C.K., Ed.; Springer: Berlin/Heidelberg, Germany, 2013.
49. Parrilla, M.; Cánovas, R.; Andrade, F.J. Enhanced potentiometric detection of hydrogen peroxide using a platinum electrode coated with nafion. *Electroanalysis* **2017**, *29*, 223–230. [CrossRef]
50. Zheng, X.; Guo, Z. Potentiometric determination of hydrogen peroxide at MnO_2-doped carbon paste electrode. *Talanta* **2000**, *50*, 1157–1162. [CrossRef]
51. Zhao, J.; Yan, Y.; Zhu, L.; Li, X.; Li, G. An amperometric biosensor for the detection of hydrogen peroxide released from human breast cancer cells. *Biosens. Bioelectron.* **2013**, *41*, 815–819. [CrossRef]
52. Li, J.; Tan, S.N.; Ge, H. Silica sol-gel immobilized amperometric biosensor for hydrogen peroxide. *Anal. Chim. Acta* **1996**, *335*, 137–145. [CrossRef]
53. Wang, B.; Dong, S. Sol–gel-derived amperometric biosensor for hydrogen peroxide based on methylene green incorporated in Nafion film. *Talanta* **2000**, *51*, 565–572. [CrossRef]
54. Tripathi, V.S.; Kandimalla, V.B.; Ju, H. Amperometric biosensor for hydrogen peroxide based on ferrocene-bovine serum albumin and multiwall carbon nanotube modified ormosil composite. *Biosens. Bioelectron.* **2006**, *21*, 1529–1535. [CrossRef]
55. Yu, J.; Ju, H. Amperometric biosensor for hydrogen peroxide based on hemoglobin entrapped in titania sol–gel film. *Anal. Chim. Acta* **2003**, *486*, 209–216. [CrossRef]
56. Povedano, E.; Montiel, V.R.-V.; Gamella, M.; Serafín, V.; Pedrero, M.; Moranova, L.; Bartosik, M.; Montoya, J.J.; Yáñez-Sedeño, P.; Campuzano, S. A novel zinc finger protein–based amperometric biosensor for miRNA determination. *Anal. Bioanal. Chem.* **2019**, *412*, 5031–5041. [CrossRef] [PubMed]
57. Li, Y.; Huan, K.; Deng, D.; Tang, L.; Wang, J.; Luo, L. Facile synthesis of $ZnMn_2O_4$@ rGO microspheres for ultrasensitive electrochemical detection of hydrogen peroxide from human breast cancer cells. *ACS Appl. Mater. Interfaces* **2019**, *12*, 3430–3437. [CrossRef] [PubMed]
58. Dong, W.; Ren, Y.; Bai, Z.; Yang, Y.; Chen, Q. Fabrication of hexahedral Au-Pd/graphene nanocomposites biosensor and its application in cancer cell H_2O_2 detection. *Bioelectrochemistry* **2019**, *128*, 274–282. [CrossRef] [PubMed]
59. Jiao, J.; Pan, M.; Liu, X.; Li, B.; Liu, J.; Chen, Q. A non-enzymatic sensor based on trimetallic nanoalloy with poly (diallyldimethylammonium chloride)-capped reduced graphene oxide for dynamic monitoring hydrogen peroxide production by cancerous cells. *Sensors* **2020**, *20*, 71. [CrossRef]
60. Shu, Y.; Zhang, L.; Cai, H.; Yang, Y.; Zeng, J.; Ma, D.; Gao, Q. Hierarchical mo2c@ MoS_2 nanorods as electrochemical sensors for highly sensitive detection of hydrogen peroxide and cancer cells. *Sens. Actuators B Chem.* **2020**, 127863. [CrossRef]
61. Thiruppathi, M.; Lin, P.-Y.; Chou, Y.-T.; Ho, H.-Y.; Wu, L.-C.; Ho, J.-A.A. Simple aminophenol-based electrochemical probes for non-enzymatic, dual amperometric detection of NADH and hydrogen peroxide. *Talanta* **2019**, *200*, 450–457. [CrossRef]
62. Maji, S.K. Plasmon-enhanced electrochemical biosensing of hydrogen peroxide from cancer cells by gold nanorods. *ACS Appl. Nano Mater.* **2019**, *2*, 7162–7169. [CrossRef]
63. Du, H.; Zhang, X.; Liu, Z.; Qu, F. A supersensitive biosensor based on MoS_2 nanosheet arrays for the real-time detection of H_2O_2 secreted from living cells. *Chem. Commun.* **2019**, *55*, 9653–9656. [CrossRef] [PubMed]
64. Li, L.; Zhang, Y.; Zhang, L.; Ge, S.; Liu, H.; Ren, N.; Yan, M.; Yu, J. Based device for colorimetric and photoelectrochemical quantification of the flux of H_2O_2 releasing from MCF-7 cancer cells. *Anal. Chem.* **2016**, *88*, 5369–5377. [CrossRef]
65. Zhang, L.-N.; Deng, H.-H.; Lin, F.-L.; Xu, X.-W.; Weng, S.-H.; Liu, A.-L.; Lin, X.-H.; Xia, X.-H.; Chen, W. In situ growth of porous platinum nanoparticles on graphene oxide for colorimetric detection of cancer cells. *Anal. Chem.* **2014**, *86*, 2711–2718. [CrossRef]
66. Ge, S.; Liu, W.; Liu, H.; Liu, F.; Yu, J.; Yan, M.; Huang, J. Colorimetric detection of the flux of hydrogen peroxide released from living cells based on the high peroxidase-like catalytic performance of porous PtPd nanorods. *Biosens. Bioelectron.* **2015**, *71*, 456–462. [CrossRef] [PubMed]

67. Ye, X.; Shi, H.; He, X.; Wang, K.; He, D.; Yan, L.A.; Xu, F.; Lei, Y.; Tang, J.; Yu, Y. Iodide-responsive Cu–Au nanoparticle-based colorimetric platform for ultrasensitive detection of target cancer cells. *Anal. Chem.* **2015**, *87*, 7141–7147. [CrossRef] [PubMed]
68. Kim, J.-H.; Patra, C.R.; Arkalgud, J.R.; Boghossian, A.A.; Zhang, J.; Han, J.-H.; Reuel, N.F.; Ahn, J.-H.; Mukhopadhyay, D.; Strano, M.S. Single-molecule detection of H_2O_2 mediating angiogenic redox signaling on fluorescent single-walled carbon nanotube array. *ACS Nano* **2011**, *5*, 7848–7857. [CrossRef] [PubMed]
69. Maji, S.K.; Sreejith, S.; Mandal, A.K.; Ma, X.; Zhao, Y. Immobilizing gold nanoparticles in mesoporous silica covered reduced graphene oxide: A hybrid material for cancer cell detection through hydrogen peroxide sensing. *ACS Appl. Mater. Interfaces* **2014**, *6*, 13648–13656. [CrossRef] [PubMed]
70. Wang, Y.; Tang, L.; Li, Z.; Lin, Y.; Li, J. In situ simultaneous monitoring of ATP and GTP using a graphene oxide nanosheet–based sensing platform in living cells. *Nat. Protoc.* **2014**, *9*, 1944. [CrossRef]
71. McKibbin, P.L.; Kobori, A.; Taniguchi, Y.; Kool, E.T.; David, S.S. Surprising repair activities of nonpolar analogs of 8-oxoG expose features of recognition and catalysis by base excision repair glycosylases. *J. Am. Chem. Soc.* **2012**, *134*, 1653–1661. [CrossRef] [PubMed]
72. Li, L.; Lin, H.; Lei, C.; Nie, Z.; Huang, Y.; Yao, S. Label-free fluorescence assay for thrombin based on unmodified quantum dots. *Biosens. Bioelectron.* **2014**, *54*, 42–47. [CrossRef] [PubMed]
73. Ren, D.; Wong, N.T.; Handoko, A.D.; Huang, Y.; Yeo, B.S. Mechanistic insights into the enhanced activity and stability of agglomerated Cu nanocrystals for the electrochemical reduction of carbon dioxide to n-propanol. *J. Phys. Chem. Lett.* **2016**, *7*, 20–24. [CrossRef]
74. Zhou, Y.; Zhang, Y.; Lau, C.; Lu, J. Sequential determination of two proteins by temperature-triggered homogeneous chemiluminescent immunoassay. *Anal. Chem.* **2006**, *78*, 5920–5924. [CrossRef] [PubMed]
75. Liu, M.; Lin, Z.; Lin, J.-M. A review on applications of chemiluminescence detection in food analysis. *Anal. Chim. Acta* **2010**, *670*, 1–10. [CrossRef]
76. Kong, H.; Liu, D.; Zhang, S.; Zhang, X. Protein sensing and cell discrimination using a sensor array based on nanomaterial-assisted chemiluminescence. *Anal. Chem.* **2011**, *83*, 1867–1870. [CrossRef] [PubMed]
77. Ji, D.; Mohsen, M.G.; Harcourt, E.M.; Kool, E.T. ATP-Releasing Nucleotides: Linking DNA Synthesis to Luciferase Signaling. *Angew. Chem. Int. Ed.* **2016**, *55*, 2087–2091. [CrossRef]
78. Liu, B.-F.; Ozaki, M.; Hisamoto, H.; Luo, Q.; Utsumi, Y.; Hattori, T.; Terabe, S. Microfluidic chip toward cellular ATP and ATP-conjugated metabolic analysis with bioluminescence detection. *Anal. Chem.* **2005**, *77*, 573–578. [CrossRef] [PubMed]
79. Ju, J.; Chen, W. In situ growth of surfactant-free gold nanoparticles on nitrogen-doped graphene quantum dots for electrochemical detection of hydrogen peroxide in biological environments. *Anal. Chem.* **2015**, *87*, 1903–1910. [CrossRef] [PubMed]
80. Bai, J.; Jiang, X. A facile one-pot synthesis of copper sulfide-decorated reduced graphene oxide composites for enhanced detecting of H_2O_2 in biological environments. *Anal. Chem.* **2013**, *85*, 8095–8101. [CrossRef]
81. Lu, Y.; Liu, Y.; Zhang, S.; Wang, S.; Zhang, S.; Zhang, X. Aptamer-Based Plasmonic Sensor Array for Discrimination of Proteins and Cells with the Naked Eye. *Anal. Chem.* **2013**, *85*, 6571–6574. [CrossRef]
82. Gong, Y.; Chen, X.; Lu, Y.; Yang, W. Self-assembled dipeptide–gold nanoparticle hybrid spheres for highly sensitive amperometric hydrogen peroxide biosensors. *Biosens. Bioelectron.* **2015**, *66*, 392–398. [CrossRef] [PubMed]
83. Wang, Y.; Gu, H. Core-Shell-Type Magnetic Mesoporous Silica Nanocomposites for Bioimaging and Therapeutic Agent Delivery. *Adv. Mater.* **2015**, *27*, 576–585. [CrossRef]
84. Yang, J.; Shen, D.; Zhou, L.; Li, W.; Li, X.; Yao, C.; Wang, R.; El-Toni, A.M.; Zhang, F.; Zhao, D. Spatially confined fabrication of core–shell gold nanocages@mesoporous silica for near-infrared controlled photothermal drug release. *Chem. Mater.* **2013**, *25*, 3030–3037. [CrossRef]
85. Lee, D.; Khaja, S.D.; Velasquez-Castano, J.C.; Dasari, M.; Sun, C.; A Petros, J.; Taylor, W.R.; Murthy, N. In vivo imaging of hydrogen peroxide with chemiluminescent nanoparticles. *Nat. Mater.* **2007**, *6*, 765–769. [CrossRef] [PubMed]
86. Arnous, A.; Petrakis, C.; Makris, D.P.; Kefalas, P. A peroxyoxalate chemiluminescence-based assay for the evaluation of hydrogen peroxide scavenging activity employing 9,10-diphenylanthracene as the fluorophore. *J. Pharmacol. Toxicol. Methods* **2002**, *48*, 171–177. [CrossRef]
87. Koike, R.; Kato, Y.; Motoyoshiya, J.; Nishii, Y.; Aoyama, H. Unprecedented chemiluminescence behaviour during peroxyoxalate chemiluminescence of oxalates with fluorescent or electron-donating aryloxy groups. *Luminescence* **2006**, *21*, 164–173. [CrossRef]
88. Stevani, C.V.; Silva, S.M.; Baader, W.J. Studies on the Mechanism of the Excitation Step in Peroxyoxalate Chemiluminescence. *Eur. J. Org. Chem.* **2000**, *2000*, 4037–4046. [CrossRef]
89. Matsumoto, M. Advanced chemistry of dioxetane-based chemiluminescent substrates originating from bioluminescence. *J. Photochem. Photobiol. C* **2004**, *5*, 27–53. [CrossRef]
90. Lee, D.; Dasari, M.; Erigala, V.; Murthy, N.; Yu, J.; Dickson, R. Detection of hydrogen peroxide with chemiluminescent micelles. *Int. J. Nanomed.* **2008**, *3*, 471–476. [CrossRef]
91. Weissleder, R.; Pittet, M.J. Imaging in the era of molecular oncology. *Nature* **2008**, *452*, 580–589. [CrossRef]
92. Zhang, K.; Kaufman, R.J. From endoplasmic-reticulum stress to the inflammatory response. *Nature* **2008**, *454*, 455–462. [CrossRef] [PubMed]
93. Peer, D.; Karp, J.M.; Hong, S.; Farokhzad, O.C.; Margalit, R.; Langer, R. Nanocarriers as an emerging platform for cancer therapy. *Nat. Nanotechnol.* **2007**, *2*, 751–760. [CrossRef] [PubMed]

94. Nan, Y.; Zhao, W.; Li, N.; Liang, Z.; Xu, X. Chemiluminescence-triggered fluorophore release: Approach for in vivo fluorescence imaging of hydrogen peroxide. *Sens. Actuators B Chem.* **2019**, *281*, 296–302. [CrossRef]
95. Lee, Y.-D.; Lim, C.-K.; Singh, A.; Koh, J.; Kim, J.; Kwon, I.C.; Kim, S. Dye/Peroxalate Aggregated Nanoparticles with Enhanced and Tunable Chemiluminescence for Biomedical Imaging of Hydrogen Peroxide. *ACS Nano* **2012**, *6*, 6759–6766. [CrossRef] [PubMed]
96. Lee, E.S.; Deepagan, V.G.; Gil You, D.; Jeon, J.; Yi, G.-R.; Lee, J.Y.; Lee, D.S.; Suh, Y.D.; Park, J.H. Nanoparticles based on quantum dots and a luminol derivative: Implications for in vivo imaging of hydrogen peroxide by chemiluminescence resonance energy transfer. *Chem. Commun.* **2016**, *52*, 4132–4135. [CrossRef] [PubMed]
97. Geng, J.; Li, K.; Qin, W.; Tang, B.Z.; Liu, B. Red-Emissive Chemiluminescent Nanoparticles with Aggregation-Induced Emission Characteristics for In Vivo Hydrogen Peroxide Imaging. *Part. Part. Syst. Charact.* **2014**, *31*, 1238–1243. [CrossRef]
98. Jia, Y.; Sun, S.; Cui, X.; Wang, X.; Yang, L. Enzyme-like catalysis of polyoxometalates for chemiluminescence: Application in ultrasensitive detection of H_2O_2 and blood glucose. *Talanta* **2019**, *205*, 120139. [CrossRef] [PubMed]
99. Klassen, N.V.; Marchington, D.; McGowan, H.C. H_2O_2 Determination by the I_3^- Method and by $KMnO_4$ Titration. *Anal. Chem.* **1994**, *66*, 2921–2925. [CrossRef]
100. Kieber, R.J.; Helz, G.R. Two-method verification of hydrogen peroxide determinations in natural waters. *Anal. Chem.* **1986**, *58*, 2312–2315. [CrossRef]
101. Putt, K.S.; Pugh, R.B. A High-throughput microtiter plate based method for the determination of peracetic acid and hydrogen peroxide. *PLoS ONE* **2013**, *8*, e79218. [CrossRef] [PubMed]
102. Zaribafan, A.; Haghbeen, K.; Fazli, M.; Akhondali, A. Spectrophotometric method for hydrogen peroxide determination through oxidation of organic dyes. *Environ. Stud. Persian Gulf* **2014**, *1*, 93–101.
103. Matsubara, C.; Kudo, K.; Kawashita, T.; Takamura, K. Spectrophotometric determination of hydrogen peroxide with titanium 2-((5-bromopyridyl)azo)-5-(N-propyl-N-sulfopropylamino)phenol reagent and its application to the determination of serum glucose using glucose oxidase. *Anal. Chem.* **1985**, *57*, 1107–1109. [CrossRef]
104. Clapp, P.A.; Evans, D.F.; Sheriff, T.S. Spectrophotometric determination of hydrogen peroxide after extraction with ethyl acetate. *Anal. Chim. Acta* **1989**, *218*, 331–334. [CrossRef]
105. Mukhopadhyay, D.; Dasgupta, P.; Roy, D.S.; Palchoudhuri, S.; Chatterjee, I.; Ali, S.; Dastidar, S.G. A Sensitive In vitro Spectrophotometric Hydrogen Peroxide Scavenging Assay using 1,10-Phenanthroline. *Free Radic. Antioxid.* **2016**, *6*, 124–132. [CrossRef]
106. Elnemma, E.M. Spectrophotometric Determination of Hydrogen Peroxide by a Hydroquinone-Aniline System Catalyzed by Molybdate. *Bull. Korean Chem. Soc.* **2004**, *25*, 127–129.
107. Zhang, L.-S.; Wong, G.T. Spectrophotometric determination of H_2O_2 in marine waters with leuco crystal violet. *Talanta* **1994**, *41*, 2137–2145. [CrossRef]
108. Huang, Y.; Cai, R.; Mao, L.; LIU, Z.; HUANG, H. Spectrophotometric determination of hydrogen peroxide using β-CD-Hemin as a mimetic enzyme of peroxidase. *Anal. Sci.* **1999**, *15*, 889–894. [CrossRef]
109. Zhang, Q.; Fu, S.; Li, H.; Liu, Y. A novel method for the determination of hydrogen peroxide in bleaching effluents by spectroscopy. *BioResources* **2013**, *8*, 3699–3705. [CrossRef]
110. Eisenberg, G. Industrial and engineering chemistry. *Ind. Eng. Chem. Anal. Ed.* **1943**, *15*, 327–328. [CrossRef]
111. Graf, E.; Penniston, J.T. Method for determination of hydrogen peroxide, with its application illustrated by glucose assay. *Clin. Chem.* **1980**, *26*, 658–660. [CrossRef]
112. Pick, E.; Keisari, Y. A simple colorimetric method for the measurement of H_2O_2 produced by cells in culture *J. Immunol. Methods* **1980**, *38*, 161–170. [CrossRef]
113. Fernando, C.D.; Soysa, P. Optimized enzymatic colorimetric assay for determination of hydrogen peroxide (H_2O_2) scavenging activity of plant extracts. *MethodsX* **2015**, *2*, 283–291. [CrossRef] [PubMed]
114. Su, G.; Wei, Y.; Guo, M. Direct Colorimetric Detection of Hydrogen Peroxide Using 4-Nitrophenyl Boronic Acid or Its Pinacol Ester. *Am. J. Anal. Chem.* **2011**, *2*, 879–884. [CrossRef]
115. Nitinaivinij, K.; Parnklang, T.; Thammacharoen, C.; Ekgasit, S.; Wongravee, K. Colorimetric determination of hydrogen peroxide by morphological decomposition of silver nanoprisms coupled with chromaticity analysis. *Anal. Methods* **2014**, *6*, 9816–9824. [CrossRef]
116. Takahashi, A.; Hashimoto, K.; Kumazawa, S.; Nakayama, T. Determination of Hydrogen Peroxide by High-Performance Liquid Chromatography with a Cation-Exchange Resin Gel Column and Electrochemical Detector. *Anal. Sci.* **1999**, *15*, 481–483. [CrossRef]
117. Wada, M.; Inoue, K.; Ihara, A.; Kishikawa, N.; Nakashima, K.; Kuroda, N. Determination of organic peroxides by liquid chromatography with on-line post-column ultraviolet irradiation and peroxyoxalate chemiluminescence detection. *J. Chromatogr. A* **2003**, *987*, 189–195. [CrossRef]
118. Nepomnyashchikh, Y.V.; Borkina, G.G.; Karavaeva, A.V.; Perkel', A.L. Photometric and Gas-Chromatographic Determination of Hydrogen Peroxide and Peroxybutanoic Acid in Oxidized Butanoic Acid. *J. Anal. Chem.* **2005**, *60*, 1024–1028. [CrossRef]
119. Magara, K.; Ikeda, T.; Sugimoto, T.; Hosoya, S. Quantitative Analysis of Hydrogen Peroxide by High Performance Liquid Chromatography. *Jpn. TAPPI J.* **2007**, *61*, 1481–1493. [CrossRef]
120. Tarno, H.; Qi, H.; Endoh, R.; Kobayashi, M.; Goto, H.; Futai, K. Types of frass produced by the ambrosia beetle Platypus quercivorus during gallery construction, and host suitability of five tree species for the beetle. *J. For. Res.* **2011**, *16*, 68–75. [CrossRef]

121. Wielandt, H. On the eigenvalues of A + B and AB. *J. Res. Natl. Bur. Stand. Sect. B Math. Sci.* **1973**, *77B*, 61. [CrossRef]
122. Xu, K.; Tang, B.; Huang, H.; Yang, G.; Chen, Z.; Li, P.; An, L. Strong red fluorescent probes suitable for detecting hydrogen peroxide generated by mice peritoneal macrophages. *Chem. Commun.* **2005**, *48*, 5974–5976. [CrossRef]
123. Paździoch-Czochra, M.; Wideńska, A. Spectrofluorimetric determination of hydrogen peroxide scavenging activity. *Anal. Chim. Acta* **2002**, *452*, 177–184. [CrossRef]
124. Miller, E.W.; Albers, A.E.; Pralle, A.; Isacoff, E.Y.; Chang, C.J. Boronate-Based Fluorescent Probes for Imaging Cellular Hydrogen Peroxide. *J. Am. Chem. Soc.* **2005**, *127*, 16652–16659. [CrossRef]
125. Qian, P.; Qin, Y.; Lyu, Y.; Li, Y.; Wang, L.; Wang, S.; Liu, Y. A hierarchical cobalt/carbon nanotube hybrid nanocomplex-based ratiometric fluorescent nanosensor for ultrasensitive detection of hydrogen peroxide and glucose in human serum. *Anal. Bioanal. Chem.* **2019**, *411*, 1517–1524. [CrossRef] [PubMed]
126. Onoda, M.; Uchiyama, T.; Mawatari, K.-I.; Kaneko, K.; Nakagomi, K. Simple and Rapid Determination of Hydrogen Peroxide Using Phosphine-based Fluorescent Reagents with Sodium Tungstate Dihydrate. *Anal. Sci.* **2006**, *22*, 815–817. [CrossRef] [PubMed]
127. Lyublinskaya, O.; Antunes, F. Measuring intracellular concentration of hydrogen peroxide with the use of genetically encoded H_2O_2 biosensor HyPer. *Redox Biol.* **2019**, *24*, 101200. [CrossRef]
128. Belousov, V.V.; Fradkov, A.F.; Lukyanov, K.; Staroverov, D.; Shakhbazov, K.S.; Terskikh, A.V.; Lukyanov, S. Genetically encoded fluorescent indicator for intracellular hydrogen peroxide. *Nat. Methods* **2006**, *3*, 281–286. [CrossRef]
129. Zheng, M.; Åslund, F.; Storz, G. Activation of the OxyR Transcription Factor by Reversible Disulfide Bond Formation. *Science* **1998**, *279*, 1718–1722. [CrossRef]
130. Markvicheva, K.N.; Bilan, D.; Mishina, N.; Gorokhovatsky, A.Y.; Vinokurov, L.M.; Lukyanov, S.; Belousov, V.V. A genetically encoded sensor for H_2O_2 with expanded dynamic range. *Bioorganic Med. Chem.* **2011**, *19*, 1079–1084. [CrossRef]
131. Bilan, D.S.; Pase, L.; Joosen, L.; Gorokhovatsky, A.Y.; Ermakova, Y.G.; Gadella, T.W.J.; Grabher, C.; Schultz, C.; Lukyanov, S.; Belousov, V.V. HyPer-3: A Genetically Encoded H_2O_2 Probe with Improved Performance for Ratiometric and Fluorescence Lifetime Imaging. *ACS Chem. Biol.* **2013**, *8*, 535–542. [CrossRef]
132. Pak, V.V.; Ezerina, D.; Lyublinskaya, O.; Pedre, B.; Tyurin-Kuzmin, P.A.; Mishina, N.M.; Thauvin, M.; Young, D.; Wahni, K.; Gache, S.A.M.; et al. Ultrasensitive Genetically Encoded Indicator for Hydrogen Peroxide Identifies Roles for the Oxidant in Cell Migration and Mitochondrial Function. *Cell Metab.* **2020**, *31*, 642–653.e6. [CrossRef] [PubMed]
133. Ermakova, Y.; Bilan, D.; Matlashov, M.; Mishina, N.; Markvicheva, K.N.; Subach, O.M.; Subach, F.V.; Bogeski, I.; Hoth, M.; Enikolopov, G.; et al. Red fluorescent genetically encoded indicator for intracellular hydrogen peroxide. *Nat. Commun.* **2014**, *5*, 5222. [CrossRef]
134. Choi, H.-J.; Kim, S.-J.; Mukhopadhyay, P.; Cho, S.; Woo, J.-R.; Storz, G.; Ryu, S.-E. Structural Basis of the Redox Switch in the OxyR Transcription Factor. *Cell* **2001**, *105*, 103–113. [CrossRef]
135. Xu, J.; Zhang, Y.; Yu, H.; Gao, X.; Shao, S. Mitochondria-Targeted Fluorescent Probe for Imaging Hydrogen Peroxide in Living Cells. *Anal. Chem.* **2015**, *88*, 1455–1461. [CrossRef] [PubMed]
136. Xiao, H.; Li, P.; Hu, X.; Shi, X.; Zhang, W.; Tang, B. Simultaneous fluorescence imaging of hydrogen peroxide in mitochondria and endoplasmic reticulum during apoptosis. *Chem. Sci.* **2016**, *7*, 6153–6159. [CrossRef] [PubMed]
137. Shen, R.; Liu, P.; Zhang, Y.; Yu, Z.; Chen, X.; Zhou, L.; Nie, B.; Żaczek, A.; Chen, J.; Liu, J. Sensitive Detection of Single-Cell Secreted H_2O_2 by Integrating a Microfluidic Droplet Sensor and Au Nanoclusters. *Anal. Chem.* **2018**, *90*, 4478–4484. [CrossRef] [PubMed]
138. Chen, Y.; Ye, J.; Lv, G.; Liu, W.; Jiang, H.; Liu, X.; Wang, X. Hydrogen Peroxide and Hypochlorite Responsive Fluorescent Nanoprobes for Sensitive Cancer Cell Imaging. *Biosensors* **2022**, *12*, 111. [CrossRef]
139. Hu, L.; Yuan, Y.; Zhang, L.; Zhao, J.; Majeed, S.; Xu, G. Copper nanoclusters as peroxidase mimetics and their applications to H_2O_2 and glucose detection. *Anal. Chim. Acta* **2013**, *762*, 83–86. [CrossRef]
140. Liu, H.; Gu, C.; Xiong, W.; Zhang, M. A sensitive hydrogen peroxide biosensor using ultra-small $CuInS_2$ nanocrystals as peroxidase mimics. *Sens. Actuators B Chem.* **2015**, *209*, 670–676. [CrossRef]
141. Su, L.; Qin, W.; Zhang, H.; Rahman, Z.U.; Ren, C.; Ma, S.; Chen, X. The peroxidase/catalase-like activities of MFe_2O_4 (M = Mg, Ni, Cu) MNPs and their application in colorimetric biosensing of glucose. *Biosens. Bioelectron.* **2015**, *63*, 384–391. [CrossRef]
142. Regalado, C.; García-Almendárez, B.E.; Duarte-Vázquez, M.A. Biotechnological applications of peroxidases. *Phytochem. Rev.* **2004**, *3*, 243–256. [CrossRef]
143. Hamid, M. Potential applications of peroxidases. *Food Chem.* **2009**, *115*, 1177–1186. [CrossRef]
144. Chekin, F.; Gorton, L.; Tapsobea, I. Direct and mediated electrochemistry of peroxidase and its electrocatalysis on a variety of screen-printed carbon electrodes: Amperometric hydrogen peroxide and phenols biosensor. *Anal. Bioanal. Chem.* **2014**, *407*, 439–446. [CrossRef]
145. Yang, H.; Liu, B.; Ding, Y.; Li, L.; Ouyang, X. Fabrication of cuprous oxide nanoparticles-graphene nanocomposite for determination of acetaminophen. *J. Electroanal. Chem.* **2015**, *757*, 88–93. [CrossRef]
146. Chulkova, I.; Derina, K.; Taishibekova, Y. The modified electrode for the determination of cholesterol. In Proceedings of the Chemistry and Chemical Technology in the XXI Century: Materials of the XVI International Scientific-Practical Conference of Students and Young Scientists Dedicated to the 115th Anniversary of Professor L.P. Kuleva, Tomsk, Russia, 25–29 May 2015; pp. 195–197.

147. Jelikić-Stankov, M.D.; Djurdjevic, P.; Stankov, D. Determination of uric acid in human serum by an enzymatic method using N-methyl-N-(4-aminophenyl)-3-methoxyaniline reagent. *J. Serb. Chem. Soc.* **2003**, *68*, 691–698. [CrossRef]
148. Zhou, B.; Wang, J.; Guo, Z.; Tan, H.; Zhu, X. A simple colorimetric method for determination of hydrogen peroxide in plant tissues. *Plant Growth Regul.* **2006**, *49*, 113–118. [CrossRef]
149. Chinnadayyala, S.R.; Kakoti, A.; Santhosh, M.; Goswami, P. A novel amperometric alcohol biosensor developed in a 3rd generation bioelectrode platform using peroxidase coupled ferrocene activated alcohol oxidase as biorecognition system. *Biosens. Bioelectron.* **2014**, *55*, 120–126. [CrossRef]
150. Yu, F.; Huang, Y.; Cole, A.J.; Yang, V.C. The artificial peroxidase activity of magnetic iron oxide nanoparticles and its application to glucose detection. *Biomaterials* **2009**, *30*, 4716–4722. [CrossRef]
151. Mu, J.; Zhang, L.; Zhao, M.; Wang, Y. Co_3O_4 nanoparticles as an efficient catalase mimic: Properties, mechanism and its electrocatalytic sensing application for hydrogen peroxide. *J. Mol. Catal. A Chem.* **2013**, *378*, 30–37. [CrossRef]
152. Yoon, J.; Lee, T.; Bapurao G., B.; Jo, J.; Oh, B.-K.; Choi, J.-W. Electrochemical H_2O_2 biosensor composed of myoglobin on MoS_2 nanoparticle-graphene oxide hybrid structure. *Biosens. Bioelectron.* **2017**, *93*, 14–20. [CrossRef]
153. Wen, Z.; Ci, S.; Li, J. Pt Nanoparticles Inserting in Carbon Nanotube Arrays: Nanocomposites for Glucose Biosensors. *J. Phys. Chem. C* **2009**, *113*, 13482–13487. [CrossRef]
154. Pingarrón, J.M.; Yáñez-Sedeño, P.; González-Cortés, A. Gold nanoparticle-based electrochemical biosensors. *Electrochim. Acta* **2008**, *53*, 5848–5866. [CrossRef]
155. Zhou, M.; Zhai, Y.M.; Dong, S.J. Electrochemical sensing and biosensing platform based on chemically reduced graphene oxide. *Anal. Chem.* **2009**, *81*, 5603–5613. [CrossRef]
156. Xu, X.; Jiang, S.; Hu, Z.; Liu, S. Nitrogen-doped carbon nanotubes: High electrocatalytic activity toward the oxidation of hydrogen peroxide and its application for biosensing. *ACS Nano* **2010**, *4*, 4292–4298. [CrossRef]
157. Luo, Y.; Liu, H.; Rui, Q.; Tian, Y. Detection of Extracellular H_2O_2 Released from Human Liver Cancer Cells Based on TiO_2 Nanoneedles with Enhanced Electron Transfer of Cytochrome c. *Anal. Chem.* **2009**, *81*, 3035–3041. [CrossRef] [PubMed]
158. Shao, Y.; Wang, J.; Wu, H.; Liu, J.; Aksay, I.A.; Lin, Y. Graphene based electrochemical sensors and biosensors: A review. *Electroanal. Int. J. Devoted Fundam. Pract. Asp. Electroanal.* **2010**, *22*, 1027–1036. [CrossRef]
159. Kim, G.; Lee, Y.-E.K.; Xu, H.; Philbert, M.A.; Kopelman, R. Nanoencapsulation method for high selectivity sensing of hydrogen peroxide inside live cells. *Anal. Chem.* **2010**, *82*, 2165–2169. [CrossRef] [PubMed]
160. Sun, X.; Guo, S.; Liu, Y.; Sun, S. Dumbbell-like PtPd–Fe_3O_4 nanoparticles for enhanced electrochemical detection of H_2O_2. *Nano Lett.* **2012**, *12*, 4859–4863. [CrossRef] [PubMed]
161. Wang, T.; Zhu, H.; Zhuo, J.; Zhu, Z.; Papakonstantinou, P.; Lubarsky, G.; Lin, J.; Li, M. Biosensor Based on Ultrasmall MoS_2 Nanoparticles for Electrochemical Detection of H_2O_2 Released by Cells at the Nanomolar Level. *Anal. Chem.* **2013**, *85*, 10289–10295. [CrossRef] [PubMed]
162. Dou, B.; Yang, J.; Yuan, R.; Xiang, Y. Trimetallic Hybrid Nanoflower-Decorated MoS_2 Nanosheet Sensor for Direct in Situ Monitoring of H_2O_2 Secreted from Live Cancer Cells. *Anal. Chem.* **2018**, *90*, 5945–5950. [CrossRef] [PubMed]
163. Chang, H.-C.; Ho, J.-A.A. Gold nanocluster-assisted fluorescent detection for hydrogen peroxide and cholesterol based on the inner filter effect of gold nanoparticles. *Anal. Chem.* **2015**, *87*, 10362–10367. [CrossRef] [PubMed]
164. Cui, H.; Wang, W.; Duan, C.-F.; Dong, Y.-P.; Guo, J.-Z. Synthesis, characterization, and electrochemiluminescence of luminol-reduced gold nanoparticles and their application in a hydrogen peroxide sensor. *Chem.–A Eur. J.* **2007**, *13*, 6975–6984. [CrossRef] [PubMed]
165. Liu, Q.; Yang, Y.; Lv, X.; Ding, Y.; Zhang, Y.; Jing, J.; Xu, C. One-step synthesis of uniform nanoparticles of porphyrin functionalized ceria with promising peroxidase mimetics for H_2O_2 and glucose colorimetric detection. *Sens. Actuators B Chem.* **2017**, *240*, 726–734. [CrossRef]
166. Xiong, X.; You, C.; Cao, X.; Pang, L.; Kong, R.; Sun, X. Ni_2P nanosheets array as a novel electrochemical catalyst electrode for non-enzymatic H_2O_2 sensing. *Electrochim. Acta* **2017**, *253*, 517–521. [CrossRef]
167. Chen, L.; Wang, N.; Wang, X.; Ai, S. Protein-directed in situ synthesis of platinum nanoparticles with superior peroxidase-like activity, and their use for photometric determination of hydrogen peroxide. *Mikrochim. Acta* **2013**, *180*, 1517–1522. [CrossRef]
168. Ma, B.; Kong, C.; Hu, X.; Liu, K.; Huang, Q.; Lv, J.; Lu, W.; Zhang, X.; Yang, Z.; Yang, S. A sensitive electrochemical nonenzymatic biosensor for the detection of H_2O_2 released from living cells based on ultrathin concave Ag nanosheets. *Biosens. Bioelectron.* **2018**, *106*, 29–36. [CrossRef]
169. Asif, M.; Liu, H.; Aziz, A.; Wang, H.; Wang, Z.; Ajmal, M.; Xiao, F.; Liu, H. Core-shell iron oxide-layered double hydroxide: High electrochemical sensing performance of H_2O_2 biomarker in live cancer cells with plasma therapeutics. *Biosens. Bioelectron.* **2017**, *97*, 352–359. [CrossRef]
170. Li, Z.; Xin, Y.; Wu, W.; Fu, B.; Zhang, Z. Topotactic Conversion of Copper(I) Phosphide Nanowires for Sensitive Electrochemical Detection of H_2O_2 Release from Living Cells. *Anal. Chem.* **2016**, *88*, 7724–7729. [CrossRef]
171. Su, S.; Han, X.; Lu, Z.; Liu, W.; Zhu, D.; Chao, J.; Fan, C.; Wang, L.; Song, S.; Weng, L.; et al. Facile Synthesis of a MoS_2–Prussian Blue Nanocube Nanohybrid-Based Electrochemical Sensing Platform for Hydrogen Peroxide and Carcinoembryonic Antigen Detection. *ACS Appl. Mater. Interfaces* **2017**, *9*, 12773–12781. [CrossRef] [PubMed]
172. Yuan, L.; Lin, W.; Xie, Y.; Chen, B.; Zhu, S. Single fluorescent probe responds to H_2O_2, NO, and H_2O_2/NO with three different sets of fluorescence signals. *J. Am. Chem. Soc.* **2012**, *134*, 1305–1315. [CrossRef] [PubMed]

173. Boero, C.; Casulli, M.A.; Olivo, J.; Foglia, L.; Orso, E.; Mazza, M.; Carrara, S.; De Micheli, G. Design, development, and validation of an in-situ biosensor array for metabolite monitoring of cell cultures. *Biosens. Bioelectron.* **2014**, *61*, 251–259. [CrossRef]
174. Shi, B.-X.; Wang, Y.; Zhang, K.; Lam, T.-L.; Chan, H.L.-W. Monitoring of dopamine release in single cell using ultrasensitive ITO microsensors modified with carbon nanotubes. *Biosens. Bioelectron.* **2011**, *26*, 2917–2921. [CrossRef] [PubMed]
175. Li, D.-W.; Qin, L.-X.; Li, Y.; Nia, R.P.; Long, Y.-T.; Chen, H.-Y. CdSe/ZnS quantum dot–Cytochrome c bioconjugates for selective intracellular O 2–Sensing. *Chem. Commun.* **2011**, *47*, 8539–8541. [CrossRef] [PubMed]
176. Han, M.; Liu, S.; Bao, J.; Dai, Z. Pd nanoparticle assemblies—As the substitute of HRP, in their biosensing applications for H_2O_2 and glucose. *Biosens. Bioelectron.* **2012**, *31*, 151–156. [CrossRef]
177. Wang, Y.; Hasebe, Y. Carbon felt-based bioelectrocatalytic flow-through detectors: Highly sensitive amperometric determination of H_2O_2 based on a direct electrochemistry of covalently modified horseradish peroxidase using cyanuric chloride as a linking agent. *Sens. Actuators B Chem.* **2011**, *155*, 722–729. [CrossRef]
178. Wang, Z.; Yang, Y.; Leng, K.; Li, J.; Zheng, F.; Shen, G.; Yu, R. A Sequence-Selective Electrochemical DNA Biosensor Based on HRP-Labeled Probe for Colorectal Cancer DNA Detection. *Anal. Lett.* **2008**, *41*, 24–35. [CrossRef]
179. Crulhas, B.P.; Ramos, N.P.; Castro, G.R.; Pedrosa, V.A. Detection of hydrogen peroxide releasing from prostate cancer cell using a biosensor. *J. Solid State Electrochem.* **2016**, *20*, 2427–2433. [CrossRef]
180. Zhou, J.; Liao, C.; Zhang, L.; Wang, Q.; Tian, Y. Molecular Hydrogel-Stabilized Enzyme with Facilitated Electron Transfer for Determination of H_2O_2 Released from Live Cells. *Anal. Chem.* **2014**, *86*, 4395–4401. [CrossRef]
181. Goenka, S.; Sant, V.; Sant, S. Graphene-based nanomaterials for drug delivery and tissue engineering. *J. Control. Release* **2014**, *173*, 75–88. [CrossRef]
182. Pumera, M.; Ambrosi, A.; Bonanni, A.; Chng, E.L.K.; Poh, H.L. Graphene for electrochemical sensing and biosensing. *TrAC Trends Anal. Chem.* **2010**, *29*, 954–965. [CrossRef]
183. Gao, H.; Duan, H. 2D and 3D graphene materials: Preparation and bioelectrochemical applications. *Biosens. Bioelectron.* **2015**, *65*, 404–419. [CrossRef] [PubMed]
184. Favero, G.; Fusco, G.; Mazzei, F.; Tasca, F.; Antiochia, R. Electrochemical Characterization of Graphene and MWCNT Screen-Printed Electrodes Modified with AuNPs for Laccase Biosensor Development. *Nanomaterials* **2015**, *5*, 1995–2006. [CrossRef] [PubMed]
185. Song, Y.; Luo, Y.; Zhu, C.; Li, H.; Du, D.; Lin, Y. Recent advances in electrochemical biosensors based on graphene two-dimensional nanomaterials. *Biosens. Bioelectron.* **2016**, *76*, 195–212. [CrossRef] [PubMed]
186. Chen, Y.; Tan, C.; Zhang, H.; Wang, L. Two-dimensional graphene analogues for biomedical applications. *Chem. Soc. Rev.* **2015**, *44*, 2681–2701. [CrossRef]
187. Tan, S.M.; Sofer, Z.; Pumera, M. Biomarkers Detection on Hydrogenated Graphene Surfaces: Towards Applications of Graphane in Biosensing. *Electroanalysis* **2013**, *25*, 703–705. [CrossRef]
188. Yang, G.; Zhu, C.; Du, D.; Zhu, J.; Lin, Y. Graphene-like two-dimensional layered nanomaterials: Applications in biosensors and nanomedicine. *Nanoscale* **2015**, *7*, 14217–14231. [CrossRef] [PubMed]
189. Gupta, A.; Sakthivel, T.; Seal, S. Recent development in 2D materials beyond graphene. *Prog. Mater. Sci.* **2015**, *73*, 44–126. [CrossRef]
190. Lin, Y.; Connell, J.W. Advances in 2D boron nitride nanostructures: Nanosheets, nanoribbons, nanomeshes, and hybrids with graphene. *Nanoscale* **2012**, *4*, 6908–6939. [CrossRef] [PubMed]
191. Zhu, C.; Du, D.; Lin, Y. Graphene and graphene-like 2D materials for optical biosensing and bioimaging: A review. *2D Mater.* **2015**, *2*, 032004. [CrossRef]
192. McCreery, R.L. Advanced Carbon Electrode Materials for Molecular Electrochemistry. *Chem. Rev.* **2008**, *108*, 2646–2687. [CrossRef] [PubMed]
193. Takahashi, S.; Abiko, N.; Anzai, J.-I. Redox Response of Reduced Graphene Oxide-Modified Glassy Carbon Electrodes to Hydrogen Peroxide and Hydrazine. *Materials* **2013**, *6*, 1840–1850. [CrossRef] [PubMed]
194. Hamilton, C.E.; Lomeda, J.R.; Sun, Z.; Tour, J.M.; Barron, A.R. High-Yield Organic Dispersions of Unfunctionalized Graphene. *Nano Lett.* **2009**, *9*, 3460–3462. [CrossRef]
195. Lv, W.; Guo, M.; Liang, M.-H.; Jin, F.-M.; Cui, L.; Zhi, L.; Yang, Q.-H. Graphene-DNA hybrids: Self-assembly and electrochemical detection performance. *J. Mater. Chem.* **2010**, *20*, 6668–6673. [CrossRef]
196. Woo, S.; Kim, Y.-R.; Chung, T.D.; Piao, Y.; Kim, H. Synthesis of a graphene–carbon nanotube composite and its electrochemical sensing of hydrogen peroxide. *Electrochim. Acta* **2012**, *59*, 509–514. [CrossRef]
197. Wang, Y.; Shao, Y.; Matson, D.W.; Li, J.; Lin, Y. Nitrogen-Doped Graphene and Its Application in Electrochemical Biosensing. *ACS Nano* **2010**, *4*, 1790–1798. [CrossRef]
198. Li, M.; Wu, Z.-S.; Ren, W.; Cheng, H.-M.; Tang, N.; Wu, W.; Zhong, W.; Du, Y. The doping of reduced graphene oxide with nitrogen and its effect on the quenching of the material's photoluminescence. *Carbon* **2012**, *50*, 5286–5291. [CrossRef]
199. Yeh, M.-H.; Li, Y.-S.; Chen, G.-L.; Lin, L.-Y.; Li, T.-J.; Chuang, H.-M.; Hsieh, C.-Y.; Lo, S.-C.; Chiang, W.-H.; Ho, K.-C. Facile Synthesis of Boron-doped Graphene Nanosheets with Hierarchical Microstructure at Atmosphere Pressure for Metal-free Electrochemical Detection of Hydrogen Peroxide. *Electrochim. Acta* **2015**, *172*, 52–60. [CrossRef]

200. Yang, G.-H.; Zhou, Y.-H.; Wu, J.-J.; Cao, J.-T.; Li, L.-L.; Liu, H.-Y.; Zhu, J.-J. Microwave-assisted synthesis of nitrogen and boron co-doped graphene and its application for enhanced electrochemical detection of hydrogen peroxide. *RSC Adv.* **2013**, *3*, 22597–22604. [CrossRef]
201. Zor, E.; Saglam, M.E.; Akin, I.; Saf, A.O.; Bingol, H.; Ersoz, M. Green synthesis of reduced graphene oxide/nanopolypyrrole composite: Characterization and H_2O_2 determination in urine. *RSC Adv.* **2014**, *4*, 12457–12466. [CrossRef]
202. Luo, J.; Chen, Y.; Ma, Q.; Liu, R.; Liu, X. Layer-by-layer assembled ionic-liquid functionalized graphene–polyaniline nanocomposite with enhanced electrochemical sensing properties. *J. Mater. Chem. C* **2014**, *2*, 4818–4827. [CrossRef]
203. Wang, Q.; Li, M.; Szunerits, S.; Boukherroub, R. Environmentally Friendly Reduction of Graphene Oxide Using Tyrosine for Nonenzymatic Amperometric H_2O_2 Detection. *Electroanalysis* **2014**, *26*, 156–163. [CrossRef]
204. Nguyen, V.H.; Tran, T.H.; Shim, J.-J. Glassy carbon electrode modified with a graphene oxide/poly(o-phenylenediamine) composite for the chemical detection of hydrogen peroxide. *Mater. Sci. Eng. C* **2014**, *44*, 144–150. [CrossRef] [PubMed]
205. Huang, Y.; Li, S.F.Y. Electrocatalytic performance of silica nanoparticles on graphene oxide sheets for hydrogen peroxide sensing. *J. Electroanal. Chem.* **2013**, *690*, 8–12. [CrossRef]
206. Kong, F.-Y.; Li, W.-W.; Wang, J.-Y.; Fang, H.-L.; Fan, D.-H.; Wang, W. Direct electrolytic exfoliation of graphite with hemin and single-walled carbon nanotube: Creating functional hybrid nanomaterial for hydrogen peroxide detection. *Anal. Chim. Acta* **2015**, *884*, 37–43. [CrossRef] [PubMed]
207. Lei, W.; Wu, L.; Huang, W.; Hao, Q.; Zhang, Y.; Xia, X. Microwave-assisted synthesis of hemin–graphene/poly(3,4-ethylenedioxythiophene) nanocomposite for a biomimetic hydrogen peroxide biosensor. *J. Mater. Chem. B* **2014**, *2*, 4324–4330. [CrossRef]
208. Zhang, T.; Gu, Y.; Li, C.; Yan, X.; Lu, N.; Liu, H.; Zhang, Z.; Zhang, H. Fabrication of Novel Electrochemical Biosensor Based on Graphene Nanohybrid to Detect H_2O_2 Released from Living Cells with Ultrahigh Performance. *ACS Appl. Mater. Interfaces* **2017**, *9*, 37991–37999. [CrossRef]
209. Xi, F.; Zhao, D.; Wang, X.; Chen, P. Non-enzymatic detection of hydrogen peroxide using a functionalized three-dimensional graphene electrode. *Electrochem. Commun.* **2013**, *26*, 81–84. [CrossRef]
210. Zhang, J.; Zhao, M.; Yang, J.; Wu, G.; Wu, H.; Chen, C.; Liu, A. Metal-free rGO/GO hybrid microelectrode array for sensitive and in-situ hydrogen peroxide sensing. *Electrochim. Acta* **2019**, *326*, 134967. [CrossRef]
211. Tian, Y.; Wei, Z.; Zhang, K.; Peng, S.; Zhang, X.; Liu, W.; Chu, K. Three-dimensional phosphorus-doped graphene as an efficient metal-free electrocatalyst for electrochemical sensing. *Sens. Actuators B Chem.* **2017**, *241*, 584–591. [CrossRef]
212. Radhakrishnan, S.; Kim, S.J. An enzymatic biosensor for hydrogen peroxide based on one-pot preparation of CeO_2-reduced graphene oxide nanocomposite. *RSC Adv.* **2015**, *5*, 12937–12943. [CrossRef]
213. Wang, S.; Zhu, Y.; Yang, X.; Li, C. Photoelectrochemical detection of H_2O_2 based on flower-like $CuInS_2$-graphene hybrid. *Electroanalysis* **2014**, *26*, 573–580. [CrossRef]
214. Song, H.; Ni, Y.; Kokot, S. Investigations of an electrochemical platform based on the layered MoS_2–graphene and horseradish peroxidase nanocomposite for direct electrochemistry and electrocatalysis. *Biosens. Bioelectron.* **2014**, *56*, 137–143. [CrossRef]
215. Liu, Y.; Liu, X.; Guo, Z.; Hu, Z.; Xue, Z.; Lu, X. Horseradish peroxidase supported on porous graphene as a novel sensing platform for detection of hydrogen peroxide in living cells sensitively. *Biosens. Bioelectron.* **2017**, *87*, 101–107. [CrossRef]
216. Fan, Z.; Lin, Q.; Gong, P.; Liu, B.; Wang, J.; Yang, S. A new enzymatic immobilization carrier based on graphene capsule for hydrogen peroxide biosensors. *Electrochim. Acta* **2015**, *151*, 186–194. [CrossRef]
217. Wu, P.; Cai, Z.; Chen, J.; Zhang, H.; Cai, C. Electrochemical measurement of the flux of hydrogen peroxide releasing from RAW 264.7 macrophage cells based on enzyme-attapulgite clay nanohybrids. *Biosens. Bioelectron.* **2011**, *26*, 4012–4017. [CrossRef] [PubMed]
218. Wang, Y.; Zhang, H.; Yao, D.; Pu, J.; Zhang, Y.; Gao, X.; Sun, Y. Direct electrochemistry of hemoglobin on graphene/Fe_3O_4 nanocomposite-modified glass carbon electrode and its sensitive detection for hydrogen peroxide. *J. Solid State Electrochem.* **2013**, *17*, 881–887. [CrossRef]
219. Cheng, Y.; Feng, B.; Yang, X.; Yang, P.; Ding, Y.; Chen, Y.; Fei, J. Electrochemical biosensing platform based on carboxymethyl cellulose functionalized reduced graphene oxide and hemoglobin hybrid nanocomposite film. *Sens. Actuators B Chem.* **2013**, *182*, 288–293. [CrossRef]
220. Xie, L.; Xu, Y.; Cao, X. Hydrogen peroxide biosensor based on hemoglobin immobilized at graphene, flower-like zinc oxide, and gold nanoparticles nanocomposite modified glassy carbon electrode. *Colloids Surf. B Biointerfaces* **2013**, *107*, 245–250. [CrossRef] [PubMed]
221. Li, M.; Xu, S.; Tang, M.; Liu, L.; Gao, F.; Wang, Y. Direct electrochemistry of horseradish peroxidase on graphene-modified electrode for electrocatalytic reduction towards H_2O_2. *Electrochim. Acta* **2011**, *56*, 1144–1149. [CrossRef]
222. Zhang, L.; Han, G.; Liu, Y.; Tang, J.; Tang, W. Immobilizing haemoglobin on gold/graphene–chitosan nanocomposite as efficient hydrogen peroxide biosensor. *Sens. Actuators B Chem.* **2014**, *197*, 164–171. [CrossRef]
223. Liu, H.; Su, X.; Duan, C.; Dong, X.; Zhou, S.; Zhu, Z. Microwave-assisted hydrothermal synthesis of Au NPs–Graphene composites for H_2O_2 detection. *J. Electroanal. Chem.* **2014**, *731*, 36–42. [CrossRef]
224. Vilian, A.T.E.; Chen, S.-M. Simple approach for the immobilization of horseradish peroxidase on poly-l-histidine modified reduced graphene oxide for amperometric determination of dopamine and H_2O_2. *RSC Adv.* **2014**, *4*, 55867–55876. [CrossRef]

225. Xiong, W.; Qu, Q.; Liu, S. Self-assembly of ultra-small gold nanoparticles on an indium tin oxide electrode for the enzyme-free detection of hydrogen peroxide. *Mikrochim. Acta* **2014**, *181*, 983–989. [CrossRef]
226. Sheng, Q.; Wang, M.; Zheng, J. A novel hydrogen peroxide biosensor based on enzymatically induced deposition of polyaniline on the functionalized graphene–carbon nanotube hybrid materials. *Sens. Actuators B Chem.* **2011**, *160*, 1070–1077. [CrossRef]
227. Zhou, K.; Zhu, Y.; Yang, X.; Luo, J.; Li, C.; Luan, S. A novel hydrogen peroxide biosensor based on Au–graphene–HRP–chitosan biocomposites. *Electrochim. Acta* **2010**, *55*, 3055–3060. [CrossRef]
228. Wang, T.; Liu, J.; Ren, J.; Wang, J.; Wang, E. Mimetic biomembrane–AuNPs–graphene hybrid as matrix for enzyme immobilization and bioelectrocatalysis study. *Talanta* **2015**, *143*, 438–441. [CrossRef] [PubMed]
229. Nandini, S.; Manjunatha, R.; Shanmugam, S.; Melo, J.S.; Suresh, G.S. Electrochemical biosensor for the selective determination of hydrogen peroxide based on the co-deposition of palladium, horseradish peroxidase on functionalized-graphene modified graphite electrode as composite. *J. Electroanal. Chem.* **2013**, *689*, 233–242. [CrossRef]
230. Nalini, S.; Shanmugam, S.; Neelagund, S.E.; Melo, J.S.; Suresh, G.S.; Nandini, S. Amperometric hydrogen peroxide and cholesterol biosensors designed by using hierarchical curtailed silver flowers functionalized graphene and enzymes deposits. *J. Solid State Electrochem.* **2014**, *18*, 685–701. [CrossRef]
231. Huang, K.-J.; Niu, D.-J.; Liu, X.; Wu, Z.-W.; Fan, Y.; Chang, Y.-F.; Wu, Y.-Y. Direct electrochemistry of catalase at amine-functionalized graphene/gold nanoparticles composite film for hydrogen peroxide sensor. *Electrochim. Acta* **2011**, *56*, 2947–2953. [CrossRef]
232. Dinesh, B.; Mani, V.; Saraswathi, R.; Chen, S.-M. Direct electrochemistry of cytochrome c immobilized on a graphene oxide–carbon nanotube composite for picomolar detection of hydrogen peroxide. *RSC Adv.* **2014**, *4*, 28229–28237. [CrossRef]
233. Mani, V.; Dinesh, B.; Chen, S.-M.; Saraswathi, R. Direct electrochemistry of myoglobin at reduced graphene oxide-multiwalled carbon nanotubes-platinum nanoparticles nanocomposite and biosensing towards hydrogen peroxide and nitrite. *Biosens. Bioelectron.* **2014**, *53*, 420–427. [CrossRef]
234. Liu, F.; Xu, Q.; Huang, W.; Zhang, Z.; Xiang, G.; Zhang, C.; Liang, C.; Lian, H.; Peng, J. Green synthesis of porous graphene and its application for sensitive detection of hydrogen peroxide and 2,4-dichlorophenoxyacetic acid. *Electrochim. Acta* **2019**, *295*, 615–623. [CrossRef]
235. Ebrahimi, A.; Zhang, K.; Dong, C.; Subramanian, S.; Butler, D.; Bolotsky, A.; Goodnight, L.; Cheng, Y.; Robinson, J.A. FeSx-graphene heterostructures: Nanofabrication-compatible catalysts for ultra-sensitive electrochemical detection of hydrogen peroxide. *Sens. Actuators B Chem.* **2019**, *285*, 631–638. [CrossRef]
236. Hu, Z.; Dai, Z.; Hu, X.; Yang, B.; Liu, Q.; Gao, C.; Zheng, X.; Yu, Y. A facile preparation of FePt-loaded few-layer MoS_2 nanosheets nanocomposites (F-MoS_2-FePt NCs) and their application for colorimetric detection of H_2O_2 in living cells. *J. Nanobiotechnol.* **2019**, *17*, 38. [CrossRef] [PubMed]
237. Dai, H.; Chen, Y.; Niu, X.; Pan, C.; Chen, H.; Chen, X. High-performance electrochemical biosensor for nonenzymatic H_2O_2 sensing based on Au@C-Co3O4 heterostructures. *Biosens. Bioelectron.* **2018**, *118*, 36–43. [CrossRef]
238. Liu, J.; Bo, X.; Yang, J.; Yin, D.; Guo, L. One-step synthesis of porphyrinic iron-based metal-organic framework/ordered mesoporous carbon for electrochemical detection of hydrogen peroxide in living cells. *Sens. Actuators B Chem.* **2017**, *248*, 207–213. [CrossRef]
239. Shu, Y.; Xu, J.; Chen, J.; Xu, Q.; Xiao, X.; Jin, D.; Pang, H.; Hu, X. Ultrasensitive electrochemical detection of H_2O_2 in living cells based on ultrathin MnO_2 nanosheets. *Sens. Actuators B Chem.* **2017**, *252*, 72–78. [CrossRef]
240. Ensafi, A.A.; Jafari-Asl, M.; Rezaei, B. A novel enzyme-free amperometric sensor for hydrogen peroxide based on Nafion/exfoliated graphene oxide–Co_3O_4 nanocomposite. *Talanta* **2013**, *103*, 322–329. [CrossRef]
241. Sarkar, A.; Ghosh, A.B.; Saha, N.; Bhadu, G.R.; Adhikary, B. Newly Designed Amperometric Biosensor for Hydrogen Peroxide and Glucose Based on Vanadium Sulfide Nanoparticles. *ACS Appl. Nano Mater.* **2018**, *1*, 1339–1347. [CrossRef]
242. Li, S.-J.; Du, J.-M.; Zhang, J.-P.; Zhang, M.-J.; Chen, J. A glassy carbon electrode modified with a film composed of cobalt oxide nanoparticles and graphene for electrochemical sensing of H_2O_2. *Mikrochim. Acta* **2014**, *181*, 631–638. [CrossRef]
243. Zheng, L.; Ye, D.; Xiong, L.; Xu, J.; Tao, K.; Zou, Z.; Huang, D.; Kang, X.; Yang, S.; Xia, J. Preparation of cobalt-tetraphenylporphyrin/reduced graphene oxide nanocomposite and its application on hydrogen peroxide biosensor. *Anal. Chim. Acta* **2013**, *768*, 69–75. [CrossRef]
244. Hosu, I.S.; Wang, Q.; Vasilescu, A.; Peteu, S.F.; Raditoiu, V.; Railian, S.; Zaitsev, V.; Turcheniuk, K.; Wang, Q.; Li, M.; et al. Cobalt phthalocyanine tetracarboxylic acid modified reduced graphene oxide: A sensitive matrix for the electrocatalytic detection of peroxynitrite and hydrogen peroxide. *RSC Adv.* **2015**, *5*, 1474–1484. [CrossRef]
245. Liu, X.; Zhu, H.; Yang, X. An amperometric hydrogen peroxide chemical sensor based on graphene-Fe_3O_4 multilayer films modified ITO electrode. *Talanta* **2011**, *87*, 243–248. [CrossRef] [PubMed]
246. Yang, S.; Li, G.; Wang, G.; Zhao, J.; Hu, M.; Qu, L. A novel nonenzymatic H_2O_2 sensor based on cobalt hexacyanoferrate nanoparticles and graphene composite modified electrode. *Sens. Actuators B Chem.* **2015**, *208*, 593–599. [CrossRef]
247. Kubendhiran, S.; Thirumalraj, B.; Chen, S.-M.; Karuppiah, C. Electrochemical co-preparation of cobalt sulfide/reduced graphene oxide composite for electrocatalytic activity and determination of H_2O_2 in biological samples. *J. Colloid Interface Sci.* **2018**, *509*, 153–162. [CrossRef]

248. Karimi, M.A.; Banifatemeh, F.; Hatefi-Mehrjardi, A.; Tavallali, H.; Eshaghia, Z.; Deilamy-Rad, G. A novel rapid synthesis of Fe_2O_3/graphene nanocomposite using ferrate (VI) and its application as a new kind of nanocomposite modified electrode as electrochemical sensor. *Mater. Res. Bull.* **2015**, *70*, 856–864. [CrossRef]
249. Li, Z.; Zheng, X.; Zheng, J. A non-enzymatic sensor based on Au@Ag nanoparticles with good stability for sensitive detection of H_2O_2. *New J. Chem.* **2016**, *40*, 2115–2120. [CrossRef]
250. Zhu, S.; Guo, J.; Dong, J.; Cui, Z.; Lu, T.; Zhu, C.; Zhang, D.; Ma, J. Sonochemical fabrication of Fe_3O_4 nanoparticles on reduced graphene oxide for biosensors. *Ultrason. Sonochem.* **2013**, *20*, 872–880. [CrossRef]
251. Zhang, P.; Huang, Y.; Lu, X.; Zhang, S.; Li, J.; Wei, G.; Su, Z. One-Step Synthesis of Large-Scale Graphene Film Doped with Gold Nanoparticles at Liquid–Air Interface for Electrochemistry and Raman Detection Applications. *Langmuir* **2014**, *30*, 8980–8989. [CrossRef] [PubMed]
252. Ye, Y.; Kong, T.; Yu, X.; Wu, Y.; Zhang, K.; Wang, X. Enhanced nonenzymatic hydrogen peroxide sensing with reduced graphene oxide/ferroferric oxide nanocomposites. *Talanta* **2012**, *89*, 417–421. [CrossRef]
253. Yang, X.; Wang, L.; Zhou, G.; Sui, N.; Gu, Y.; Wan, J. Electrochemical Detection of H_2O_2 Based on Fe_3O_4 Nanoparticles with Graphene Oxide and Polyamidoamine Dendrimer. *J. Clust. Sci.* **2014**, *26*, 789–798. [CrossRef]
254. Zhu, M.; Li, N.; Ye, J. Sensitive and Selective Sensing of Hydrogen Peroxide with Iron-Tetrasulfophthalocyanine-Graphene-Nafion Modified Screen-Printed Electrode. *Electroanalysis* **2012**, *24*, 1212–1219. [CrossRef]
255. Palanisamy, S.; Chen, S.-M.; Sarawathi, R. A novel nonenzymatic hydrogen peroxide sensor based on reduced graphene oxide/ZnO composite modified electrode. *Sens. Actuators B Chem.* **2012**, *166–167*, 372–377. [CrossRef]
256. Jiang, B.-B.; Wei, X.-W.; Wu, F.-H.; Wu, K.-L.; Chen, L.; Yuan, G.-Z.; Dong, C.; Ye, Y. A non-enzymatic hydrogen peroxide sensor based on a glassy carbon electrode modified with cuprous oxide and nitrogen-doped graphene in a nafion matrix. *Mikrochim. Acta* **2014**, *181*, 1463–1470. [CrossRef]
257. Xu, F.; Deng, M.; Li, G.; Chen, S.; Wang, L. Electrochemical behavior of cuprous oxide–reduced graphene oxide nanocomposites and their application in nonenzymatic hydrogen peroxide sensing. *Electrochim. Acta* **2013**, *88*, 59–65. [CrossRef]
258. Liu, M.; Liu, R.; Chen, W. Graphene wrapped Cu_2O nanocubes: Non-enzymatic electrochemical sensors for the detection of glucose and hydrogen peroxide with enhanced stability. *Biosens. Bioelectron.* **2013**, *45*, 206–212. [CrossRef] [PubMed]
259. Li, L.; Du, Z.; Liu, S.; Hao, Q.; Wang, Y.; Li, Q.; Wang, T. A novel nonenzymatic hydrogen peroxide sensor based on MnO_2/graphene oxide nanocomposite. *Talanta* **2010**, *82*, 1637–1641. [CrossRef] [PubMed]
260. Dong, S.; Xi, J.; Wu, Y.; Liu, H.; Fu, C.; Liu, H.; Xiao, F. High loading MnO_2 nanowires on graphene paper: Facile electrochemical synthesis and use as flexible electrode for tracking hydrogen peroxide secretion in live cells. *Anal. Chim. Acta* **2015**, *853*, 200–206. [CrossRef] [PubMed]
261. Feng, X.; Zhang, Y.; Song, J.; Chen, N.; Zhou, J.; Huang, Z.; Ma, Y.; Zhang, L.; Wang, L. MnO_2/graphene nanocomposites for nonenzymatic electrochemical detection of hydrogen peroxide. *Electroanalysis* **2015**, *27*, 353–359. [CrossRef]
262. Hassan, M.; Jiang, Y.; Bo, X.; Zhou, M. Sensitive nonenzymatic detection of hydrogen peroxide at nitrogen-doped graphene supported-CoFe nanoparticles. *Talanta* **2018**, *188*, 339–348. [CrossRef] [PubMed]
263. Lu, N.; Zheng, B.; Gu, Y.; Yan, X.; Zhang, T.; Liu, H.; Xu, H.; Xu, Z.; Li, X.; Zhang, Z. Fabrication of CoNPs-embedded porous carbon composites based on morphochemical imprinting strategy for detection of H_2O_2 released from living cells. *Electrochim. Acta* **2019**, *321*, 134717. [CrossRef]
264. Benvidi, A.; Nafar, M.T.; Jahanbani, S.; Tezerjani, M.D.; Rezaeinasab, M.; Dalirnasab, S. Developing an electrochemical sensor based on a carbon paste electrode modified with nano-composite of reduced graphene oxide and $CuFe_2O_4$ nanoparticles for determination of hydrogen peroxide. *Mater. Sci. Eng. C* **2017**, *75*, 1435–1447. [CrossRef] [PubMed]
265. Dang, W.; Sun, Y.; Jiao, H.; Xu, L.; Lin, M. AuNPs-NH_2/Cu-MOF modified glassy carbon electrode as enzyme-free electrochemical sensor detecting H_2O_2. *J. Electroanal. Chem.* **2020**, *856*, 113592. [CrossRef]
266. Wang, W.; Tang, H.; Wu, Y.; Zhang, Y.; Li, Z. Highly electrocatalytic biosensor based on Hemin@ AuNPs/reduced graphene oxide/chitosan nanohybrids for non-enzymatic ultrasensitive detection of hydrogen peroxide in living cells. *Biosens. Bioelectron.* **2019**, *132*, 217–223. [CrossRef]
267. Sun, Y.; Luo, M.; Meng, X.; Xiang, J.; Wang, L.; Ren, Q.; Guo, S. Graphene/Intermetallic PtPb Nanoplates Composites for Boosting Electrochemical Detection of H_2O_2 Released from Cells. *Anal. Chem.* **2017**, *89*, 3761–3767. [CrossRef] [PubMed]
268. Zhang, Y.; Liu, Y.; He, J.; Pang, P.; Gao, Y.; Hu, Q. Electrochemical behavior of graphene/Nafion/Azure I/Au nanoparticles composites modified glass carbon electrode and its application as nonenzymatic hydrogen peroxide sensor. *Electrochim. Acta* **2013**, *90*, 550–555. [CrossRef]
269. Wang, L.; Dong, Y.; Zhang, Y.; Zhang, Z.; Chi, K.; Yuan, H.; Zhao, A.; Ren, J.; Xiao, F.; Wang, S. PtAu alloy nanoflowers on 3D porous ionic liquid functionalized graphene-wrapped activated carbon fiber as a flexible microelectrode for near-cell detection of cancer. *NPG Asia Mater.* **2016**, *8*, e337. [CrossRef]
270. Yuan, B.; Xu, C.; Liu, L.; Shi, Y.; Li, S.; Zhang, R.; Zhang, D. Polyethylenimine-bridged graphene oxide–gold film on glassy carbon electrode and its electrocatalytic activity toward nitrite and hydrogen peroxide. *Sens. Actuators B* **2014**, *198*, 55–61. [CrossRef]
271. Lorestani, F.; Shahnavaz, Z.; Mn, P.; Alias, Y.; Manan, N.S.A. One-step hydrothermal green synthesis of silver nanoparticle-carbon nanotube reduced-graphene oxide composite and its application as hydrogen peroxide sensor. *Sens. Actuators B Chem.* **2015**, *208*, 389–398. [CrossRef]

272. Chang, H.-C.; Wang, X.; Shiu, K.-K.; Zhu, Y.; Wang, J.; Li, Q.; Chen, B.; Jiang, H. Layer-by-layer assembly of graphene, Au and poly (toluidine blue O) films sensor for evaluation of oxidative stress of tumor cells elicited by hydrogen peroxide. *Biosens. Bioelectron.* **2013**, *41*, 789–794. [CrossRef] [PubMed]
273. Bai, X.; Shiu, K.-K. Investigation of the optimal weight contents of reduced graphene oxide–gold nanoparticles composites and theirs application in electrochemical biosensors. *J. Electroanal. Chem.* **2014**, *720–721*, 84–91. [CrossRef]
274. Liu, R.; Li, S.; Zhang, G.; Dolbecq, A.; Mialane, P.; Keita, B. Polyoxometalate-Mediated Green Synthesis of Graphene and Metal Nanohybrids: High-Performance Electrocatalysts. *J. Clust. Sci.* **2014**, *25*, 711–740. [CrossRef]
275. Yu, B.; Feng, J.; Liu, S.; Zhang, T. Preparation of reduced graphene oxide decorated with high density Ag nanorods for non-enzymatic hydrogen peroxide detection. *RSC Adv.* **2013**, *3*, 14303–14307. [CrossRef]
276. Fang, Y.; Guo, S.; Zhu, C.; Zhai, Y.; Wang, E. Self-Assembly of Cationic Polyelectrolyte-Functionalized Graphene Nanosheets and Gold Nanoparticles: A Two-Dimensional Heterostructure for Hydrogen Peroxide Sensing. *Langmuir* **2010**, *26*, 11277–11282. [CrossRef] [PubMed]
277. Xu, F.; Sun, Y.; Zhang, Y.; Shi, Y.; Wen, Z.; Li, Z. Graphene–Pt nanocomposite for nonenzymatic detection of hydrogen peroxide with enhanced sensitivity. *Electrochem. Commun.* **2011**, *13*, 1131–1134. [CrossRef]
278. Gao, C.-H.; Zhu, X.-Z.; Zhang, L.; Zhou, D.-Y.; Wang, Z.-K.; Liao, L.-S. Comparative studies on the inorganic and organic p-type dopants in organic light-emitting diodes with enhanced hole injection. *Appl. Phys. Lett.* **2013**, *102*, 153301. [CrossRef]
279. Pang, P.; Yang, Z.; Xiao, S.; Xie, J.; Zhang, Y.; Gao, Y. Nonenzymatic amperometric determination of hydrogen peroxide by graphene and gold nanorods nanocomposite modified electrode. *J. Electroanal. Chem.* **2014**, *727*, 27–33. [CrossRef]
280. Li, X.-R.; Xu, M.-C.; Chen, H.-Y.; Xu, J.-J. Bimetallic Au@Pt@Au core–shell nanoparticles on graphene oxide nanosheets for high-performance H_2O_2 bi-directional sensing. *J. Mater. Chem. B* **2015**, *3*, 4355–4362. [CrossRef]
281. Yao, L.; Yan, Y.; Lee, J.-M. Synthesis of Porous Pd Nanostructure and Its Application in Enzyme-Free Sensor of Hydrogen Peroxide. *ACS Sustain. Chem. Eng.* **2017**, *5*, 1248–1252. [CrossRef]
282. Fu, L.; Lai, G.; Jia, B.; Yu, A. Preparation and Electrocatalytic Properties of Polydopamine Functionalized Reduced Graphene Oxide-Silver Nanocomposites. *Electrocatalysis* **2014**, *6*, 72–76. [CrossRef]
283. Liu, S.; Tian, J.; Wang, L.; Sun, X. Microwave-assisted rapid synthesis of Ag nanoparticles/graphene nanosheet composites and their application for hydrogen peroxide detection. *J. Nanopart. Res.* **2011**, *13*, 4539–4548. [CrossRef]
284. Shan, L.; Liu, H.; Wang, G. Preparation of tungsten-doped $BiVO_4$ and enhanced photocatalytic activity. *J. Nanopart. Res.* **2015**, *17*, 181. [CrossRef]
285. Liu, J.; Bo, X.; Zhao, Z.; Guo, L. Highly exposed Pt nanoparticles supported on porous graphene for electrochemical detection of hydrogen peroxide in living cells. *Biosens. Bioelectron.* **2015**, *74*, 71–77. [CrossRef]
286. Zhang, C.; Zhang, Y.; Miao, Z.; Ma, M.; Du, X.; Lin, J.; Han, B.; Takahashi, S.; Anzai, J.-I.; Chen, Q. Dual-function amperometric sensors based on poly (diallydimethylammoniun chloride)-functionalized reduced graphene oxide/manganese dioxide/gold nanoparticles nanocomposite. *Sens. Actuators B Chem.* **2016**, *222*, 663–673. [CrossRef]
287. Liu, S.; Wang, L.; Tian, J.; Luo, Y.; Zhang, X.; Sun, X. Aniline as a dispersing and stabilizing agent for reduced graphene oxide and its subsequent decoration with Ag nanoparticles for enzymeless hydrogen peroxide detection. *J. Colloid Interface Sci.* **2011**, *363*, 615–619. [CrossRef] [PubMed]
288. Zhu, J.; Kim, K.; Liu, Z.; Feng, H.; Hou, S. Electroless Deposition of Silver Nanoparticles on Graphene Oxide Surface and Its Applications for the Detection of Hydrogen Peroxide. *Electroanalysis* **2014**, *26*, 2513–2519. [CrossRef]
289. Zhang, P.; Zhang, X.; Zhang, S.; Lu, X.; Li, Q.; Su, Z.; Wei, G. One-pot green synthesis, characterizations, and biosensor application of self-assembled reduced graphene oxide–gold nanoparticle hybrid membranes. *J. Mater. Chem. B* **2013**, *1*, 6525–6531. [CrossRef]
290. Li, S.-J.; Shi, Y.-F.; Liu, L.; Song, L.-X.; Pang, H.; Du, J.-M. Electrostatic self-assembly for preparation of sulfonated graphene/gold nanoparticle hybrids and their application for hydrogen peroxide sensing. *Electrochim. Acta* **2012**, *85*, 628–635. [CrossRef]
291. Zhang, Y.; Wang, Z.; Ji, Y.; Liu, S. Synthesis of Ag nanoparticle–carbon nanotube–reduced graphene oxide hybrids for highly sensitive non-enzymatic hydrogen peroxide detection. *RSC Adv.* **2015**, *5*, 39037–39041. [CrossRef]
292. Nia, P.M.; Lorestani, F.; Meng, W.P.; Alias, Y. A novel non-enzymatic H_2O_2 sensor based on polypyrrole nanofibers–silver nanoparticles decorated reduced graphene oxide nano composites. *Appl. Surf. Sci.* **2015**, *332*, 648–656.
293. Lu, D.; Zhang, Y.; Lin, S.; Wang, L.; Wang, C. Synthesis of PtAu bimetallic nanoparticles on graphene–carbon nanotube hybrid nanomaterials for nonenzymatic hydrogen peroxide sensor. *Talanta* **2013**, *112*, 111–116. [CrossRef] [PubMed]
294. Liu, P.; Li, J.; Liu, X.; Li, M.; Lu, X. One-pot synthesis of highly dispersed PtAu nanoparticles–CTAB–graphene nanocomposites for nonenzyme hydrogen peroxide sensor. *J. Electroanal. Chem.* **2015**, *751*, 1–6. [CrossRef]
295. Cui, X.; Wu, S.; Li, Y.; Wan, G. Sensing hydrogen peroxide using a glassy carbon electrode modified with in-situ electrodeposited platinum-gold bimetallic nanoclusters on a graphene surface. *Mikrochim. Acta* **2014**, *182*, 265–272. [CrossRef]
296. Sun, Y.; He, K.; Zhang, Z.; Zhou, A.; Duan, H. Real-time electrochemical detection of hydrogen peroxide secretion in live cells by Pt nanoparticles decorated graphene–carbon nanotube hybrid paper electrode. *Biosens. Bioelectron.* **2015**, *68*, 358–364. [CrossRef]
297. Xu, C.; Zhang, L.; Liu, L.; Shi, Y.; Wang, H.; Wang, X.; Wang, F.; Yuan, B.; Zhang, D. A novel enzyme-free hydrogen peroxide sensor based on polyethylenimine-grafted graphene oxide-Pd particles modified electrode. *J. Electroanal. Chem.* **2014**, *731*, 67–71. [CrossRef]
298. Liu, H.; Chen, X.; Huang, L.; Wang, J.; Pan, H. Palladium Nanoparticles Embedded into Graphene Nanosheets: Preparation, Characterization, and Nonenzymatic Electrochemical Detection of H_2O_2. *Electroanalysis* **2014**, *26*, 556–564. [CrossRef]

299. Chen, X.-M.; Cai, Z.-X.; Huang, Z.-Y.; Oyama, M.; Jiang, Y.-Q.; Chen, X. Ultrafine palladium nanoparticles grown on graphene nanosheets for enhanced electrochemical sensing of hydrogen peroxide. *Electrochim. Acta* **2013**, *97*, 398–403. [CrossRef]
300. Sun, W.; Cai, X.; Wang, Z.; Zhao, H.; Lan, M. A novel modification method via in-situ reduction of AuAg bimetallic nanoparticles by polydopamine on carbon fiber microelectrode for H_2O_2 detection. *Microchem. J.* **2020**, *154*, 104595. [CrossRef]
301. Xie, Y.; Zhan, Y. Electrochemical capacitance of porous reduced graphene oxide/nickel foam. *J. Porous Mater.* **2015**, *22*, 403–412. [CrossRef]
302. Zhao, Y.; Hu, Y.; Hou, J.; Jia, Z.; Zhong, D.; Zhou, S.; Huo, D.; Yang, M.; Hou, C. Electrochemical biointerface based on electrodeposition AuNPs on 3D graphene aerogel: Direct electron transfer of Cytochrome c and hydrogen peroxide sensing. *J. Electroanal. Chem.* **2019**, *842*, 16–23. [CrossRef]
303. Fu, Y.; Huang, D.; Li, C.; Zou, L.; Ye, B. Graphene blended with SnO_2 and Pd-Pt nanocages for sensitive non-enzymatic electrochemical detection of H_2O_2 released from living cells. *Anal. Chim. Acta* **2018**, *1014*, 10–18. [CrossRef] [PubMed]
304. Lu, J.; Hu, Y.; Wang, P.; Liu, P.; Chen, Z.; Sun, D. Electrochemical biosensor based on gold nanoflowers-encapsulated magnetic metal-organic framework nanozymes for drug evaluation with in-situ monitoring of H_2O_2 released from H9C2 cardiac cells. *Sens. Actuators B Chem.* **2020**, *311*, 127909. [CrossRef]
305. Hu, J.; Wisetsuwannaphum, S.; Foord, J.S. Glutamate biosensors based on diamond and graphene platforms. *Faraday Discuss.* **2014**, *172*, 457–472. [CrossRef] [PubMed]
306. Chang, Q.; Tang, H. Optical determination of glucose and hydrogen peroxide using a nanocomposite prepared from glucose oxidase and magnetite nanoparticles immobilized on graphene oxide. *Mikrochim. Acta* **2014**, *181*, 527–534. [CrossRef]
307. Zhang, J.; Zhang, F.; Yang, H.; Huang, X.; Liu, H.; Zhang, J.; Guo, S. Graphene Oxide as a Matrix for Enzyme Immobilization. *Langmuir* **2010**, *26*, 6083–6085. [CrossRef]
308. Dresselhaus, M.; Dresselhaus, G.; Jorio, A. Unusual properties and structure of carbon nanotubes. *Annu. Rev. Mater. Res.* **2004**, *34*, 247–278. [CrossRef]
309. Primo, E.N.; Gutiérrez, F.; Rubianes, M.D.; Ferreyra, N.F.; Rodríguez, M.C.; Pedano, M.L.; Gasnier, A.; Gutierrez, A.; Eguílaz, M.; Dalmasso, P.; et al. Electrochemistry in one dimension: Applications of carbon nanotubes. *Electrochem. Carbon Electrodes* **2015**, 83–120. [CrossRef]
310. Yu, D.; Wang, P.; Zhao, Y.; Fan, A. Iodophenol blue-enhanced luminol chemiluminescence and its application to hydrogen peroxide and glucose detection. *Talanta* **2016**, *146*, 655–661. [CrossRef] [PubMed]
311. Georgakilas, V.; Perman, J.A.; Tucek, J.; Zboril, R. Broad Family of Carbon Nanoallotropes: Classification, Chemistry, and Applications of Fullerenes, Carbon Dots, Nanotubes, Graphene, Nanodiamonds, and Combined Superstructures. *Chem. Rev.* **2015**, *115*, 4744–4822. [CrossRef]
312. Yáñez-Sedeño, P.; González-Cortés, A.; Agüí, L.; Pingarrón, J.M. Uncommon carbon nanostructures for the preparation of electrochemical immunosensors. *Electroanalysis* **2016**, *28*, 1679–1691. [CrossRef]
313. Jacobs, C.B.; Peairs, M.J.; Venton, B.J. Review: Carbon nanotube based electrochemical sensors for biomolecules. *Anal. Chim. Acta* **2010**, *662*, 105–127. [CrossRef] [PubMed]
314. Yang, J.; Xu, Y.; Zhang, R.; Wang, Y.; He, P.; Fang, Y. Direct Electrochemistry and Electrocatalysis of the Hemoglobin Immobilized on Diazonium-Functionalized Aligned Carbon Nanotubes Electrode. *Electroanal. Int. J. Devoted Fundam. Pract. Asp. Electroanal.* **2009**, *21*, 1672–1677. [CrossRef]
315. Esplandiu, M.J.; Pacios, M.; Cyganek, L.; Bartroli, J.; Del Valle, M. Enhancing the electrochemical response of myoglobin with carbon nanotube electrodes. *Nanotechnology* **2009**, *20*, 355502. [CrossRef]
316. Zhang, H.; Ruan, J.; Liu, W.; Jiang, X.; Du, T.; Jiang, H.; Alberto, P.; Gottschalk, K.-E.; Wang, X. Monitoring dynamic release of intracellular hydrogen peroxide through a microelectrode based enzymatic biosensor. *Anal. Bioanal. Chem.* **2018**, *410*, 4509–4517. [CrossRef] [PubMed]
317. Wang, S.; Xie, F.; Liu, G. Direct electrochemistry and electrocatalysis of heme proteins on SWCNTs-CTAB modified electrodes. *Talanta* **2009**, *77*, 1343–1350. [CrossRef]
318. Chen, L.; Lu, G. Novel amperometric biosensor based on composite film assembled by polyelectrolyte-surfactant polymer, carbon nanotubes and hemoglobin. *Sens. Actuators B Chem.* **2007**, *121*, 423–429. [CrossRef]
319. Zong, S.; Cao, Y.; Ju, H. Amperometric biosensor for hydrogen peroxide based on myoglobin doped multiwalled carbon nanotube enhanced grafted collagen matrix. *Anal. Lett.* **2007**, *40*, 1556–1568. [CrossRef]
320. Nagaraju, D.; Pandey, R.K.; Lakshminarayanan, V. Electrocatalytic studies of Cytochrome c functionalized single walled carbon nanotubes on self-assembled monolayer of 4-ATP on gold. *J. Electroanal. Chem.* **2009**, *627*, 63–68. [CrossRef]
321. Chen, S.; Yuan, R.; Chai, Y.; Yin, B.; Xu, Y. Multilayer Assembly of Hemoglobin and Colloidal Gold Nanoparticles on Multiwall Carbon Nanotubes/Chitosan Composite for Detecting Hydrogen Peroxide. *Electroanal. Int. J. Devoted Fundam. Pract. Asp. Electroanal.* **2008**, *20*, 2141–2147. [CrossRef]
322. Ren, Q.-Q.; Wu, J.; Zhang, W.-C.; Wang, C.; Qin, X.; Liu, G.-C.; Li, Z.-X.; Yu, Y. Real-time in vitro detection of cellular H_2O_2 under camptothecin stress using horseradish peroxidase, ionic liquid, and carbon nanotube-modified carbon fiber ultramicroelectrode. *Sens. Actuators B Chem.* **2017**, *245*, 615–621. [CrossRef]
323. Pandey, R.R.; Guo, Y.; Gao, Y.; Chusuei, C.C. A Prussian blue ZnO carbon nanotube composite for chronoamperometrically assaying H_2O_2 in BT20 and 4T1 breast cancer cells. *Anal. Chem.* **2019**, *91*, 10573–10581. [CrossRef] [PubMed]

324. Bai, J.; Sun, C.; Jiang, X. Carbon dots-decorated multiwalled carbon nanotubes nanocomposites as a high-performance electrochemical sensor for detection of H_2O_2 in living cells. *Anal. Bioanal. Chem.* **2016**, *408*, 4705–4714. [CrossRef]
325. Sahraoui, Y.; Chaliaa, S.; Maaref, A.; Haddad, A.; Bessueille, F.; Jaffrezic-Renault, N. Synergistic effect of polyoxometalate and single walled carbon nanotubes on peroxidase-like mimics and highly sensitive electrochemical detection of hydrogen peroxide. *Electroanalysis* **2020**, *32*, 683–689. [CrossRef]
326. Liu, J.-X.; Ding, S.-N. Non-enzymatic amperometric determination of cellular hydrogen peroxide using dendrimer-encapsulated Pt nanoclusters/carbon nanotubes hybrid composites modified glassy carbon electrode. *Sens. Actuators B Chem.* **2017**, *251*, 200–207. [CrossRef]
327. Roushani, M.; Bakyas, K.; Dizajdizi, B.Z. Development of sensitive amperometric hydrogen peroxide sensor using a CuNPs/MB/MWCNT-C60-Cs-IL nanocomposite modified glassy carbon electrode. *Mater. Sci. Eng. C* **2016**, *64*, 54–60. [CrossRef] [PubMed]
328. Zhao, P.; Zhao, Y.; Jiang, L.; Chen, S.; Ji, Z.; Hou, C.; Huo, D.; Yang, M. 3D carbon nanotubes spaced graphene aerogel incorporated with prussian blue nanoparticles for real-time detection of H_2O_2 released from living cells. *J. Electrochem. Soc.* **2020**, *167*, 047511. [CrossRef]
329. Zhang, Y.; Xiao, J.; Sun, Y.; Wang, L.; Dong, X.; Ren, J.; He, W.; Xiao, F. Flexible nanohybrid microelectrode based on carbon fiber wrapped by gold nanoparticles decorated nitrogen doped carbon nanotube arrays: In situ electrochemical detection in live cancer cells. *Biosens. Bioelectron.* **2018**, *100*, 453–461. [CrossRef] [PubMed]
330. Guo, X.; Cao, Q.; Liu, Y.; He, T.; Liu, J.; Huang, S.; Tang, H.; Ma, M. Organic electrochemical transistor for in situ detection of H_2O_2 released from adherent cells and its application in evaluating the in vitro cytotoxicity of nanomaterial. *Anal. Chem.* **2019**, *92*, 908–915. [CrossRef]
331. Tavakkoli, H.; Akhond, M.; Ghorbankhani, G.A.; Absalan, G. Electrochemical sensing of hydrogen peroxide using a glassy carbon electrode modified with multiwalled carbon nanotubes and zein nanoparticle composites: Application to HepG2 cancer cell detection. *Mikrochim. Acta* **2020**, *187*, 105. [CrossRef]
332. Wayu, M.B.; Spidle, R.T.; Devkota, T.; Deb, A.K.; Delong, R.K.; Ghosh, K.C.; Wanekaya, A.K.; Chusuei, C.C. Morphology of hydrothermally synthesized ZnO nanoparticles tethered to carbon nanotubes affects electrocatalytic activity for H_2O_2 detection. *Electrochim. Acta* **2013**, *97*, 99–104. [CrossRef]
333. Goran, J.M.; Phan, E.N.H.; Favela, C.A.; Stevenson, K.J. H_2O_2 Detection at carbon nanotubes and nitrogen-doped carbon nanotubes: Oxidation, reduction, or disproportionation? *Anal. Chem.* **2015**, *87*, 5989–5996. [CrossRef] [PubMed]
334. Li, X.; Li, H.; Liu, T.; Hei, Y.; Hassan, M.; Zhang, S.; Lin, J.; Wang, T.; Bo, X.; Wang, H.-L.; et al. The biomass of ground cherry husks derived carbon nanoplates for electrochemical sensing. *Sens. Actuators B Chem.* **2018**, *255*, 3248–3256. [CrossRef]
335. Wu, Z.-L.; Li, C.-K.; Yu, J.-G.; Chen, X.-Q. MnO_2/reduced graphene oxide nanoribbons: Facile hydrothermal preparation and their application in amperometric detection of hydrogen peroxide. *Sens. Actuators B Chem.* **2017**, *239*, 544–552. [CrossRef]
336. Zhang, M.; Zheng, J.; Wang, J.; Xu, J.; Hayat, T.; Alharbi, N.S. Direct electrochemistry of cytochrome c immobilized on one dimensional Au nanoparticles functionalized magnetic N-doped carbon nanotubes and its application for the detection of H_2O_2. *Sens. Actuators B Chem.* **2019**, *282*, 85–95. [CrossRef]
337. Chen, S.; Yuan, R.; Chai, Y.; Hu, F. Electrochemical sensing of hydrogen peroxide using metal nanoparticles: A review. *Mikrochim. Acta* **2013**, *180*, 15–32. [CrossRef]
338. Chou, T.-C.; Wu, K.-Y.; Hsu, F.-X.; Lee, C.-K. Pt-MWCNT modified carbon electrode strip for rapid and quantitative detection of H_2O_2 in food. *J. Food Drug Anal.* **2018**, *26*, 662–669. [CrossRef]
339. Tabrizi, M.A.; Shamsipur, M.; Saber, R.; Sarkar, S. Flow injection amperometric sandwich-type aptasensor for the determination of human leukemic lymphoblast cancer cells using MWCNTs-Pdnano/PTCA/aptamer as labeled aptamer for the signal amplification. *Anal. Chim. Acta* **2017**, *985*, 61–68. [CrossRef]
340. Werner, E. Determination of cellular H_2O_2 production. *Sci. Signal.* **2003**, *2003*, PL3. [CrossRef]
341. Ma, Z.; Jiang, M.; Zhu, Q.; Luo, Y.; Chen, G.; Pan, M.; Xie, T.; Huang, X.; Chen, D. A porous hollow fiber sensor for detection of cellular hydrogen peroxide release based on cell-in-lumen configuration. *Sens. Actuators B Chem.* **2020**, *321*, 128516. [CrossRef]
342. Zhao, Y.; Huo, D.; Bao, J.; Yang, M.; Chen, M.; Hou, J.; Fa, H.; Hou, C. Biosensor based on 3D graphene-supported Fe_3O_4 quantum dots as biomimetic enzyme for in situ detection of H_2O_2 released from living cells. *Sens. Actuators B Chem.* **2017**, *244*, 1037–1044. [CrossRef]
343. Xi, J.; Zhang, Y.; Wang, Q.; Xiao, J.; Chi, K.; Duan, X.; Chen, J.; Tang, C.; Sun, Y.; Xiao, F.; et al. Multi-element doping design of high-efficient carbocatalyst for electrochemical sensing of cancer cells. *Sens. Actuators B Chem.* **2018**, *273*, 108–117. [CrossRef]
344. Zhao, A.; She, J.; Manoj, D.; Wang, T.; Sun, Y.; Zhang, Y.; Xiao, F. Functionalized graphene fiber modified by dual nanoenzyme: Towards high-performance flexible nanohybrid microelectrode for electrochemical sensing in live cancer cells. *Sens. Actuators B Chem.* **2020**, *310*, 127861. [CrossRef]
345. Yang, Y.; Ohnoutek, L.; Ajmal, S.; Zheng, X.; Feng, Y.; Li, K.; Wang, T.; Deng, Y.; Liu, Y.; Xu, D.; et al. "Hot edges" in an inverse opal structure enable efficient CO_2 electrochemical reduction and sensitive in situ Raman characterization. *J. Mater. Chem. A* **2019**, *7*, 11836–11846. [CrossRef]
346. Sinha, A.; Dhanjai; Zhao, H.; Huang, Y.; Lu, X.; Chen, J.; Jain, R. MXene: An emerging material for sensing and biosensing. *TrAC Trends Anal. Chem.* **2018**, *105*, 424–435. [CrossRef]

347. Chen, J.; Tong, P.; Huang, L.; Yu, Z.; Tang, D. Ti_3C_2 MXene nanosheet-based capacitance immunoassay with tyramine-enzyme repeats to detect prostate-specific antigen on interdigitated micro-comb electrode. *Electrochim. Acta* **2019**, *319*, 375–381. [CrossRef]
348. Chen, X.; Sun, X.; Xu, W.; Pan, G.; Zhou, D.; Zhu, J.; Wang, H.; Bai, X.; Dong, B.; Song, H. Ratiometric photoluminescence sensing based on Ti3C2MXene quantum dots as an intracellular pH sensor. *Nanoscale* **2018**, *10*, 1111–1118. [CrossRef]
349. Lin, H.; Chen, Y.; Shi, J. Insights into 2D MXenes for versatile biomedical applications: Current advances and challenges ahead. *Adv. Sci.* **2018**, *5*, 1800518. [CrossRef]
350. Dai, C.; Lin, H.; Xu, G.; Liu, Z.; Wu, R.; Chen, Y. Biocompatible 2D titanium carbide (MXenes) composite nanosheets for pH-responsive MRI-guided tumor hyperthermia. *Chem. Mater.* **2017**, *29*, 8637–8652. [CrossRef]
351. Liu, J.; Jiang, X.; Zhang, R.; Zhang, Y.; Wu, L.; Lu, W.; Li, J.; Li, Y.; Zhang, H. MXene-Enabled Electrochemical Microfluidic Biosensor: Applications toward Multicomponent Continuous Monitoring in Whole Blood. *Adv. Funct. Mater.* **2019**, *29*, 1807326. [CrossRef]
352. Guan, Q.; Ma, J.; Yang, W.; Zhang, R.; Zhang, X.; Dong, X.; Fan, Y.; Cai, L.; Cao, Y.; Zhang, Y.; et al. Highly fluorescent Ti_3C_2 MXene quantum dots for macrophage labeling and Cu^{2+} ion sensing. *Nanoscale* **2019**, *11*, 14123–14133. [CrossRef]

MDPI
St. Alban-Anlage 66
4052 Basel
Switzerland
Tel. +41 61 683 77 34
Fax +41 61 302 89 18
www.mdpi.com

Nanomaterials Editorial Office
E-mail: nanomaterials@mdpi.com
www.mdpi.com/journal/nanomaterials

www.ingramcontent.com/pod-product-compliance
Lightning Source LLC
Chambersburg PA
CBHW040317270926
51850CB00007BC/1504
9783036559711